# Cohesive Sediments

# Cohesive Sediments

4th Nearshore and Estuarine
Cohesive Sediment Transport Conference
INTERCOH '94
11–15 July 1994: Wallingford, England, UK

*Edited by*

**NEVILLE BURT**

**REG PARKER**

**JACQUELINE WATTS**

JOHN WILEY & SONS

Chichester · New York · Weinheim · Brisbane · Singapore · Toronto

*Other Wiley Editorial Offices*

John Wiley & Sons, Inc., 605 Third Avenue,
New York, NY 10158-0012, USA

VCH Verlagsgesellschaft mbH, Pappelallee 3,
D-69469 Weinheim, Germany

Jacaranda Wiley Ltd, 33 Park Road, Milton,
Queensland 4064, Australia

John Wiley & Sons (Asia) Pte Ltd, Clementi Loop #02-01,
Jin Xing Distripark, Singapore 129809

John Wiley & Sons (Canada) Ltd, 22 Worcester Road,
Rexdale, Ontario M9W 1L1, Canada

*British Library Cataloguing in Publication Data*

A catalogue record for this book is available from the British Library

ISBN 0-471-97098-0

Produced from authors' camera-ready copy.
Printed and bound in Great Britain by Bookcraft (Bath) Ltd.
This book is printed on acid-free paper responsibly manufactured from sustainable forestation,
for which at least two trees are planted for each one used for paper production.

# Contents

# List of contributors

Dr J. N. Aldridge,
*MAFF,*
*Pakefield Road,*
*Lowestoft,*
*Suffolk, UK*

Mr K. S. Black,
*School of Biological and Medical Sciences,*
*University of St Andrews,*
*Getty Marine Laboratory,*
*St Andrews,*
*Fife KY16 8LB, UK*

Mr T. J. Chesher,
*HR Wallingford,*
*Howbery Park,*
*Wallingford,*
*Oxon OX10 8BA, UK*

Mr J. Cornelisse,
*Delft Hydraulics,*
*PO Box 177,*
*2600 MH Delft,*
*The Netherlands*

Dr M. Crapper,
*Department of Civil Engineering,*
*University of Edinburgh,*
*Riccarton Campus,*
*Edinburgh EH14 4AS, UK*

Dr M. Dearnaley,
*HR Wallingford,*
*Howbery Park,*
*Wallingford,*
*Oxon OX10 8BA, UK*

Mr P. J. de Wit,
*Delft University of Technology,*
*PO Box 5048,*
*2600 GA Delft,*
*The Netherlands*

Professor D. Eisma,
*Netherlands Institute for Sea Research,*
*PO Box 59,*
*1790 AB Den Burg,*
*The Netherlands*

Dr M. J. Fennessy,
*Institute of Marine Studies,*
*University of Plymouth,*
*Drake Circus,*
*Plymouth PL4 8AA, UK*

Mr J. C. Galland,
*EDFLNH,*
*6 Quai Watier,*
*BP49,*
*78401 Chatou Cedex,*
*France*

Professor G. Gust,
*Technical University of Hamburg-Harburg,*
*Lauenbruch ost 1,*
*21079 Hamburg,*
*Germany*

Mr L. Hamm,
*SOGREAH Ingénierie,*
*BP 172,*
*38042 Grenoble,*
*Cedex 9,*
*France*

Mr C. Johansen,
*Department of Civil Engineering,*
*Aalborg University,*
*Sohngaardsholmsvej 57,*
*DK 9000 Aalborg, Denmark*

Professor T. E. R. Jones,
*Department of Mathematics,*
*University of Plymouth,*
*Drake Circus,*
*Plymouth, UK*

Dr R. Kirby,
*Dredging Research Ltd,*
*6 Queens Drive,*
*Taunton,*
*Somerset TA1 4XW, UK*

Dr B. G. Krishnappan,
*National Water Research Institute,*
*867 Lakeshore Road,*
*Burlington,*
*Ontario, Canada*

Professor T. Kusuda,
*Department of Civil Engineering,*
*Kyushu University 36,*
*6-10-1 Hakozaki Higashi-ku,*
*Fukuoka 812,*
*Japan*

Dr P. le Hir,
*IFREMER—Centre de BREST,*
*DEL–Laboratoire Chimie des*
*Contaminants et Modelisation (CCM),*
*BP 70,*
*29280 Plouzané,*
*France*

Professor A. J. Mehta,
*Coastal and Ocean Engineering*
*Department,*
*University of Florida,*
*336 Weil Hall, Gainesville,*
*Florida 32611,*
*USA*

Dr R. Parker,
*Blackdown Consultants Ltd,*
*Unit 1, Georges Farm,*
*West Buckland,*
*Wellington TA21 9LE, UK*

Dr D. Paterson,
*School of Biological and Medical Sciences,*
*University of St Andrews,*
*Getty Marine Laboratory,*
*St Andrews,*
*Fife KY16 8LB, UK*

Dr G. S. Sills,
*Department of Engineering,*
*University of Oxford,*
*Department of Engineering,*
*Parks Road,*
*Oxford OX1 3PJ, UK*

Mr A. M. Teeter,
*Department of the Army,*
*Waterways Experiment Station,*
*3909 Halls Ferry Road,*
*Vicksburg,*
*Mississippi 39180-6199,*
*USA*

Dr W. B. M. ten Brinke,
*Department of Physical Geography,*
*Utrecht University,*
*PO Box 80115,*
*3508 TC Utrecht,*
*The Netherlands*

Dr C. Teisson,
*Laboratoire d'Hydraulique EDF,*
*6 Quai Watier, BP49,*
*78401 Chatou Cedex, France*

Dr E. A. Toorman,
*Katholieke Universiteit Leuven,*
*de Croylaan 2,*
*Hydraulic Laboratory,*
*B-3001 Heverlee,*
*Belgium*

Ms H. Torfs,
*Katholieke Universiteit Leuven,*
*Hydraulics Laboratory,*
*B-3001 Heverlee,*
*Belgium*

Dr W. van Leussen,
*National Institute for Coastal and Marine*
*Management,*
*Ministry of Transport & Public Works,*
*PO Box 20907,*
*2500 EX The Hague,*
*The Netherlands*

Mr H. Verbeek,
*Rikswaterstaat – RIZA,*
*van Leauwenhoekwej 20,*
*3316 AV Dordrecht,*
*The Netherlands*

Dr J. R. West,
*School of Civil Engineering,*
*The University of Birmingham,*
*Edgbaston,*
*Birmingham B15 2TT, UK*

Mr D. H. Willis,
*National Research Council Canada,*
*Coastal Engineering, M-32,*
*Montreal Road,*
*Ottawa,*
*Ontario*
*K1A OR6,*
*Canada*

Mr H. Winterwerp,
*Delft Hydraulics*
*PO Box 177,*
*2600 MH Delft*
*The Netherlands*

# Preface

Cohesive sediments contribute to a wide range of design, maintenance and management problems in estuaries and coastal regions. Accumulation of sediment in navigation channels and berths often results in the need for expensive dredging operations. New developments can cause significant changes to the sediment transport patterns, so a good understanding of the likely changes is necessary before any engineering project can proceed. In addition, many pollutants are preferentially adsorbed on to the fine cohesive fraction of the sediment and therefore for environmental reasons it is important to be able to predict the movement of the sediment.

The behaviour of cohesive sediment is affected by many physical, chemical and biological factors. As yet it is not possible to predict the behaviour from physical properties alone. In addition, there are many different methods for measuring the physical properties so that results from researchers around the world cannot always be easily compared.

Under the general umbrella of the Steering Committee of the Nearshore and Estuarine Cohesive Sediment Workshops, Intercoh '94 was the 4th of the series of workshops started in Florida by Ashish J. Mehta and held approximately every three years. It brought together engineers and scientists working on cohesive sediment transport problems to share and discuss their research and experience and to initiate the process of harmonising amongst the international research community the characterisation of cohesive sediment behaviour and properties for transport modelling.

The conference provided a forum for presentation and discussion of many research projects, one of which is the EC-funded MAST G8M-Coastal Morphodynamics, Project 4 (Cohesive Sediments). This continuing project involved participants from 12 European Institutes working on improving the understanding of the basic processes and their contribution to cohesive sediment transport.

Generous support from the Commission of the European Communities, Directorate General for Science, Research and Development MAST programme allowed the deliberations of a Review Coordination Committee whose role was to undertake reviews of the selected topic areas of the conference (Particles/Aggregates, Settling and consolidation, Rheology, In situ settling velocity, Deposition and erosion, Biological influences, Sediment/water column exchange, Modelling), assist in paper selection and, where needed, in peer review.

The organisation and running of the conference was also generously supported by the Commission of the European Communities, Directorate General for Science, Research and Development.

It is with great pleasure that we acknowledge this support and the support, advice and guidance of Dr Boissonnais and Dr Fragakis.

The time, freely given by the Technical Steering Committee is greatly appreciated.

The contribution of the many referees who gave of their time for peer review cannot be overvalued. The great practical and social success of the conference is almost entirely due to the efforts of Jacqueline Watts and the staff of HR Wallingford.

Neville Burt and Reg Parker

# CONFERENCE OVERVIEW

# 1    On the characterisation of cohesive sediment for transport modelling

**W. R. PARKER**
*Blackdown Consultants Ltd, UK*

Introduction.

Whether it is the transport and impact of the sediment itself or the role of the sediment as the transport agency for a wide spectrum of micro-contaminants, the environmental management of almost all of the water cycle , including estuaries and coastal waters , involves cohesive sediments.  Management depends on prediction and prediction needs models ; models of processes, models of behaviour and models of systems.

The Oxford English Dictionary  includes in its definitions of the term "model"  the following:-
> : representation in 3 dimensions, especially on a smaller scale.
> : a simplified (often mathematical) description of a system to assist calculations and predictions.

Models, or numerical simulations, of natural processes are both an exploratory scientific tool and, especially in the form of predictive simulations, an essential engineering and environmental management facility.

As exploratory tools they allow examination of the interactions between the model components or the sensitivity of process to component parameters.  As management tools they provide essential guidance in assessing the optimum solutions to problems or the consequences of proposed actions.

However unlike "models" of static objects, or even moving objects, models or simulations of cohesive sediment transport attempt to represent natural systems of great complexity: systems which respond to phenomena operating on the one hand on submicron scales to, on the other hand, scales of kilometres: on time scales from fractions of seconds to tens of years.

The development of reliable numerical simulations as  management tools rests upon the degree to which the relevant natural biochemical, biological, chemical, physical and physicochemical processes and mechanisms involved are understood and can be represented in the simulation.

The G6M publication (Ref G6M; Berlamont et al 1993) lists some 32 variables and other parameters as descriptors necessary to be determined for the physico-chemical characterisation of cohesive sediments.  These descriptors mostly refer to physical parameters (e.g.particle size, bed or suspension concentration, specific surface), physico-chemical parameters (e.g. cation exchange, sodium absorption, pH, Eh) and physical processes (e.g.settling, deposition, consolidation, erosion).

It is hardly surprising therefore that those who devote themselves to developing numerical simulations look with  bewilderment, concern  and even despair at the complexity of cohesive sediment systems and the disparity in views and results concerning how they work.

Many characteristic parameters and concepts widely used in non-cohesive systems such as size, density or strength have, in cohesive sediment systems, time, energy, chemistry and biologically influenced character.  For example particle size is a unique descriptor of the solid transport unit of sand, but not so for cohesive sediment.

*Cohesive Sediments.*
Edited by Neville Burt, Reg Parker and Jacqueline Watts
©1997 John Wiley & Sons Ltd.

If the ultimate objective of modelling is to allow successful and harmonious management of our environment then it is essential that careful attention is paid to understanding how the natural system to be simulated operates, (e.g.Teisson et al 1993). The natural laws and rules must be identified and understood.   As Francis Bacon wrote

"We cannot command nature except by obeying her"

And we cannot obey, unless we understand!

Needs of Models.

Successful simulation of the transport of cohesive sediment requires the natural laws which govern its behaviour are identified, quantified and represented in mathematical form.  In so far as this behaviour involves the complex interaction of fluids and charged particles, it involves some of the more intractable areas of natural science: the biology , microbiology, chemistry and physics of fine particle systems in fluids of varying  physical and chemical character. Many aspects of this behaviour are not realistically scalable and for this, and reasons of cost and flexibility, numerical simulation is the standard modelling approach.

There now exists  a generally powerful capability to simulate the behaviour of water flow in estuaries in two or three dimensions.

To simulate cohesive sediment transport it is necessary to add to this hydrodynamic base the appropriate algorithms describing the properties and behaviour of cohesive sediment, their response to hydrodynamic stresses and their effect on fluid properties.

FIGURE 1   A COHESIVE SEDIMENT SYSTEM

TRANSPORT

EROSION                              DEPOSITION

MATERIAL OF TIME DEPENDANT
PROPERTIES

It is the objective of this conference to focus attention on the need for some degree of unity in the views of how cohesive sediment systems work and on the need to unify the methods and  concepts used to provide the information on which simulations depend.


System Concepts.

Many diagrams, of varying complexity, have been prepared representing the behaviour of fine sediment.There have been many attempts to represent the interlinked network of mechanisms, processes and phenomena which constitute a natural cohesive sediment system (e.g. Postma and Kalle 1955; Allen et al 1977; Parker and Kirby 1977; Mehta et al 1982) and there seems to be a general consensus regarding the physical framework.  At their most simple they may be  represented as in Fig 1.  Models need the functional relationships describing both the particulate and mass behaviour of the fine sediment moved within this framework by processes driven by hydrodynamic and gravitational energy, (Fig 2).

FIGURE 2    SYSTEM CHARACTERISATION

FLUID PROPERTIES        MATERIAL PROPERTIES

SYSTEM BEHAVIOUR

FLUID BEHAVIOUR        MATERIAL BEHAVIOUR

What is less clear is the degree to which the physical manifestations are in fact mediated by mechanisms having biological or biochemical origins. Are the conclusions, rates or coefficients arising from supposedly abiotic experiments, or experiments which ignore microbiological impacts vulnerable to alternative formulation? Do the data sets used to describe the workings of a fine sediment population, to prepare hierarchies of the relative importance of various processes and mechanisms really give the whole picture, particularly when they are of shorter duration than many biological mechanisms?

Characterisation : A Necessary Impediment to Progress?.
Appropriate characterisation of cohesive sediment must encompass the relevant physical, physicochemical, hydrodynamic and fluid mechanical interactions between particles and between arrays of particles and fluids forming the behavioural network (Fig 2). Increasingly, as awareness grows, the mediation of what are generally regarded as physical processes by microbiological agencies, is regarded as being of the first order importance rather than a second order modification, (e.g. Montague 1986; van Gemeerden 1993). How long will it be before we see included in the constitutive equations describing fine sediment behaviour, terms or groups balancing bacterial metabolism and mucous production, microbiological grazing and digestion, bioroughness production and pelletisation.
Across the range of factors which are required for a complete characterisation of cohesive sediment for transport modelling, the level of knowledge and scientific security is highly variable. The G6M Report lists 4 main aspects necessary to evaluate the physico-chemical characteristics of Cohesive Sediments.

        * Physico-chemical properties of the suspension
        * Physico-chemical properties of the mud
        * Characteristics of the sediment bed
        * Water-bed exchange processes.

This pioneering effort to present a coherent picture begs many questions, not least of which, from a practical viewpoint, are

"Why are we characterising the sediment?" and
"Which characteristics are most pertinent to our interest?".

Are all muds so different in their properties that it is necessary to execute all the 32 determinations? . One can point to scatter in global plots of erosion function to support this view, but how much of this is scatter is due to the different experimental conditions, especially chemical and microbiological?. Whilst ever more detailed understanding is rightly sought, should the practical uses of science not also be a spur to identify the order of importance of these variables.
Thus although "characterisation" of cohesive sediments must examine Fluid Properties and Material Properties, and through their interaction Behavioural Properties, the key aspects of this characterisation lie not solely in precise description or quantification of the individual components but in determining their interaction. In coupling, for example, fluid chemistry, microbiology and energy with particle size, shape and charge to describe aggregation in a manner that will yield realistic information about the aggregate. Alternatively should interactions of macro and micro flow structure (streamlines, turbulence intensities) with aggregate properties to describe vertical mass transfer, resultant concentration fields, or hydrodynamic effects of aggregate arrays on fluid flow, transfer of stress, energy dissipation and changes in fluid properties.

The necessity or viability of characterising various sectors of the cohesive sediment system must also be considered. If all are equally important they are not all equally tractable! Aggregation presents formidable problems, not only of characterising the system components, but the simulation of the processes of aggregate growth and breakup in natural systems involving natural macromolecules. Can we translate such models to give realistic results in a cohesive system simulation or should the focus of attention be on direct measurement of the resultant parameters of interest (size, settling velocity).

For convenience the conference will consider both material and behavioural characterisation together, moving from the transport unit (particles and aggregates), through settling, deposition, consolidation and then erosion: processes, mechanisms, apparatus will feature in presentation and discussion.

Particles and Aggregates.
The grouping of individual mineral grains and organic debris to produce composite particles (e.g. aggregates or flocs) is known to depend upon the size and mineralogy and hence the surface properties of the constituents and to be achieved by a combination of electrochemical and hydrodynamic energy (interparticle potentials, Brownian motion, fluid shear) and, some-times to predominance, by microbiological products. The size, density and other physical characteristics of these composite particles are a temporally and spatially varying product of kinetic energy dissipation inducing either growth or breakup, (e.g. Hunt 1986; Lick et al 1993). At present, characterisation covers the size, mineralogy and surface properties of the mineral grains and the essential chemistry (e.g.ionic strength and pH) of the fluid. Should considera-tion not also be extended to include the macromolecular components of the fluid and the organic coatings of the particles?

Perhaps also it would prove useful to agree some overall consistent terminology. Do the terms "floc" or "aggregates" mean the same thing? Is the discrimination between flocculation, coagulation and aggregation clear in a mechanistic context or are these all related to "composite particles" in a global sense.

Settling.
The "settling velocity" of the material in suspension is a crucial input parameter to simulations, delimiting vertical mass fluxes. Burt (1986) noted the differences between laboratory and field determinations of settling velocity. Others have discussed the effects of water chemistry and concentration It is self evident that as the character (size, density, etc) of flocs, aggregates or other compos ite particles is locally a function of number concentration and turbulence, so might the settling velocity be expected to vary. The great efforts to measure in situ settling velocity bear witness to this. These efforts reveal much about spatial and temporal variability in unit behaviour. However, in view of the apparently key role of Kolmogorov scales in determining the size and density of aggregates, does the apparatus which is used to determine the settling velocity actually influence it by
   a) influencing the turbulence spectrum and thereby the growth or breakup of the transport unit
   b) removing or modifying the distribution of energy within the turbulence spectrum and, as suggested by recent publication e.g. Maxey (1987); Fung (1993) or Wang and Maxey (1993), removing the influence of the turbulence itself on vertical transfer rates as compared to still water ? (see Berlamont et al 1993).

In circumstances where the pattern of suspended solids in the flow is anisotropic, what is the effect of the macrostructure of the flow on the vertical transfer of solids mass and on deposition? Is settling velocity the key parameter?

Deposition.

The incorporation of material into the substrate below the flow might seem to be the least indistinct part of both the behavioural and numerical models. Deposition occurs at a stress which characterises the inability of the flow to support the material units or detach them from the substrate once attached, (Lau and Krishnappan 1994). The role of flow structure in transporting units to the bed seems unclear.

With respect to Fig 3, two extreme end member situations may be proposed: low concentration circumstances of unit attachment to the substrate or high concentration conditions where deposition is less explicit and where, perhaps, it is more a matter of phase transformation from discontinuous to continuous structure, (e.g. Parker 1989).

FIGURE 3    DEPOSITION

SETTLING AND / OR VERTICAL FLOW TRANSFER

| LOW CONCENTRATION | HIGH CONCENTRATION |
|---|---|
| | HYPERCONCENTRATED LAYER FORMATION |
| SETTLING | HINDERED SETTLING |
| UNIT ATTACHMENT | PHASE TRANSFORMATION FROM UNSTRUCTURED TO STRUCTURED |

The work of Sills and Thomas(1984) indicates that slow deposition would lead to a more open and porous substrate compared to that produced by "phase transformation" at high concentration. The principal consequence of this appears to lie in the characteristics of flow deposited, experimental substrates used as analogues to investigate the erosion of natural muds. Rather than depositing a convenient mass of sediment it appears necessary to reproduce the natural flux of sediment onto the bed in order to produce the correct structural starting point for the subsequent erosion experiment.

Consolidation.

The self weight consolidation of fine sediments has been, and continues to be, a topic of intensive study. The focus of attention hitherto has been to develop descriptions of the processes involved and to derive mathematical formulations of them. The experimental basis for these are quasi one dimensional experiments generally conducted in 100mm diameter settling columns. Widely varying materials, introduction methods and introduction schemes have been tested. The most common factor in these experiments is that they involve quiescent fluid conditions.

Is this consistent with the field situation? Sediment deposition arises because fluid stresses are too low to prevent deposition or to create erosion. Are stresses at these levels significant in affecting the selfweight consolidation of soft materials? Observations reported by Parker (1989) suggest that material aparently having a continuous structure can also be deformed by fluid stresses. What effect does this have on near surface consolidation and the development of erosion resistance? What may be of greater importance is the effect of subcritical stress arising from surface waves on the rate and paths of selfweight consolidation.

Erosion.

It appears to be a matter of general consensus that the resistance of a cohesive bed to erosion arises as a consequence of both interparticle bonds and the mechanical interaction of groups of particles. The degree to which either or both of these predominates may well be related to the state of structural evolution of the material and its history.

It is also well established that the response of materials, such as cohesive sediment, to stresses of various origins is a product of both the amplitude and duration of the stress, the total strain which results and the ability of the material to dissipate stresses induced by local structural adjustment.

Appreciation of the links between interparticle and mechanical forces and rheological properties has led to a rapid expansion in the use of rheological techniques and models in relation to predicting erosion thresholds and rates, (e.g. Dade et al 1992; Otsubo and Muraoka 1988) and the interaction of waves with cohesive substrates (Foda et al 1993; Maa and Mehta 1988, 1989; Sukakiyama and Bijker 1989). However there is great disparity between rheological measurement methods, interpretation of rheological data and its incorporation into theories for erosion and deposition. It is in this area that detailed characterisation of materials appears to be most crucial and the need for agreement of methods and definitions amongst the most urgent. There is a need for a clear view on the conditions wherein controlled shear as opposed to controlled stress techniques are appropriate (James et al 1982-1987; Williams and Williams 1989, 1991),the interpretation of the data derived from controlled or applied shear rate measurements compared with those from controlled stress techniques, and the application of other techniques (e.g. Buscall et al 1988).

Unconfined, non analytical geometries (e.g. bobs in beakers) and concentric cylinder viscometry require very careful use. Buscall et al (1993) demonstrate the occurrence of slip phenomena and the dependence of low shear apparent viscosity results on methodology in weakly flocculated, concentrated colloidal dispersions. James et al (1986) have demonstrated the limitations of such methods for high volume fraction natural muds. Whether controlled shear rate methods are applicable at solids volume fractions greater than the space filling volume fraction appears open to grave doubt.

However, the spectroscopic approach to evaluating the structural level of sediments (Williams and Williams 1992) appears to offer the first line of approach for assessing the likely response of cohesive substrates to the spectral characteristics of the fluid forcing, allowing comparison of the temporal mean stress approach with ones based on energy spectra.

Methodologies such as controlled stress rheometry and shear wave spectroscopy open the way to evaluating cohesive sediment responses at subcritical stresses and strains, those below amplitudes involving structural breakdown and flow, with the prospect of methodologically independent results.

There is also a need for increased attention to biological effects. If erosion resistance relates to interparticle forces, the role of pH in influencing the total interaction potential is well known and pH effects on structural properties are reported (e.g. Williams and Williams 1991). The microbiological mediation of fluid chemistry, within cohesive substrates, (? and composite particles) is very well documented. It gives a quite different kinetics to the development of erosion resistance and its modification. It extends beyond the binding effects of biopolymers to influence the basic chemical environment of the material. Are the effects of these processes represented, or even representable, in models?


Material Quantities.

In many practical, as well as experimental, situations the quantity of sediment involved becomes of substantial importance; in the population of sediment in a turbidity maximum and the water quality impact of resuspension; in the mass flux onto "the bed". Measurements of concentration form a crucial input to experiment and simulation.

Five aspects of concentration interact in both a physical and metrological sense:

Mass Concentration $C$

Volumetric Concentration $\phi$

Effective Volume Concentration $\phi_E$

Hydrodynamic Volume Concentration $\phi_H$

Number Concentration Number per unit volume $Cn$

Although most simulations are concerned with mass fluxes, the majority of measuring systems (other than samplers and nuclear systems) respond to some combination of volume and number concentrations (e.g. Gibbs and Wolanski 1992). The nature of the materials and the fluid characteristics mean that a given mass of sediment can, in principle, comprise a temporally varying spectrum of aggregate size and density and thus number ($Cn$) and volumetric ($\phi$, $\phi_E$) concentration. This presents challenges to the calibration of acoustic, optical or other indirect concentration measuring devices, placing a premium on in situ calibration. Rheological properties are predominantly influenced by volume and number concentrations, though results are often presented in terms of mass concentration or bulk density.


In Situ Versus Laboratory Determinations.

Although laboratory investigation of basic principles by study of ideal or, so called, model systems will remain a cornerstone of cohesive sediment research, it is apparent that for practical and other purposes, in situ measurements appear generally more desirable. Whereas laboratory determinations of individual mineral and other clast size are made on treated material, in situ sizing of transport units is a major priority and research area. Other targets for in situ determination must eventually include

    \* Solute, biopolymer and colloid constituents of fluids.

    \* Vertical transfer in free field flow.

    \* Reliable determination of both mass, volume and number concentration.

    \* Rheological and rheospectral characteristics of aggregated and structured materials.

Deposition, consolidation and erosion can already be measured in situ with an accuracy and resolution appropriate for model timestep calibration. However the methods used for erosion measurements require more refinement. From rheological considerations it may be suggested that erosion functions have some dependence on the spectral character of turbulent energy and that the erosion function for a bed may be related to the stress path to which it is subjected, (e.g. Amos et al 1992). It may be that temporally compressed, in situ, erosion experiments,

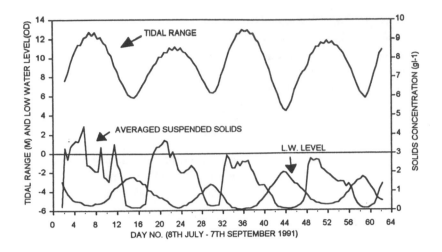

Fig. 4 Decreasing tidally averaged suspended solids during July to September

using apparatus which limit the turbulence scales occurring or do not reproduce the appropriate distribution of energy within the turbulence spectrum, whilst avoiding the problems of an unrealistic substrate, may still not provide a realistic entrainment scenario. The substantial variations in hydrodynamics consequent upon changes in suspended load and bedload (Gust 1976, Gust and Walger 1976, Wang and Larsen 1994) are a significant warning in this direction.

System Evaluation.

The detailed study of materials and mechanisms allows evaluation of processes and phenome na. However many of the processes studied (e.g. aggregation, erosion) are of short duration (seconds to hours). The context of these processes may change as the character of a cohesive sediment system changes because the driving functions are modulated over longer timescales e.g. tides over spring/neap or seasonal cycles; microbiological activity over seasonal cycles.
As longer and more representative data sets are acquired it becomes evident that the essentially physically dominated mechanisms are themselves modulated on longer periods by "non-physical" forces. Fig 4 shows the long term summer decline in tidally averaged suspended solids levels in a macro-tidal estuary. Note also the repeated phase advance of suspended solids levels compared to tidal range rather than the more usually described phase lag .This is shown in more detail in Fig 5. The dynamics of this sediment population has a crucial effect on the modulation of oxygen levels (Parker et al. 1994).Modelling in this estuary has progressed from simulating individual spring and neap tide ,through spring-neap-spring simulations, (Olesen et al 1992; Larsen et al 1992), to annual simulations including seasonality in terms related to the total solids population.
What is of great interest is to know with some certainty where the sediment goes. It returns in October or November and it is suspected that it is stored intertidally, (c.f. Frostick and McCave 1979).
Characterisation of the system appears to be profoundly affected by the length of the data set. There is still a need for long data sets of basic descriptive parameters.

10

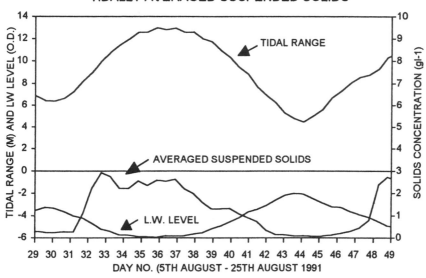

Fig 5 Neap-Spring-Neap Cycle of tidal range, low water level and suspended solids

Modelling.
The focus of this conference considers the concepts, criteria and means whereby input informa tion is derived for models.  Some consideration is also given to the models themselves. As  research tools model evolution is toward greater complexity, greater apparent sophistication and the need for ever more powerful platforms.
Is there not also an equal requirement for more rigorous parameterisation of more simple or robust models?  It is a seductive argument but is the situation sufficiently clear to allow a decision?  Is the data available to say whether, or not, the simulations accurately represent the natural system?


Conclusions.
The characterisation of cohesive sediment for transport modelling needs to cover
        Fluid and Material Properties
        Fluid and Material Behaviour
        System Behaviour
It needs to consider extremes of physical and temporal scales.
The nature of the materials and the processes which operate place a premium on experimental rigour requiring the combined efforts of multidisciplinary teams.  This conference is aimed at  furthering the causes of scientific excellence which is the only means for real progress, the cause of common methodology and approach which will foster collaboration and exchange of  results and the cause of communication between specialists in the fields which need to be brought to bear to understand natural cohesive sediment systems.

References.

Allen, G.P., P.Castaing and J.M.Jouanneau. 1977. Mechanisms de remise en suspension et de dispersion des sediments fins dans l'estuaire de la Gironde. Bull.Geol.Soc.France, 7 (19) 2. 167-176.

Amos, C.L., G.R.Daborn, H.A.Christian, A.Atkinson and A.Robertson. 1992. In situ erosion measurements on fine-grained sediments from the Bay of Fundy. Marine Geology. 108. 175-196.

Berlamont, J., M.Ockenden, E.Toorman and J. Winterwerp. 1993. The characterisation of cohesive sediment properties. Coastal Engineering. 21. 105-128.

Burt, T.N. 1986. Field settling velocities of estuary muds. in A.J.Mehta(Ed). Estuarine Cohesive Sediment Dynamics. Springer-Verlag. 126-150.

Buscall, R., P.D.A.Mills, J.W.Goodwin and D.W.Lawson. 1988. Scaling behaviour of the rheology of aggregate networks formed from colloidal particles. J.Chem.Soc., Faraday Trans. 1. 84(12), 4249-4260.

Buscall, R., J.I.McGowan and A.J.Morton-Jones. 1993. The rheology of concentrated dispersions of weakly attracting colloidal particles with and without wall slip. J.Rheol. 37(4). 621-641.

Dade, W.B., A.R.M.Nowell and P.A.Jumars. 1992. Predicting erosion resistance of muds. Marine Geology. 105. 285-297.

Foda, M.A., J.R.Hunt and H-T Chou. 1993. A non-linear model for the fluidisation of marine mud by waves. J.Geophys.Res. 98(C4). 7039-7047.

Frostick, L.E. and I.N.McCave. 1979. Seasonal shifts within an estuary mediated by algal growth. Estuarine and Coastal Marine Science. 9(5). 569-576.

Fung, J.C.H. 1993. Gravitational settling of particles and bubbles in homogeneous turbulence. J.Geophys. Res. 98(C11). 20287-20297.

Gibbs, R.J. and E.Wolanski. 1992. The effect of flocs on optical backscattering measurements of suspended material concentration. Marine Geology. 107. 289-291.

Gust, G. 1976. Observations on turbulent drag reduction in a dilute suspension of clay in sea water. Jour.Fluid Mech. 75(8). 29-47.

Gust, G. and E.Walger. 1976. The influence of suspended cohesive sediments on boundary layer structure and erosive activity of turbulent seawater flow. Marine Geology. 22. 189-206.

G6M. On the methodology and accuracy of measuring physico-chemical properties to characterise cohesive sediments. Report sponsored by C.E.C. Directorate General XII. July 1993.

Hunt, J.R. 1986. Particle aggregate breakup by fluid shear. in Estuarine Cohesive Sediment Dynamics. A.J.Mehta (Ed). Springer-Verlag. 85-109.

James, A.E. and D.J.A.Williams. 1982. Particle interactions and rheological effects in kaolinite suspensions. Adv.Colloid and Interface Science. 17. 219-232.

James, A.E. and D.J.A.Williams. 1982. Flocculation and rheology of kaolinite/quartz suspensions. Rheol.Acta. 21. 176-183.

James, A.E., D.J.A.Williams and P.R.Williams. 1986. Small strain low shear rate rheometry of cohesive sediments. In J.Dronkers and W.van Leussen (Ed). Physical Processes in Estuaries. Springer. 488-502.

James, A.E., D.J.A.Williams and P.R.Williams. 1987. Direct measurement of static yield properties of cohesive suspensions. Rheol.Acta. 26. 437-446.

Larsen, H.G.H., K.W.Olesen, A.J.Parfitt and W.R.Parker. 1992. Water quality modelling in a hypertidal, muddy estuary. in Hydraulic and Environmental Modelling: Estuarine and River Waters. Falconer, Shiono and Matthews (Eds). Ashgate. 223-240.

Lau, Y.L. and B.G.Krishnappan. 1994. Does re-entrainment occur during cohesive sediment settling? J. Hydraulic Engineering. 120(2). 236-244.

Lick, W., H.Huang and R.Jepsen. 1993. Flocculation of fine grained sediments due to differential settling. J. Geophys. Res. 98(C6). 10279-10288.

Maa, J.P-Y. and A.J.Mehta. 1988. Soft mud properties : Voigt Model. J. Waterways, Port, Coastal and Ocean Engineering. 114(6). 765-770.

Maa, J.P-Y. and A.J.Mehta. 1989. Soft mud response to water waves. J. Waterways, Port, Coastal and Ocean Engineering. 116(5). 634-

Maxey, M.R. 1987. The gravitational settling of aerosol particles in homogeneous turbulence and random flow fields. J.Fluid Mech. 174. 441-465.

Mehta, A.J., T.M.Parchure, J.G.Dixit and R.Ariathurai. 1982. Resuspension potential of deposited cohesive sediment beds. in Estuarine Comparisons. V.S.Kennedy (Ed). Academic Press. 591-609.

Montague, C.L. 1986. Influence of biota on erodibility of sediments. in Mehta, A.J.(Ed). Estuarine Cohesive Sediment Dynamics. Springer-Verlag. 251-269.

Olesen, K.W., W.R.Parker, A.J.Parfitt and H.Engrobb. 1992. Field, laboratory and model investigations of the feasibility of a tidal barrage in a muddy hypertidal estuary. in Tidal Power: Trends and Developments. R.Clare (Ed). Proc 4th Conference on Tidal Power. Thomas Telford. 279-295.

Otsubo, K. and K.Muraoka. 1988. Critical shear stress of cohesive bottom sediments. J. Hydraulic Engrg. 114 (10). 1241-1256.

Parker, W.R. and R.Kirby. 1977. Fine sediment studies relevant to dredging practice and control. Proc. 2nd Int.Symp. Dredging Technology. BHRA. Cranfield. 15-26.

Parker, W.R. 1989. Definition and determination of the bed in high concentration fine sediment regimes. J. Coastal res. Special Issue 5. 175-184.

Parker, W.R., L.D.Marshall and A.J.Parfitt. 1994. Modulation of dissolved oxygen levels in a hypertidal estuary by sediment resuspension. Proc. ECSA 23. Neth.J. Aquatic Ecology

Postma, H. and K.Kalle. 1955. Die entetehung von trubing zonen im unterlauf der flusse, speziell im hinblick auf die verhaltnisse der unterelbe. Deutsche Hydr. Zeit. 8. 137-144.

Sills, G.C. and R.C.Thomas. 1984. Settlement and consolidation in the laboratory of steadily deposited sediment. in Sea Bed Mechanics. B.Denness (Ed). Graham & Trotman. London. 41-49.

Sukakiyama, T. and E.W.Bijker. 1989. Mass transport velocity in mud layer due to progressive waves. J. Waterways, Port, Coastal and Ocean Engineering.115(5). 614-

Teisson, C., M.Ockenden, P.Lehir, C.Kranenburg and L.Hamm. 1993. Cohesive sediment transport processes. Coastal Engineering. 21. 129-162.

van Gemeerden, H. 1993. Microbial mats: A joint venture. Marine Geology. 113. 3-25.

Wang, L-P. and M.R.Maxey. 1993. Settling velocity and concentration distribution of heavy particles in homogeneous isotropic turbulence. J. Fluid Mech. 256. 27-68.

Wang, Z. and P.Larsen. 1994. Turbulent structure of water and clay suspensions with bed-
    load. J. Hydraulic Engineering. 120(6). 577-600.

Williams, D.J.A. and P.R.Williams. 1989. Rheology of concentrated cohesive sediments.
    J. Coastal Res. (Special Issue). 5. 165-174.

Williams, P.R. and D.J.A. Williams. 1989. Rheometry for concentrated cohesive
    sediments. J. Coastal Res. (Special Issue). 5. 151-164.

Williams, D.J.A. and P.R.Williams. 1991. Rheology and microstructure of concentrated
    illite suspensions. Ch 28 in Bennet et al (Eds). Microstructure of Fine Grained
    Terrigenous Marine Sediments. Springer-Verlag. 267-271.

Williams, P.R. and D.J.A. Williams. 1992. The determination of dynamic moduli at high
    frequencies . J. Non-Newtonian Fluid Mech. 42. 267-282.

# SETTLING VELOCITY

# 2     The *in-situ* determination of the settling velocities of suspended fine-grained sediment—a review

D. EISMA[1], K. R. DYER[2] AND W. VAN LEUSSEN[3]

[1]*Netherlands Institute for Sea Research, P.O. Box 59, 1790 AB Den Burg, Texel, The Netherlands*
[2]*Institute of Marine Studies, University of Plymouth, Drake Circus, Plymouth, Devon PL4 8AA, United Kingdom*
[3]*National Institute for Coastal and Marine Management/RIKZ PO Box 20907, 2500 EX, The Hague, The Netherlands*

ABSTRACT

Seventeen instruments and methods to determine *in-situ* the settling velocity of fine-grained particles are described: seven field settling tubes based on bottom withdrawal (3) and pipette-withdrawal (4) of the sediment, four remote or automated instruments, three *in-situ* videosystems, and three miscellaneous techniques. The relation of *in-situ* settling velocity with suspended sediment concentration, size and density is discussed. An *in-situ* comparison of some of the techniques and methods shows considerable divergence of the results between the settling velocities obtained at different concentrations.

## 1. INTRODUCTION

The settling velocities of suspended particles/flocs depend on the particle/floc size (or cross section perpendicular to the direction of fall), their density and their shape, besides the hydraulic conditions, chemical parameters and biological mechanisms. In the past it was often found to be related to the particle/floc concentration: Interaction between the particles/flocs (entrainment of small ones by large ones, flocculation during settling, hindered settling) can be important. Virtually all particulate material in suspension is flocculated to some degree: where observations in situ have been made, only rarely is unflocculated material found, as off glaciers where the outflow only contains fine mineral particles without organic material and the water is fresh. In all other cases organic matter was present, or assumed to be present, in the flocs because of the presence of living plankton. Flocs, although stable in flowing turbulent water, easily break apart when sampled, because of the additional shear during

*Cohesive Sediments.*
Edited by Neville Burt, Reg Parker and Jacqueline Watts
©1997 John Wiley & Sons Ltd.

sampling (additional turbulence, shockwaves). Therefore to know the settling velocity of natural flocs, it is necessary to measure this *in-situ* in the water.

During the last two decades various instruments have been developed to measure the settling velocities of mud flocs directly in nature. At the required depth they collect a water-sediment sample, after which the settling velocity is determined in still water. It is assumed that the suspended sediment samples are taken and handled carefully in such a way that the flocs remain in their natural state.
*Bottom-withdrawal* and *pipette-withdrawal* systems can be distinguished (Fig. 1). Different approaches in taking the samples, turning the tubes to a vertical position, making provisions to prevent internal circulations as well as possible, etc., have resulted in a variety of instruments.

More recently *in situ video recording systems* have been developed. The principle here is to capture the suspended flocs very carefully so that also the large aggregates, which are very fragile, are not destroyed. Next the suspended flocs settle in a vertical tube in still water conditions. From the video recordings both floc sizes and settling velocities can be determined.

In this review the various methods will be discussed with special reference to the different properties that can contribute to the variations in the values that are found for settling velocities.

## 2. *IN SITU* INSTRUMENTS

### 2.1. Field Settling Tubes

The principle of FST's is a settling tube, which is lowered with open ends in a horizontal position to the required sampling depth in the water column. Then the valves at both ends of the tube are closed and the tube is turned to a vertical position. This may be done at the moment of taking the sample, or when the tube is taken on board of the research vessel. The assumption is that when the settling tube is in the vertical position, the sedimentation process starts with a vertically homogeneously distributed mixture of sediment and water, and that the suspended particles and flocs have the same sizes and settling velocities as when they were suspended in the water column.
The FST's can be subdivided into Bottom Withdrawal Tubes (BWT) and Pipette Withdrawal Tubes (PWT). In the case of BWT's samples are taken from the bottom (Fig. 1a) at pre-fixed time intervals. After determining the dry weight of the sediment in the samples, a cumulative weight curve is obtained of the sediment that has settled to the bottom of the tube. The time intervals for sampling are chosen in such a way that the accumulative weight curve can be fitted as well as possible, which implies successively increasing time intervals. The accumulative weight curves are analyzed according to the method described by Vanoni (1975) for the settling of polydisperse suspensions.
In the case of PWT's, the suspended sediment concentration is determined at a pre-fixed level in the settling tube (Fig. 1b). Particles and flocs, that have a settling velocity larger than the ratio between the water depth above the sampling point and the elapsed period of time, already have settled below this sampling point. Also in this case (about 8) samples are taken at

a number of pre-fixed times. From the concentration curve the distribution of settling velocities, as well as the median settling velocity are determined.

The disadvantages of the BWT are (Van Rijn & Nienhuis, 1985):

- sediment particles sticking to the inside of the tube may not be removed during withdrawal of the sample.
- The complete procedure for determining the particle settling velocity distribution is rather time consuming compared with the PWT-method.

The disadvantage of the PWT is its difficulty to obtain reliable results when suspended sediment concentrations are low (Özturgut & Lavelle, 1986), But, this depends also on the sizes of the samples as well as on the method for determining the dry weight of the samples.

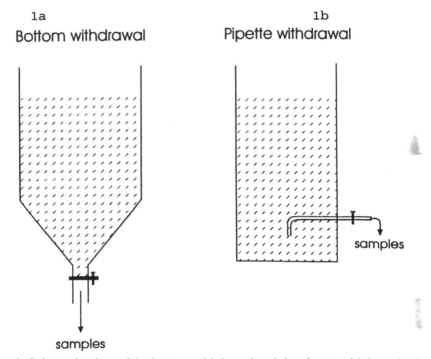

Figure 1. Schematic view of the bottom-withdrawal and the pipette-withdrawal tubes.

## 2.1.1. Bottom Withdrawal Tubes

*Owen Tube*

The well-known Owen Tube was developed in the late sixties at Hydraulics Research in the U.K. (Owen, 1971). The main part of the instument (Fig. 2) is a 1 metre long, 51 mm internal diameter perspex tube. It is so balanced that it lies horizontally in water and hangs vertically in air: after taking the sample in a horizontal position, the tube swings to the vertical position when it breaks the water surface. The sampling tube is supported by a concentric perspex tube, to which is fixed a guide vane assembly to align the tube with the flow. Closure at each end is achieved by rotating the inner tube ¼ of a turn relative to the outer one, thus twisting a sock at each end to form a seal. The rotation is carried out swiftly by a spring unit, which is released remotely from the survey vessel by a hydraulic system. A detailed description of the apparatus is given by Owen (1976). Results of measurements in the Thames, presented by Owen (1971), show important effects of the sediment concentration and turbulence (tidal range) on the settling velocities of the mud flocs. However from more extensive measurements in the Thames, Burt (1986) showed that the tidal range has no consistent effect on the settling velocities.

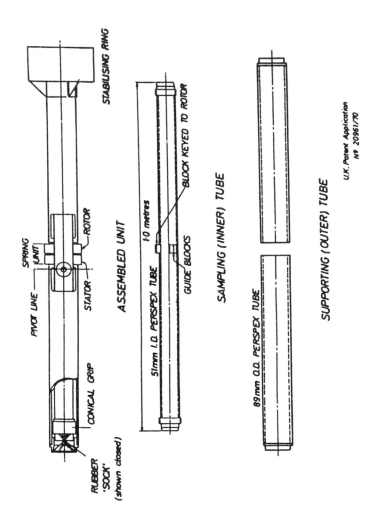

*Braystoke SK110 Settling Tube*

The Braystoke SK 110 settling tube is an improved version of the above-mentioned Owen Tube, and is commercially available at Valeport Ltd. It is sometimes called the Valeport sampler. It consists of a 1 m long perspex tube with an inner diameter of 50 mm and a volume of 2 litres (Pejrup, 1988). A tail fin is mounted on the tube so it lines up horizontally in the current direction when lowered into the water. The tube has sealing caps at both ends. These caps are closed by sending a release messenger down the suspension wire (Fig. 3). After raising the sampler to the deck of the research vessel, the tube is placed vertically in a stand and sedimentation starts. The perspex tube is divided into 10 sections of equal size, so that subsamples of 200 ml can be withdrawn at the bottom of the tube. Extensive measurements in the Elbe and Weser Estuary were presented by Puls *et al.* (1988). They show that the median settling velocity of mud flocs is a pproximately proportional to the square of the suspended sediment concentration. Estuarine flocs were shown to be destroyed and rebuilt within time scales that are smaller than the time scales of vertical mixing within the water. Pejrup (1988) applied the instrument in the Danish Wadden Sea, illustrating the importance of the suspended sediment concentration for floc growth, and of turbulent mixing for limiting the size of the flocs.

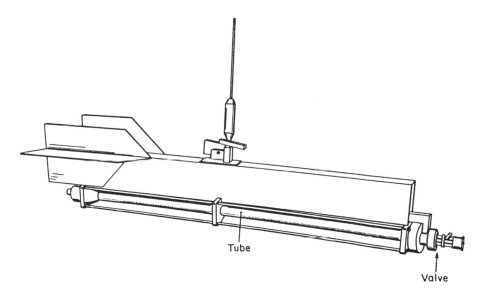

Tube

Valve

Figure 3 The Braystoke tube.

## QUISSET Tube

A more recent development is the so-called QUISSET Tube (Quasi In Situ Settling Tube), developed at Cambridge University (McCave & Atherton 1994), and improved at the Univerity of Wales, Bangor (Jones & Jago, pers. communication). It was expected that while filling the Owen tube some artificial turbulence was created around the tube's apertures and walls, resulting in some fractionation of the large flocs. To avoid this problem, as well as to be able to determine the settling velocity with sufficient accuracy also at low suspended sediment concentrations (as in shelf seas), the QUISSET tube has been developed for applications in areas with suspended sediment concentrations in the range of 1 to 100 mg/l. The settling tube is 0.9 m long with an internal diameter of 90 mm. It has a volume of 5.5 liters. In the tube an undisturbed volume of water is trapped by moving the sampling tube horizontally past a piston, and sealing it at a funnel-shaped tap end (Fig. 4). It can be released automatically at 1 m above the bed by means of a spring-release system, or manually at a specific water depth. On recovery the horizontal tube is first rolled gently to counteract any settling across the tube during recovery, then put vertically on a stand, and samples are collected by bottom withdrawal at specified intervals. Prior to withdrawal of the first sample a thermal jacket made from 25 mm thick insulating material is fitted around the tube leaving only a narrow vertical window to allow the height of the water column to be monitored. Measurements were made in the North Sea, where systematic changes within tidal cycles and during storms were observed, particularly due to the effect of resuspension (Jones *et al.* 1993; Jago *et al.* 1994). Settling velocities increased from slack water to maximum flow and decreased from maximum flow to slack water. A resuspension component and a background component of the settling velocity were distinguished, with values of of 0.1-1.0 mm/s and about 0.0001 mm/s respectively.

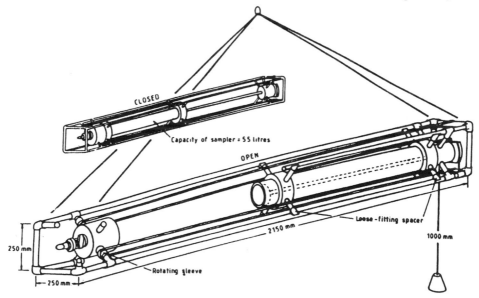

Figure 4. The Quisset tube.

## 2.1.2. Pipette Withdrawal Tubes

*Allersma Tube*

The Allersma Tube has been developed in 1962 by NEDECO, the Netherlands, for a siltation study in the Chao Phya Estuary (NEDECO, 1965). Exchangeable tubes are applied with a length of 0.28 m. The exchangeable tubes permit a large number of measurements per unit of time (for example during one tidal period). The settling tubes are mounted in a special device (Fig. 5), with which they are lowered to the desired depth and then filled and held in a vertical position when raised on board of the research vessel. Small samples (10 ml) are taken in the tubes with a pipette at a depth of 0.19 m below the water surface. More details of this instrument are given by Allersma (1980). Because the tubes come in a vertical position at the moment they are filled, the sedimentation process starts from the moment the sample is taken. Results from the Chao Phya Estuary (NEDECO, 1965) show that the settling velocity is dependent on salinity and increases exponentially with the sediment concentration to an exponent in the range of 1.2 to 1.3.

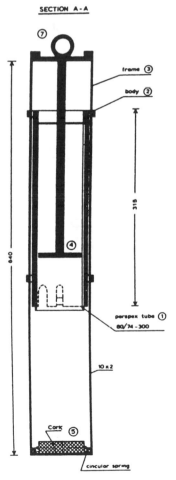

Figure 5. The Allersma tube.

23

*FIPIWITU*

Because a bottom-withdrawal tube has the disadvantage of possible incomplete removal of the fine sediment particles, accumulated at the bottom of the tube, Delft Hydraulics developed a new field instrument as shown in Fig. 6: the FIPIWITU (Field Pipette Withdrawal Tube). The instrument consists of a stainless steel tube with a length of about 0.30 m and an internal diameter of 0.12 m. The tube is used for sample collection as well as for the determination of the settling velocity by means of a settling test. It is equipped with a valve at each end and a double wall for temperature control. The tube is lowered to the sampling location in a horizontal position with opened valves. Through a messenger the valves are closed, and the tube is immediately in a vertical position. This is the starting point of the settling process (t=0). Then the tube is raised, and 200 ml samples are taken at times t= 1, 3, 6, 10, 20, 40 and 60 minutes (Van Rijn & Nienhuis 1985).

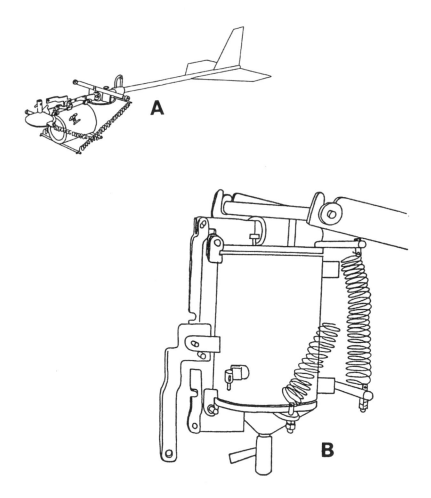

Figure 6. The FIPIWITU in lowering and sampling position (A) and in raising position (B).

24

The RWS (Rijkswaterstaat) field settling tube has been developed in the eighties by mr Van Geldermalsen of the National Institute for Coastal and Marine management / RIKZ of Rijkswaterstaat (the Netherlands). Therefore it is also called the "Van Geldermalsen Tube". The settling tube is lowered in a horizontal position and is turned to a vertical position immediately after taking the sample. Thereby the settling of the suspended sediment particles to one side of the tube is prevented. The exchangeable tubes permit a large number of measurements per unit of time (for example during one tidal period). A schematic picture of the RWS field settling tube is given in Fig. 7. On board the tube is suspended in a cardanic steel frame. Through this the settling tubes remain in a vertical position during the movements of the ship . Samples of about 200 ml, generally are taken at 1, 3, 6, 20, 40, 60 minutes after the starting point of the sedimentation process. Due to the limited time the first sample should be taken before suspending the tube in the frame. The inner diameter of the settling tube is 104 mm, whereas the initial height of the water column is about 210 mm. This relatively small height results in relatively small measuring times (generally 1 hour, sometimes 3 hours). The tube is double walled and insulated to prevent temperature-induced density flows to develop within the tube. More details can be found in Van Leussen (1994). From a series of field measurements in the Ems Estuary it became clear that the relation between the settling velocity and the suspended sediment concentration can be strongly dependent on the location within one estuary. Moreover it was shown that significant differences may occur between the median and mean settling velocities (Van Leussen, 1994).

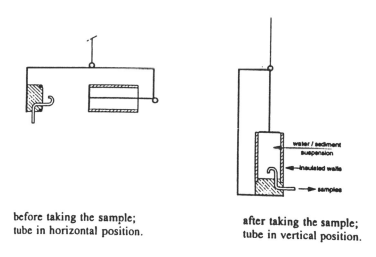

before taking the sample;
tube in horizontal position.

after taking the sample;
tube in vertical position.

Figure 7. the Rijkswaterstaat or Van Geldermalsen tube.

The BIGDAN field pipette withdrawal tube has been developed at the GKSS in Germany for the determination of settling velocities in the German Bight (Fig. 8). Because of the generally low suspended sediment concentrations (minimum values in the order of 0.5 mg/l), the settling tube has a volume of 28 litre. The tube is made of perspex, its height is 1 m and the internal diameter is 0.19 m.

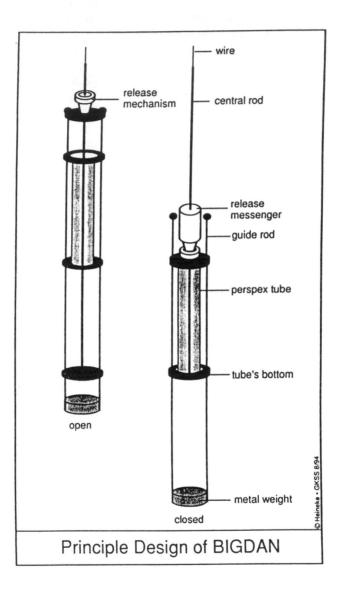

Figure 8. The Bigdan tube (the Markdan is the same but smaller).

The MARKDAN is a smaller version of the BIGDAN, designed for application in areas with higher suspended sediment concentrations, as for example estuaries. The samples are taken vertically by moving the settling tube downwards, released by a messenger. In fact it cuts out a cylindrical vertical water volume. Just after taking the sample both ends of the tube are closed, and the sedimentation process starts. Then the instrument is raised to the deck of the vessel, and is placed in a frame. To avoid convection currents the tube is insulated by a gold-coated envelope. Then a pipette-system is brought in, through which samples are taken at pre-fixed times: 3, 5, 7, 15, 30, 56, 86, 120, and 120 minutes. Accurate values for the settling velocity can be obtained in the range of 0.01 to 2 mm/s. Measurements in the German Bight by Puls and Kühl ((1995) indicated three fractions: an S-fraction ($w_S$ > 0.1 mm/s), reponsible for the variations of the suspended sediment concentrations, for example after a storm, an L-fraction ($w_S$ < 0.01 mm/s), remaining continuously in suspension, as well as an intermediate M-fraction (0.01 < $w_S$ < 0.1 mm/s). In this area there are no strong variations in the suspended sediment concentration. For measurements in the Elbe Estuary, where the suspended sediment concentration varies strongly within the tidal cycle, these fractions were defined as $w_S$ > 1 mm/s for the S-fraction and $w_S$ < 0.1 mm/s for the L-fraction.

### 2.1.3. Remote & Automated Instruments

*ROST*

The ROST (Remote Optical Settling Tube) has been developed to obtain settling velocity data for the continental shelf and the deep sea. It consists of a vertical rectangular settling column, 100 mm wide and a settling height of 0.35 m (Fig. 9). The principle of the system and the theoretical background are given by Zaneveld *et al.* (1982) and Bartz *et al.* (1985). Inside the tube the suspended sediment concentration is continuously measured by an optical beam-attenuation meter. The light source is nearly monochromatic with a wave-length in the red region, in order to reduce attenuation by dissolved materials to a minimum. Under some assumptions (cf. McCave & Gross, 1992) an accurate relation can be obtained between the suspended sediment concentration and the observed beam attenuation. The opening and closing mechanism of the settling tube consists of a low-current drain stepping motor which slowly opens the top and bottom lids against a spring. When the lids are fully open the tube is allowed to flush for a period of 2 h; at the end of this period a clutch releases and the lids snap closed (within 2 secs in water). The suspended material in the tube is then allowed to settle out for 22 h, after which the procedure is repeated. ROST is mounted on a tripod. This nondisruptive optical settling tube has been applied at a water depth of 5000 m.(Spinrad *et al.*, 1989). It was concluded that ROST is suitable for aplications in the case of fragile aggregates. The actual range of measured settling velocities was from 0.003 to 7.8 mm/s.

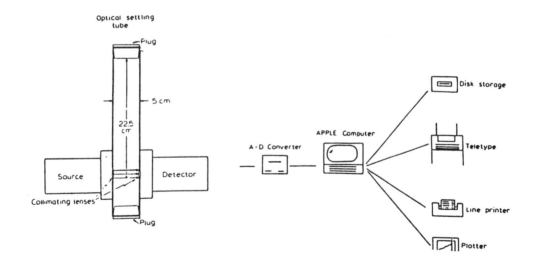

Figure 9. The ROST.
Reprinted from Zaneveld et al., *Mar. Geol.* 49, 1982, 357–376, with kind permission from Elsevier
Science – NL, Sara Burgerhartstraat 25, 1055 KV, Amsterdam, The Netherlands.

*BEAST*

BEAST (Benthic Autonomous Settling Tube) is designed to sink to the sea bed, collect a sample of water, allow its sediment to settle, and return to the surface (Bartz *et al.*, 1985; Fig. 10). Principally the system works like a Bottom Withdrawal Tube. However, rather than moving the water column down the tube, the bottom of the tube is moved up the water column. This is realized through isolating succesive volumes of water using hydraulically powered shutters. The internal diameter of the settling column is 254 mm. It consists of 8 stacked compartments with a total height of 0.819 m. Through each compartment a 5 litre sample is obtained. The concentrations in these samples are determined gravimetrically. The settling velocities are determined through the method of the Bottom Withdrawal Tube. The measurements are performed in the constant temperature environment of the sea floor, after which time the apparatus sheds ballast and returns to the surface. The system is mounted on a fibre glass tripod. Designed for, and tested in the deep sea, this instrument is too expensive to manufacture for routine use and would not work on the continental shelf because the shutter operation depends on high deep-sea pressures (McCave & Atherton, 1994).

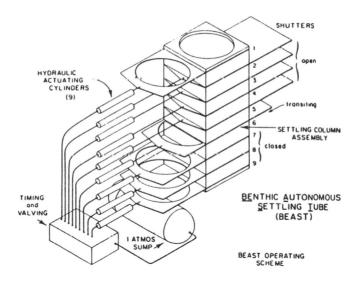

Figure 10. The BEAST.
Reprinted from Bartz et al., *Mar. Geol.* 66, 1985, 381–395, with kind permission from Elsevier Science – NL, Sara Burgerhartstraat 25, 1055 KV, Amsterdam, The Netherlands.

*Settling cylinder*

A settling cylinder was developed by Kineke & Sternberg, 1987; Kineke et al. 1989; Sternberg, 1989) for use in estuaries. This consists of a tube that is closed by a spring and contains approximately 3 liters of water and suspended sediment (Fig. 11). Between the two end plates is a vertical array of five minute optical backscatter sensors at 5 to 3 cm distance. Measuring starts as soon as the tube is closed. It is mounted on a tripod so that the lowest backscatter sensor is at approximately 20 cm from the seabed when the tripod is on the bottom. A concentration versus time graph is obtained which yields a settling velocity histogram. It has been used also in high concentration suspensions off the Amazon river mouth.

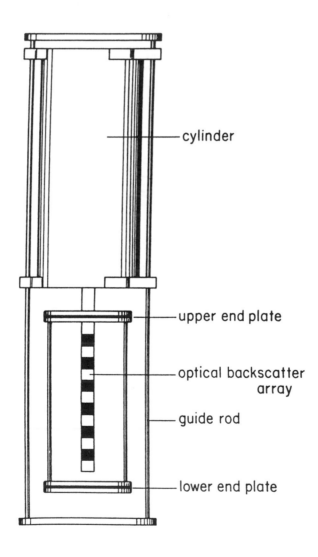

cylinder

upper end plate

optical backscatter array

guide rod

lower end plate

*Cambridge University Settling Box*

The Cambridge University Settling Box is 250 mm in diameter, 300 mm long, and captures almost 15 litres of fluid (Fig. 12). Top and bottom lids are hinged at single and diametrically opposite positions, which when fully open are 45° to the vertical. This ensures rapid flushing by both waves and a steady current. The lids close slowly in order to minimize flow disturbance, and capture a representative sample. An array of four miniature optical backscatter sensors is mounted inside the box at 100 mm, 200 mm, 250 mm, and 275 mm below the top lid, measure the vertical settling. The box is used on a quadrupod, a seabed mounted instrument frame (McCave *et al.*, pers. comm.).

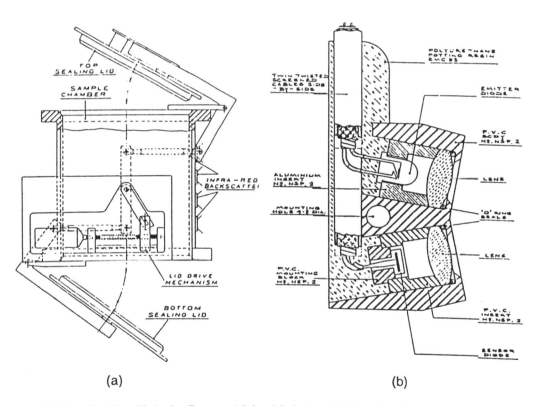

(a)         (b)

(a) The Settling Velocity Box and (b) a Miniature Optical Backscatter Sensor

Figure 12. The Cambridge settling velocity box and MOBS.

### 2.3. *In situ* video systems

*VIS*

VIS (Video In Situ) is an underwater video system, which allows the sizes and the settling velocities of suspended flocs to be determined *in situ* without disturbing the large fragile aggregates. Fig. 13 shows the principle of the system. The flocs are captured in a capture/stilling chamber, a sort of funnel (vertical pipe φ 100 mm, H=150 mm), into which a vertical settling tube (φ 26 mm ) is mounted. The settling tube contains two windows. Through one of them a light beam is introduced in the settling tube to illuminate the settling particles. The width of the beam is 0.4 mm and its height is 9 mm. Through the other window the video-recordings are made. These elements are placed in a stainless steel underwater housing (φ 600 mm, height 300 mm). The video-system is connected to a monitor, so that the settling of flocs and aggregates can directly be observed on board the research vessel and the light sheet and pumps can be monitored. The recordings are automatically marked by date and time (time step 0.01 sec); analysis of the recordings gives the sizes and settling velocities of the flocs. The difference between the movements of the capture/stilling chamber and the flowing water above should be minimized in order to capture the aggregates in the capture/stilling chamber without disturbing them Therefore the VIS is used as a floating system. To minimize the influence of the movements of the floating ship, VIS is mostly applied as a free floating submersible and only connected to the ship by the measuring cable (Van Leussen & Cornelisse, 1993a). Extensive series of measurements with VIS were carried out in the Ems estuary (Van Leussen, 1994). An abundance of macroflocs was observed shortly after maximum current velocity. Generally these macroflocs had sizes in the range of 200 to 700 μm, and sometimes more than 1 mm, and survived high current velocities (> 1 m/s); settling velocities were in the range of 0.5 to 8 mm/s. It was shown that these macroflocs play an important role in the fine-grained sediment transport in estuaries.

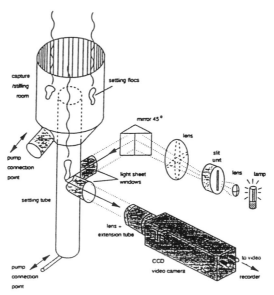

Figure 13. The VIS.

## Wallingford Video System

The Wallingford Video System involves obtaining samples of suspended sediment in the traditional manner using a Braystoke SK 110 (Valeport) bottom withdrawal tube (Fig. 14). After taking the sample and placing the tube vertically in the stand, video recordings are made of the settling flocs inside the tube. Then a video image analysis technique is applied to determine the size and settling velocity distributions of the material in suspension. A narrow light beam is used as a source of illumination. Field measurements were carried out in the Thames Estuary and in the Dover Harbour. It was concluded that in Owen tube results, median settling velocities are under-estimated because the Owen Tube analysis is based on a 30-60 minute settling period. It became also clear that the settling velocity is not dependent on the suspended matter concentration below about 50-100 mg/l and in some cases higher than previously assumed (Burt *et al.*, 1991).

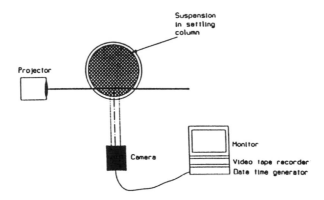

Figure 14. The Wallingford settling tube.

INSSEV (In Situ Settling Velocity Instrument) consists of a two chamber sampling device, whereby suspended particles are captured in a volume of water in the upper chamber. After a short interval, allowing turbulence in the chamber to decay, a number of flocs are allowed to fall into a still, clear water settling column below the upper chamber where a video camera observes their settling (Fig. 15). Size distributions and floc settling velocities are derived from the video recordings by image processing techniques. The instrument is mounted on a heavy tripod. When the settling column has been filled with filtered water of known density, the tripod is lowered to the bed. Some pilot experiments have been performed in the Tamar Estuary (Fennessy *èt al.*, 1994a, 1994b). These measurements were carried out in the turbidity maximum area during a mean tide. Also a Malvern Laser Particle Sizer (Bale & Morris, 1987) was mounted on the tripod to obtain comparative size data. All measurements and sampling were done at 0.5 m above the bed.

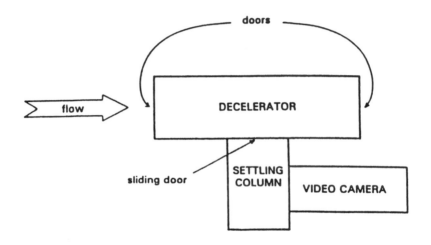

Figure 15. Scheme of the INSSEV.

## 2.4. Miscellaneous techniques

*In-situ measurements with SCUBA divers*

This has been done in clear ocean water off California by Alldredge & Gotschalk (1988) at 10 to 15 m waterdepth. Dye was injected in the water 3 cm below a particle of marine snow of 0.5 to 25.5 mm diameter (max. diameter 2.4 to 75 mm). The dye had the same density as the surrounding water and stabilized within a few seconds after injection. Turbulence was detected from the deformation of the dye and the time needed for the particle to reach the dye was measured with a stopwatch. The particle was photographed *in-situ* in the water with a 3:1

magnification extension tube mounted on a normal underwater camera. Settling velocities were in the order of 10 to 200 m.day$^{-1}$ or 0.1 to 2.6 mm.s$^{-1}$, which is of the same order as found in estuaries.

*Sediment traps*

Sediment traps measurements on the ocean floor (Sargasso Sea, Panama Basin; Deuser and Ross, 1980; Deuser, 1984,1986), extended over several years, show seasonal fluctuations in composition and in the quantity of material collected in the trap (which has a rotating set of cups so that samples are taken at weekly intervals). The fluctuations are related to seasonal variations in the production of particles (flocs) in the surface water. From the time difference between the maxima in the surface water and those in the trap, it can be estimated that the particles settle at a rate of 50 of 100 m.day$^{-1}$,, which is similar to the rates found by the *in-situ* measurements of marine snow off the California coast, described above.

*Vertical profiles of flow velocity and suspended sediment concentration*

Krone (1962) derived settling velocities from vertical profiles of flow velocity and suspended sediment concentration (d<53 µm). These were obtained at points in time, when the velocity profile corresponded very well with the logaritmic velocity profile. In this way settling velocities were obtained in the range of 3 to 5 mm/s during the decelerating flood phase.

## 3. DISCUSSION

For all settling velocity measuring instruments the following aspects should be considered:
• is there floc break-up during sampling and/or measuring ?
• is there additional flocculation inside the instrument ?
• is there an effect of water circulation in the instrument?
• is there an effect of the dimensions of the instrument on the results ?

For most instruments these questions can not be fully answered because reference data on floc settling velocity and floc size are usually not available from other (reliable) sources. The effect of the tube diameter on the settling velocities found in settling tubes was demonstrated by Puls & Kühl (1986): large diameters give more reliable results. Water circulation has often been observed in settling tubes, directly through the tube wall or with video. Taking the movement of the finest flocs as representing the water movement, the relative settling velocity of the larger flocs can be regarded as an approximation of the real settling velocity. Floc break-up or additional flocculation in the instrument can be estimated when at the same time floc size is measured *in-situ* by other instruments, giving maximum size and/or mean/median size independently. This was done during an intercalibration experiment in the Elbe estuary in 1993: the INSSEV and the VIS gave maximum and median/mean floc sizes of the same order as the NIOZ *in-situ* camera which photographed particles *in-situ* with a minimum of disturbance.

To determine the settling velocities of suspended sediment, field settling tubes have been widely used in engineering practice: both bottom withdrawal and pipette withdrawal tubes have been applied in a number of estuaries.

A large number of such measurements have been performed by Hydraulics Research in the Thames Estuary over several years.

The median settling velocity was generally found to be a logaritmic function of the suspended sediment concentration. Figure 16 contains the results of such measurements from several estuaries. They show an exponential increase of the settling velocities with the suspended sediment concentration. The following formula has been generally accepted for the settling velocity:

$$w_s = K.C^m \, ,$$

where  $C$ = suspended sediment concentration
$K$ = empirical constant
$m$ = empirical exponent

The exponent m seems to vary over a wide range: 0.6 to 3.0. Little is known about the physical, chemical, biological or instrumental background of the differences between the results from the various estuaries.

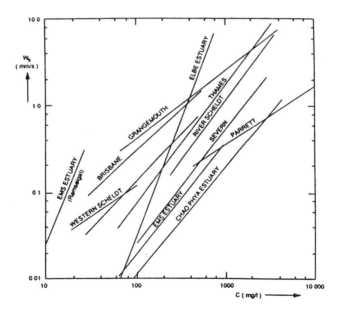

Figure 16. General relation between median settling velocity and particle concentration in different estuaries.

When the suspended sediment concentrations are higher than several grams per liter, the formula has to be modified because of hindered settling. This resuls in a decrease of $w_s$ with increasing concentration. In this case the aggregates are so closely spaced that the fluid has to be forced through the interstitial spaces between the aggregates. In that case the equation of Richardson & Zaki (1954) has been found to be applicable:

$$w_s = w_0(1 - k_2 C)^\beta ,$$

where  $k_2 =$  coefficient, depending on sediment composition
$\beta = 5$

Although these relations are widely accepted, there are also cases that show no correlation between the *in situ* settling velocity and the suspended sediment concentration. An example are the measurements in Atchafalaya Bay (Teeter, 1983; Heltzel & Teeter, 1987), the suspended sediment concentrations ranging from 13 to 560 mg/l, and the median settling velocities from 0.0005 to 0.291 mm/s. The suspended sediment had relatively low cohesiveness and was dominated by silts made up of quartz. The salinity was generally very low: smaller than a few parts per thousand.


## 4. FLOC DENSITY

The settling velocity of fine flocculated sediments is a function of both the size of the flocs and their density. For small spheres settling in the viscous Reynolds number regime these three variables can be related by Stokes' Law, and provided two are measured, the third can be calculated. Rarely are all three measured simultaneously on the same floc. Generally size and settling velocity are measured, and the effective density of 'Stokes equivalent sphere' is calculated. The effective density is the difference between the floc bulk density and that of the water. This has been also termed the excess density, the differential density and the density contract. It is important to know the density of the settling flocs in order to calculate the settling mass flux. This is the product of the settling velocity and the concentration. Though the mass concentration can be measured directly by other means, it is necessary to measure the spectral distribution of floc mass for assessment of differential settling (Lick *et al*, 1993) and for the correct calculation of the settling flux (Mehta & Lott, 1987) since the largest flocs contain the majority of the mass.

Figure 17 shows the results for one sample taken in the Elbe Estuary of settling velocity against effective diameter using the INSSEV instrument (Fennessy *et al*, 1994a). The diagonal lines are of constant effective density. The results cross the lines of constant density for spheres. This could be due to an actual increase in effective density with size, a change in the form of the floc and its roughness, or flow through the flocs (Gibbs, 1985).

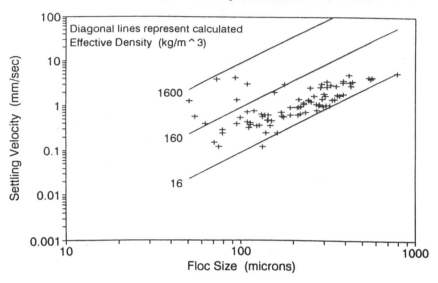

ELBE 11 June 1993 @ HW-4.44 Ebb Flow
SPM 290 mg/l Salinity: Amb 2.4 Col 8.8

Figure 17. Settling velocity against floc size for flocs observed in the Elbe estuary. The diagonal lines represent the calculated effective density (kg m$^{-3}$).

Observations have shown that the flocs are often by no means spherical. Some of the large flocs have a very irregular shape, and are often joined into 'stringers' by threads of biological origin (*eg*, Eisma, 1986; Fennessy *et al*, 1994a). However, by direct measurement Alldredge & Gotschalk (1988) found that organic aggregates whose shapes varied from spherical, including long comets, had a settling values very similar to those of nearly spherical particles. Similarly Gibbs (1985) showed that flocs from Chesapeake Bay had on the average a 1.6:1.0 length to width ratio which gave a spherical value of 0.91. Thus, the spherical approximation would appear to be reasonable under most circumstances, and the diameter normal to the direction of fall is taken as the effective diameter.

Direct measurements have shown that the effective density decreases with floc size (Gibbs, 1985; Aldredge & Gotschalk, 1988) according to : $\Delta\varphi = cD^{-n}$, where $n \approx 1\text{-}2$, though there is considerable variation in the constant c. This relationship has been used in correcting settling velocities obtained in an Owen Tube to obtain the mass distribution (Puls *et al*, 1988). Despite the decreasing density with floc size, the larger flocs have a greater settling velocity, on average, than the smaller ones, and consequently give a greater contribution to the settling mass flux.

From Fig. 17 it is apparent that a very large range of flocs have the same settling velocity. For instance, a range of particles from about 50 μm to 500 μm can have settling velocities of 1 mms$^{-1}$. Additionally, at the small sizes the effective density and settling velocity have almost a two order of magnitude range. The range in effective density is from about 30 kgm$^{-3}$, which is the same as that assumed by McCave (1975) for organic matter, and the maximum density of organic flocs obtained by Aldredge & Gotschalk (1988), to 1600 kgm$^{-3}$, which is that for quartz particles. In contrast, the effective density range at large floc size is much more uniform.

Thus it appears that at small sizes, the zero order flocs (Krone, 1962) or microflocs, range in composition from low density, primarily organic microflocs, to dominantly inorganic ones. The large size flocs which are presumably a mixture of this range of microfloc types would consequently tend to be more uniform in their effective density.

## 5. COMPARISON OF TECHNIQUES

A comparison of some of the techniques and instruments described above was carried out in June 1993 in the Elbe Estuary. Sampling was carried out hourly from an anchored pontoon at a depth of 9 m in the turbidity maximum. The experiment and the results have been described in Dyer et all (1995).

The main result is shown in Fig. 18 as a comparison between the settling velocities obtained by the systems at different concentrations. there is a considerable divergence, though this may be the result of small scale spatial and temporal patchiness in the turbidity. The Owen Tubes generally gave settling velocities at least an order of magnitude less than the direct VIS and INSSEV measurements. The degree of variation at low concentration was large, and this may have been the result of the handling techniques. A well controlled sampling protocol appeared to give reproducible results, and may give acceptible data on the median settling velocities. QUISSET and FIPIWITU appeared to break up flocs to a greater extent, though the fact that they include thermal insulation may contribute to the low settling velocities.

Comparison of *in-situ* systems is difficult because of the inherent natural variation. The direct video techniques give absolute settling velocities, rather than being derived from concentration changes. Consequently they would seem to have considerable potential and should be further developed.

# Figure 18a

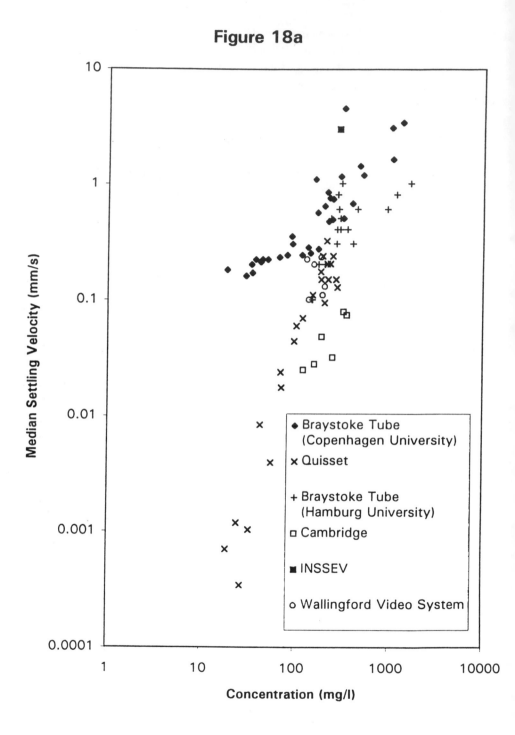

Figure 18. Relation between median settling velocity (in mm.s[-1]) and particle concentration obtained by different instruments in the Elbe estuary in June 1993.

# Figure 18b

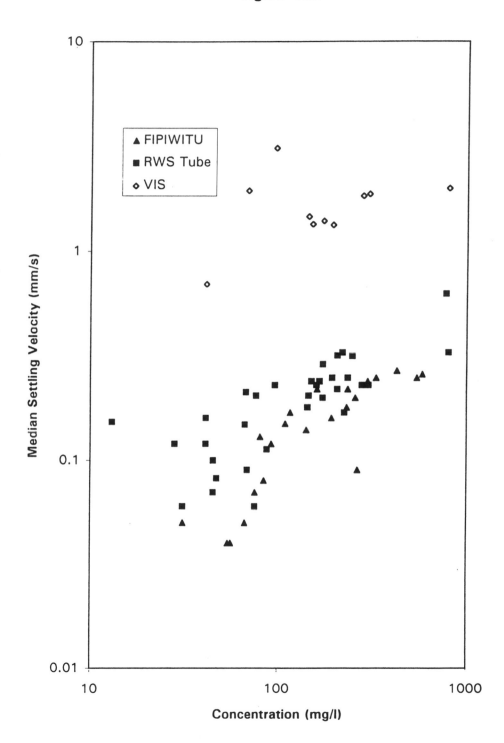

# 6. CONCLUSIONS

Seventeen instruments and methods to determine *in-situ* settling velocity of fine-grained particles are described and the results discussed. The median settling velocity is generally a logarithmic function of the suspended matter concentration. The settling velocity of individual particles is a function of floc size, density and shape. A very large size range of flocs have the same settling velocity but the large flocs tend to have more uniform settling velocities than the small sizes. Because of the variations in floc size (and presumably density and shape), the use of a single relation for size (or density) versus settling velocity is not likely to be realistic and may affect the estimation of mass settling flux. The changing content of macroflocs and microflocs may have implications for the measurement of concentrations using optical techniques calibrated in the laboratory where the floc range is constant. An *in-situ* comparison of some instruments and methods in the Elbe estuary in June 1993 indicated a considerable divergence in the relation between settling velocity and concentration, in particular at low concentrations. The comparison of *in-situ* methods is difficult because of the inherent natural variations.

# REFERENCES

ALLDREDGE, A.L. & C. GOTSCHALK, 1988. In-situ settling behaviour of marine snow. *Limnol. Oceanog.* **33**, 339-351.

ALLERSMA, E., 1980. Mud in estuaries and along coasts. Proc. *Int. Symp. Riv. Sedim.* Beijing, China: 663-685 (additional remarks: 1285-1289).

BALE, A.J. & A.W. MORRIS, 1987. In-situ measurement of particle size in estuarine waters. *Estuar. Coastal Shelf Sci.* **24**: 253-263.

BARTZ, R., J.R.V. ZANEVELD, I.N. MCCAVE & A.R.M. NOWELL, 1985 ROST and BEAST: Devices for in situ measurements of particle settling velocity. *Mar. Geol.* **66**: 381-395.

BURT, T.N., 1986. Field settling velocities of estuary muds. In: A.J. MEHTA. Estuarine Studies, **14**, *Springer Verlag*, Berlin, Heidelberg, New York: 126-150.

BURT, T.N., M.P. DEARNALEY, M.C. OCKENDEN & E.A. DELO, 1991. H.R. Wallingford's capabilities in cohesive sediment research. *Hydraulics Research Ltd, Wallingford:* 9 pp.

DEUSER, W.G., 1984. Seasonality of particulate fluxes in the Ocean's Interior. In: Proc.Workshop *Global Ocean Flux Study*, Woodshole. NSF, Natl. Acad. Press, Washington, 221-236.

DEUSER, W.G., 1986. Seasonal and interannual variations in deep-water particle fluxes in the Sargasso Sea and their relation to surface hydrography. *Deep Sea Res.* **33**: 225-246.

DEUSER, W.G. & E.H. ROSS, 1980. Seasonal change in the flux of organic carbon to the deep Sargasso Sea. *Nature* **283**: 364-365.

DYER, K.R., J. CORNELISSE, M. DEARNALEY, C. JAGS, J. KAPPENBURG, I.N. MCCAVE, M. PEJRUP, W. PULS, W. VAN LEUSSEN & K. WOLFSTEIN, in press. A comparison of in-situ techniques for estuarine floc settling velocity measurements. *Neth. J. Sea Res.*

EISMA, D., 1986. Flocculation and deflocculation of suspended matter in estuaries. *Neth. J. Sea Res.* **20**: 183-199.

FENNESSY, M.J., K.R. DYER & D.A. HUNTLEY, 1994a. INSSEV: an instrument to measure the size and settling velocity of flocs in-situ. *Mar. Geol.* **117**: 107-117.

FENNESSY, M.J., K.R. DYER & D.A. HUNTLEY, 1994b. Size and settling velocity distributions of flocs in the Tamar Estuary during a tidal cycle. *Neth. J. Aquatic Ecology* **28**: 275-282.

GIBBS, R.J., 1985. Estuarine flocs: their size, settling velocity and density. *J. Geoph. Res.* **90**: 3249-3251.

HELTZEL, S.B. & A.M. TEETER, 1987. Settling of cohesive sediments. In: N.C. KRAUS. *Coastal Sediments* 1987. ASCE: 63-70.

JAGO, C.F., A.J. BALE, M.O. GREEN, M.J. HOWARTH, S.E. JONES, I.N. MCCAVE, G.E. MILLWARD, A.W. MORRIS, A.A. ROWDEN & J.J. WILLIAMS, 1994. Resuspension processes and seston dynamics, southern North Sea. In: H.CHARNOCK, K.R.DYER, J.M. HUTHNANCE, P.S. LISS, J.H. SIMPSON & P.B. TETT. Understanding the North Sea System. *The Royal Society, Chapman & Hall, London:* 97-114.

JONES, S.E., C.F. JAGO, A.J. BALE, D. CHAPMAN, R. HOWLAND & J. JACKSON, 1993. Dynamics of suspended particulate matter in the southern North Sea II. Seasonally stratified waters. *Cont.Shelf Res.* (submitted).

KINEKE, G. & R. STERNBERG, 1987. A new instrument for measuring settling velocities in situ (abstract) *EOS Trans.: AGU* **68**: 1312.

KINEKE, G. C., R.W. STERNBERG & R. JOHNSON, 1989. A new instrument for measuring settling velocities in situ. *Mar Geol.* **90**, 149-158.

KRANCK, K., 1981. Particulate matter grain-size characteristics and flocculation in a partially mixed estuary. *Sedimentology* **28**: 107-114.

KRONE, R.B., 1962. Flume studies of the transport of sediment in estuarial shoaling processes. *Hydraul. Eng. & Sanitary Eng. Lab.*, University of California, Berkeley, Calif.: 110 pp.

LICK, W., H. HUANG & R. JEPSEN, 1993. Flocculation of fine-grained sediments due to differential settling. *J. Geoph. Res.* **98**: 10279-10288.

MCCAVE, I.N., 1975. Vertical flux of particles in the ocean. *Deep Sea Res.* **22**: 491-502.

MCCAVE, I.N. & T.F. GROSS, 1992. In-situ measurements of particle settling velocity in the deep sea. *Mar. Geol.* **99**: 403-411.

MCCAVE, I.N. & J.K. ATHERTON, 1994. Determination of particle settling velocities in the North Sea by a Quasi In Situ Settling Tube (QUISSET). *Mar. Geol.* (submitted)

MEHTA, A.J. & J.W. LOTT, 1987. Sorting of fine sediment during deposition. *Proc. Spec. Conf. Advances Understanding of Coastal Sediment Processes, Amer. Soc Civil Eng.*: 348-362.

NEDECO, 1965. A study on the siltation of the Bangkok Port Channel. Vol. II. *The Field Investigation*. Den Haag: 474 pp.

OWEN, M.W., 1971. The effect of turbulence on the settling velocities of silt flocs. *Proc 14th congres of IAHR*, Paris **4**: D4-1 - D4-5

OWEN, M.W., 1976. Determination of the settling velocities of cohesive muds. *Hydraulics Research* Report, No. IT 161

ÖZTURGUT, E. & J.W. LAVELLE, 1986. Settling analysis of fine sediment in salt water at concentrations low enough to preclude flocculation. *Mar. Geol.* **69**: 353-362.

PEJRUP, M., 1988. Suspended sediment transport across a tidal flat. *Mar. Geol.* **82**: 187-198.

PULS, W. & H. KÜHL 1986. Field measurements of settling velocities of estuarine flocs. In: WANG SY, H.W. SHEN & L.Z. DING. Proc. Third Int. Symp. River Sed. Univ. Mississippi: 525-536.

PULS, W., H. KÜHL & K. HEYMANN, 1988. Settling velocity of mud flocs: results of field measurements in the Elbe and Weser Estuaries. In: J. DRONKERS & W VAN LEUSSEN. Physical Processes in Estuaries, Springer-Verlag, Berlin: 404-424.

PULS, W., H. KÜHL, 1995. A settling tube for measuring suspended matter settling velocities in the German Bight. *Arch. f. Hydrobiol.* (Submitted).

RICHARDSON, J.F. & W.N. ZAKI, 1954. The sedimentation of suspensions of uniform spheres under conditions of viscous flow. *Chem. Eng. Sci.* **3**: 65-73.

SPINRAD, R.W., R. BARTZ & J.C. KITCHEN, 1989. In situ measurements of marine particle settling velocity and size distributions using the Remote Optical Settling Tube. *J. Geophys. Res.* **94**: 931-938.

STERNBERG, R.W., 1989. Instrumentation for estuarine research. *J. Geoph. Res.* **94**: C 10, 14.289-14.301.

TEETER, A., 1983. Investigations on Atchafalaya Bay sediments. Proc. Conf. on frontiers *Hydr. Eng. ASCE,* Cambridge, Mass. Aug. 1983: 85-90.

VAN LEUSSEN, W., 1994. Estuarine Macroflocs and their Role in Fine-grained Sediment Transport. *Ph.D. Thesis, University Utrecht*: 488 pp.

VAN LEUSSEN, W. & J.M. CORNELISSE, 1993a. The determination of the sizes and settling velocities of estuarine flocs by an underwater video system. *Neth. J. Sea Res.* **31**(3): 231-241.

VAN LEUSSEN, W. & J.M. CORNELISSE, 1993b. The role of large aggregates in estuarine fine-grained sediment dynamics. In: A.J. METHA. Nearshore and Estuarine Cohesive Sediment Transport. Coastal and Estuarine Studies No. **42**. *American Geophysical Union, Washington DC:* 75-91.

VAN RIJN, L.C. & L.E.A. NIENHUIS, 1985. In situ determination of fall velocity of suspended sediment. Proc. 21st Congress *IAHR*, Melbourne, Australia **4**: 144-148.

VANONI, V.A. ed., 1975. Sedimentation engineering *ASCE Task Comm.* New York: 745 pp.

ZANEVELD, J.R.V., R.W. SPINRAD & R. BARTZ, 1982. An optical settling tube for the determination of particle size distributions. *Mar. Geol.* **49**: 357-376.

# 3  The Kolmogorov microscale as a limiting value for the floc sizes of suspended fine-grained sediments in estuaries

W. VAN LEUSSEN

*Rijkswaterstaat, National Institute for Coastal and Marine Management/RIKZ, The Netherlands*

## ABSTRACT

Laboratory experiments and field measurements were carried out to obtain more insight into the effect of turbulence on the floc sizes of suspended fine-grained sediment in estuaries, particularly the sizes of the so-called macroflocs, which are loosely bound and very fragile, but play a significant role in the fine-grained sediment dynamics in estuaries.     In the interaction of flocs and turbulence, the turbulent eddies with dimensions in the same order of the flocs are most important. This means that the smallest turbulent eddies dominate the floc dynamics. Both laboratory experiments and field measurements demonstrate that the Kolmogorov microscale of turbulence is an adequate measure of the ultimate size of the aggregates. During the field measurements in the Ems Estuary the Kolmogorov microscales of turbulence varied in the range of 200 $\mu$m to 1 mm. In the upper part of the water column they are larger than in the lower part. In the near vicinity of the bed the smallest eddies had sizes in the range of 10 to 50 $\mu$m.

The importance of the Kolmogorov microscales of turbulence for the floc sizes and settling velocities of the macroflocs is demonstrated from results from recent turbulence research with advanced numerical simulations, where it is found that the strongest vorticity is organized in very elongated thin tubes, the so-called "worms".

Because the suspended flocs are moving continuously through zones of relatively low and high turbulence intensity, the flocs participate in a continuous process of aggregation and breakup. The distribution of these zones and their relative turbulence intensity are quantified.

## INTRODUCTION

Suspended fine-grained sediment in estuaries and coastal waters is dominantly transported in the form of flocs. It is usual to discriminate between the so-called microflocs and macroflocs (Eisma, 1986). The macroflocs have sizes up to several mm, are loosely bound and very fragile. During sampling they are often destroyed and broken up into so-called microflocs. The microflocs are more strongly bound by sticky material, mucopolysaccharides produced by organsisms, bacteria, algae and higher plants. They have sizes less than 100 $\mu$m, and are, together with the single mineral particles, the basic building units of the macroflocs.

*Cohesive Sediments.*
Edited by Neville Burt, Reg Parker and Jacqueline Watts
© 1997 John Wiley & Sons Ltd.

The properties of the flocs, such as for example the size, density and strength, are dependent on a number of parameters, including the suspended sediment concentration, turbulence, salinity, temperature, suspension residence times, the physico-chemical properties of the water/sediment mixture, coatings on the particle surfaces, dissolved organic substances, biological organisms, etc. A review is given by Van Leussen (1988).

In the light of the role of the Kolmogorov microscales -the subject of the present study- special attention will be given to the effect of turbulence on the sizes and settling velocities of the flocs. Turbulence is one of the basic parameters in the formation of mud flocs. It affects the sizes and corresponding settling velocities in two opposing ways:

1. Turbulence increases the number of collisions between the particles, thus resulting in larger flocs and larger settling velocities.
2. Turbulence results in turbulent shear stresses, thus limiting the floc size and settling velocities.

In nature, estuarine conditions and fine-grained sediment phenomena are extremely variable. a number of processes and influence factors are acting together, each on their own time scale. To unravel the relevant processes continuous measurements are needed over a long period of time. Even then it is not easy to elucidate the underlying mechanisms. In laboratory experiments the mechanisms can be studied under controlled conditions. It is possible to vary some parameter while keeping others constant and circumstances can be varied over a much wider range than that which occurs in nature.

However one should be aware of possible scale effects, through which the processes in the laboratory may differ significantly from those in nature. Therefore laboratory experiments should be carried out in association with field measurements. In fact both methods have their advantages and performing both laboratory and field measurements in a complementary way can have significant benefits.

To investigate the effect of turbulence systematically by *laboratory experiments* under controlled conditions, a special settling column has been built. Details of this facility will be given in Section 3. To obtain sufficient residence times of the particles in the settling column, before settling to the bottom, the height of the column is 4.25 m. The uniqueness of this settling column is that it is equipped with a turbulence generating facility, through which the natural turbulence properties, which are expected to govern the sizes of the macroflocs, can be simulated.

In 1990 detailed *field measurements* were carried out at five locations in the Ems Estuary. From the current velocity profiles estimates were made for the tubulent energy dissipation and the Kolmogorov microscale over a complete tidal cycle. An underwater video system VIS was applied to obtain data for the sizes and settling velocities of the suspended fine-grained sediment flocs.

After discussing both the laboratory experiments and the field measurements, the results will be compared and analyzed in terms of turbulence as a limiting values for the floc size. Special attention will be given to the theoretical background of the observed phenomena.

# FINE-GRAINED SEDIMENT PARTICLES IN A TURBULENT FLOW FIELD

Turbulence is a fluctuating velocity field, both in time and space, and should therefore be described in statistical terms. To describe the turbulent flow field, the instantaneous value of the flow velocity is usually divided into a mean value U and a turbulent fluctuating part u : $U=U+u$. In this case the root mean square value of the turbulent fluctuations u' is a measure of the turbulent intensity: $u' = \sqrt{u^2}$.

To describe the structure of the turbulent flow in a quantitive way, various scales of turbulence may be considered. For turbulent flow in an estuary, the largest eddies will be in the order of the water depth. These large eddies are stretched by local velocity gradients into smaller eddies. This process continues down to smaller scale, until the dimensions of the eddies are so small that the contained energy will be dissipated by viscosity (Tennekes and Lumley 1972; Hinze 1975). The energy of the turbulence is obtained from the mean flow field by the largest eddies and, by vortex stretching, transferred to smaller eddies.

The large energy-containing eddies are anisotropic, and in an estuary they are influenced in a complex way by topographic features, density gradients, tidal oscillations, etc. The smallest energy-dissipating eddies tend to be isotropic, and don't depend on the structure of the larger eddies. Their scale depends only on the energy flux and the viscosity.

From dimensional reasoning the size of the smallest eddies is of the order of the Kolmogorov microscale ($\eta$):

$$\eta = (\frac{\nu^3}{\varepsilon})^{\frac{1}{4}} \tag{1}$$

,where  $\nu$ = kinematic viscosity (m².s⁻¹)
$\varepsilon$ = energy dissipation per unit mass (Nm.s⁻¹.kg⁻¹)

To give an impression of the smallest eddies in a turbulent channel flow, an example is given in Fig.1. The water depth is 10 m, and calculations have been made for mean velocities of 0.25, 0.5, 1.0 and 1.5 m s⁻¹. Fig.1 shows that in this example the smallest eddies have dimensions of the order of 30 to 100 $\mu$m at the bottom and up to more than 1mm in the upper layers of the flow.

Concerning the interactions between fine sediment particles and the turbulent flow, it may be expected that these are most effective at scales where the turbulent eddies have sizes comparable with those of the sediment particles. Assuming particle sizes to be in the range a few microns (primary particles) to a few mm (large fragile aggregates), it can be shown from Fig.1 that the smallest eddies will be of utmost importance. The largest eddies are especially responsible for the mixing and dispersion of the sediment particles in the water column. The study of the interactions of particles in fluids at very small length scales, say 0.1 to 10 $\mu$m, is called microhydrodynamics (see review by Batchelor, 1976). In this theory aspects are included that are generally ignored at larger scales.

Turbulence contributes to *floc growth* because of the increase in the number of particle collisions per unit time. When $N_{ij}$ is the number of collisions occurring per unit

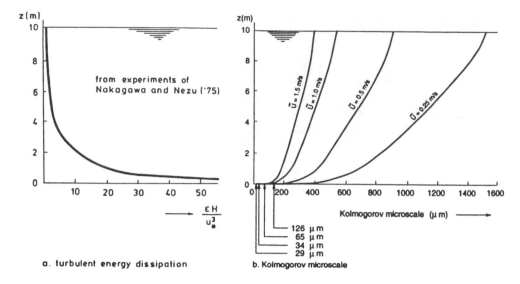

**Figure 1.** The smallest turbulent eddies in a turbulent channel kflow. Example of an open channel flow: water depth H=10 m, mean velocity U=0.25, 0.5, 1.0 and 1.5 m s⁻¹. (a) The dimensionless turbulent energy dissipation as a function of depth (from experiments of Nakagawa and Nezu (1975)). (b) Dimensions of the smallest eddies (Kolmogorov microscales); representative values for the corresponding sizes at the bottom are also indicated.

time and per unit volume between two classes of particles of diameters $d_i$ and $d_j$, the collision frequency can be written as:

$$N_{ij} = K(d_i, d_j) n_i n_j ,$$ (2)

where $n_i$ and $n_j$ are the number concentrations (volume⁻¹) of the i- and j-particles of diameter $d_i$ and $d_j$. $K(d_i, d_j)$ is the collision frequency function, which is specific for a particular mechanism and depends on the sizes of the colliding particles and on properties of the system such as temperature and pressure (cf. Friedlander 1977, Van Leussen 1988).

In this simplified model it is assumed that at a collision of two particles, they coalesce to form a third one whose volume is equal to the sum of the original two. The process where each collision leads to aggregation and the particles have been completely destabilized is called rapid flocculation. However, suspensions can exist which are still partially stable and only a fraction of the collisions is effective. Therefore a stability factor $\alpha$ is introduced, giving the percentage of effective collisions. The frequency of effective collisions will then be

$$N_{ij} = \alpha K(d_i, d_j) n_i n_j$$ (3)

where $\alpha$ takes on values between 0 and 1. A completely destabilized suspension has a stability factor of 1, which corresponds to the rapid flocculation case, whereas stable suspensions are characterized by $\alpha < < 1$.

In the formation and breakup of flocs in a turbulent flow the velocity gradient is one of the basic parameters. Generally, however, the velocity gradients are not uniform and constant, and in turbulent flows will fluctuate rapidly both in space and time. Moreover such turbulent velocity gradients are superposed upon the mean velocity gradients, which makes it difficult to determine actual gradients in real flow situations. Therefore, Camp and Stein (1943) defined an "absolute velocity gradient" G as the square root of the energy dissipation at a point divided by the fluid viscosity:

$$G = \sqrt{\frac{W}{\mu}} = \sqrt{\frac{\varepsilon}{\nu}} \qquad , \qquad (4)$$

where   $G$ = root mean square velocity gradient $(s^{-1})$
$W$ = total power dissipated per unit volume and time $(N\ m^{-2}\ s^{-1})$
$\mu$ = absolute viscosity of the fluid $(kg\ m^{-1}\ s^{-1})$
$\nu$ = kinematic viscosity of the fluid $(m^2\ s^{-1})$
$\varepsilon$ = total energy dissipation per unit mass and time $(Nm\ s^{-1}\ kg^{-1})$

Compared to our present understanding of floc formation by collisions of more or less destabilized fine particles through the perikinetic and orthokinetic mechanisms, our understanding  of floc breakup in viscous and turbulent flows is much more qualitative and speculative. Most studies on flocculation have focused on particle aggregation without considering floc breakage (Pandya and Spielman 1983). Proper characterization of floc disruption is an important problem, because in estuaries the floc growth limiting properties of a turbulent flow are fundamental in our understanding of the sizes of the suspended fragile macroflocs in estuaries.

An important factor in floc breakup is the ratio of the diameter of the floc to the size of the dissipating eddies in the turbulent flow (Kolmogorov microscale $\eta$). For $d_f > > \eta$ inertial effects are dominant, while for $d_f < < \eta$ viscous effects dominate. In the case of isotropic turbulence the floc diameter is shown to be

$$d_f :: G^{-m} , \qquad (5)$$

where the exponent m varies between 0.5 and 2, dependent on $d_f > > \eta$, or $d_f < < \eta$, breakup mechanism (surface erosion, breakage of filaments, breakage into fragments), etc. (cf. Thomas 1964, Parker et al 1972, Tambo and Hozumi 1979, Hunt 1986).

The same formula holds for the maximum floc size, being the upper size limit that can just withstand the turbulent hydrodynamic forces. For $d \approx \eta$, corresponding with floc sizes in the order of the Kolmogorov microscale, Tomi and Bagster (1978) combined viscous and inertial effects, arriving at $m=1$:

$$d_{max} :: G^{-1} \qquad (6)$$

It should be noted that almost all this research has been performed in laboratory stirred tanks, mostly with addition of aluminium sulphate, organic polymers, etc. Comparable research on floc breakup in estuaries is lacking.

Much more research has been conducted on the breakup of liquid drops in immiscible fluids (e.g. see Levich (1962), p. 454). However there is a difference between fluid droplets and suspended aggregates. The aggregates do not have a property exactly e-quivalent to surface tension and also the aggregate strength has to be included in the analysis. Levich concluded that for droplets of size d < $\eta$, in the Stokes regime of fluid motion, fragmentation does not take place. This corresponds with measurements of organic aggregates in the Ems Estuary, where Eisma (1986) determined that their maximum sizes were of the order of the smallest turbulent eddies. However no turbulence or current velocity measurements were available, from which an estimate of the Kolmogorov microscales could be made for estuarine circumstances.


## LABORATORY EXPERIMENTS

Laboratory experiments to investigate the effect of turbulence on the growth and breakup of mud flocs were carried out in a 4.25 m settling column with an inner diameter of 0.29 m. To avoid the influence of density currents as a result of temperature differences, the column was placed in a thermostatically controlled water bath. Samples from the column were taken at eleven intake points.

The turbulence inside the settling column was generated by an oscillating grid. The intensity of the turbulence is controlled by the frequency and amplitude of the oscillations and the form of the grid. The mesh-width is 0.075 m. The grids are coated with a teflon layer to prevent corrosion during tests in saline water. To verify the settling column and to determine the properties of the oscillating grid, including the vertical exchange coefficients, experiments were conducted with well-sorted non-cohesive material. The results of these experiments are given in Van Leussen (1986).

To determine the distribution of floc sizes and settling velocities, the settling column is equipped with a cuvet at the bottom, through which the settling flocs can be recorded by a film or video camera. The floc size distribution is obtained through the cuvet by a Malvern 2600 particle sizer. (Fraunhofer diffraction). The cuvet has internal dimensions of 25x25 mm and ends 80 mm above the bottom in the settling column, to avoid inflow of deposited mud from the bottom of the settling column.

Before starting the experiment the water and the fine-grained sediment are recirculated for about 20 minutes at a grid frequency of 4.0 Hz. In this way a homogeneous concentration distribution was obtained in the settling column without aggregates. At time t=0 the pump was stopped and the grid was set oscillating at the desired frequency. Grid frequencies could be chosen in a range 0.01 to 5.0 Hz.

Experiments in the settling column were conducted with kaolinite and a natural mud. Starting with kaolinite has the advantage that the results are not influenced by organic matter and the fine-grained sediment has reproducible properties. Moreover comparisons could be made with results of various other cohesive sediment research programmes, in which kaolinite has been applied (cf. Mehta 1973, Mehta and Partheniades 1975). The natural mud was obtained from the Ems Harbour at the mouth of the Ems Estuary in the northern part of the Netherlands. Before applying this sediment in the settling column, it was sieved through a 1 mm sieve. In the case of natural mud the water was obtained from the same site. The natural mud was stored in the laboratory in a dark room at a temperature of 4 to 6 °C.

A characteristic picture of the development of the floc size distribution during the flocculation process within the settling column is illustrated in Fig.2, obtained from the Malvern particle size measurements at the bottom of the settling column (pilot experiment on Ems mud; grid frequency 0.1 Hz). The experiment starts with all the particles in the smallest size class range ($< 19\mu m$). later, flocs develop in a size range 50 to 200 $\mu m$, increasing to 700 $\mu m$ over time. The figure clearly shows a bimodal distribution, consisting of fine primary particles and much larger flocs.

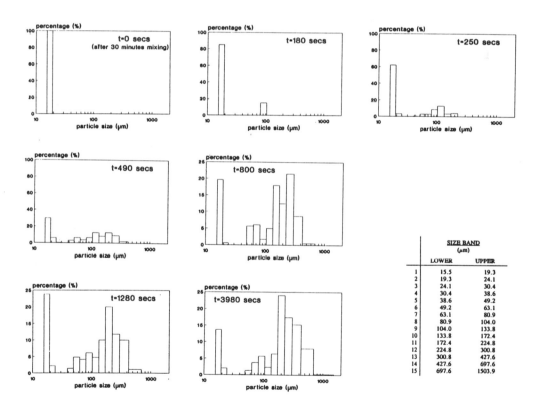

**Figure 2.** Floc size distribution during a settling column experiment, measured by a Malvern Particle Sizer. Experiment on Ems mud at a grid frequency of 0.1 Hz. $C_o = 1000$ mg/l.

Comparable experiments with non-cohesive materials have shown that in similar settling columns vertically a homogeneous distribution of vertical mixing can be obtained (Rouse, 1938; Dobbins, 1944). However detailed turbulence measurements within the column have revealed that regions of higher and lower turbulence intensity can be distinguished. This corresponds with experiences in stirred reactors. For stirred tanks with rotating blades, widely applied in chemical engineering, often three areas are discerned (cf. Cutter 1966, Tomi and Bagster 1978, Francois 1985, De Boer 1987): bulk zone, impeller zone and impeller tip zone in the near vicinity of the rotating blade. In this way a local velocity gradient G can be defined as

$$G_t = f_t . G \tag{7}$$

Table 1 gives some values for the velocity gradient intensity coefficient $f_t$ as found by Cutter (1966) and Francois (1985) with the usual geometry:

|  | volume percentage (%) | factor $f_t$ (eq. 7) (-) |
|---|---|---|
| bulk zone | 58 | 0.25 |
| impeller zone | 39.9 | 5.4 |
| impeller tip zone | 2.1 | 50 |

Table 1. Distribution of turbulent zones in a stirred reactor.

Within the settling column a similar distribution of turbulent zones can be imagined : a *grid tip zone* in the near vicinity of the grids, surrounded by a *grid zone*, and a *bulk zone* that covers the largest part of the settling column. No measurements within the settling column with sufficient detail are available . As a fair assumption the distribution of turbulent zones might be expected to conform to that in stirred reactors (Table 1).

In fact the suspended sediment particles and flocs are continuously moving through zones of relatively low and high turbulence intensity, which means that they participate in a continuous process of aggregation and breakup. After some time a quasi-equilibrium situation will result, where the growth of flocs is equal to the breakup of flocs. In this continuous process a number of time scales will be of importance: the time scale of vertical mixing, the time scale of vertical settling, the time scale of flocculation, and the time scale of floc breakup. In particular differences in these time scales between the values in the field and in the laboratory can result in different quasi-equilibrium situations for the floc size distribution.

For each zone the Kolmogorov microscales are given in Fig.3 as a function of the grid frequency. Also the mean (volume-averaged) values for $\eta$ are presented. This figure also shows the results of the settling column video recordings. These points correspond to the highest size class of floc sizes: the so-called macroflocs. The results show a good correspondence between the observed sizes of the macroflocs and the estimates for the Kolmogorov microscales bulk zone and mean values). At the very low grid frequency the dimensions of the smallest turbulent eddies were not attained due to settling out of the material (very low level of vertical mixing).

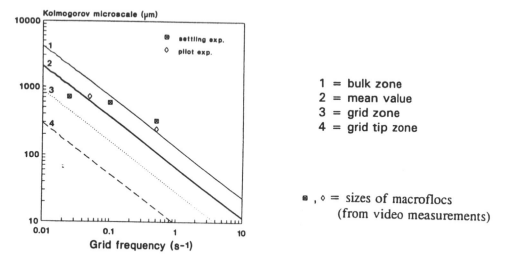

**Figure 3.** Kolmogorov microscale as a function of the grid frequency in the settling column with oscillating grids. Results of sizes of macroflocs, observed by a video system.

## FIELD MEASUREMENTS

An extensive series of field measurements were carried out in the Ems Estuary. The Ems Estuary is a partially mixed funnel-type, meso-tidal estuary in the northeastern part of the Netherlands at the border with Germany. The estuary is characterized by large areas of tidal flats, tidal channels, and salt marshes (Fig.4). The mean tidal range at the sea boundary is 2.25 m and the river discharge varies from 25 to 380 $m^3$ $s^{-1}$, with a annual mean value of 110 $m^3$ $s^{-1}$.

The measurements were conducted over a full tidal cycle at five location along the main channel of the estuary. Each half an hour vertical profiles were measured of salinity, current velocity and suspended sediment concentrations. The sizes and settling velocities of the suspended mud flocs were determined by the underwater video system VIS (Van Leussen and Cornelisse, 1993).

Similar to stirring reactor experiments, as discussed in the previous section, in real estuarine flow conditions also show regions of relatively low and high turbulence intensity. In turbulent open channel flows generally two main regions are distinguished: a region adjacent to the bottom, in which the flow is directly affected by the conditions at the bottom, called the inner region or wall region, and an outer region or core region, in which the flow is only indirectly affected by the bottom (Fig.5ᵃ). The wall region is confined to 10-20% of the water depth, while the core region covers 80-90%. About 80% of the total turbulent energy production takes place within the wall region.

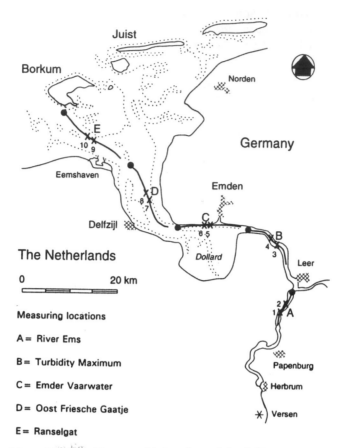

**Figure 4.** Map of the Ems Estuary with locations of the field measurements.

It appears that the deposition of aggregates is controlled by the stochastic turbulent processes in the zone near the bed. There the strongest shear and lift forces prevail and these in turn control the size and shear strength distributions of the suspended aggregates (Metha and Partheniades, 1975). In sedimentation tests in a carousel they observed that only flocs which are strong enough to withstand the maximum shearing force acting on it near the bed will settle to the bed. Flocs of lower strength will be broken up into two or more smaller units and be re-entrained into the suspension by the hydrodynamic lift forces. The broken aggregates will again participate in the aggregation process in the water column. This continuous process of aggregation and breakup is schematically presented in Fig.5[b]. It could be concluded that the properties of the suspended aggregates are significantly determined by the turbulence in the water in a narrow zone near the bed. Only aggregates that are strong enough to resist the bottom shear stresses will settle on the bed and be retained at the bed by cohesive bonds.

It can be concluded that in both the field experiment and the settling column, the fine-grained sediment particles and flocs are moving continuously through zones of high and low turbulence intensity. During these movements the flocs will grow or be broken up into smaller parts, dependent on the properties of the flocs and the physical conditions in the water. In this way the fine sediment particles participate in a continuous process of aggregation and breakup.

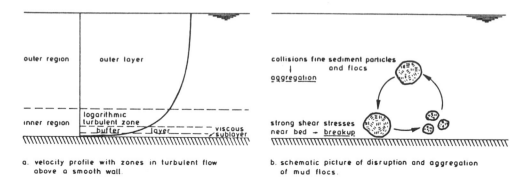

a. velocity profile with zones in turbulent flow above a smooth wall.

b. schematic picture of disruption and aggregation of mud flocs.

**Figure 5.** Schematic picture of the continuous process of aggregation and breakup of mud flocs in a turbulent flow (figures not to scale).

In estuarine flows the turbulent zones can be defined in a similar manner to that in the case of the settling column or stirred reactors. This is given in Table 2, where an outer zone, an inner zone, and a near bed zone are distinguished, as well as the very narrow zone of the viscous sublayer. Because of the much larger dimensions in nature, the outer areas are much larger, and the near bed zone relatively much smaller than in the case of the stirring experiments. It should be noticed that the value for the viscous sublayer applies for the case of a sort of "standard situation": water depth $H=10$ m and mean current velocity $U=1.0$ m s$^{-1}$. Smaller current velocities or water depths (e.g. flume experiments) will result in quite different values.

|  | volume percentage (%) | factor $f_t$ (eq. 7) (-) |
|---|---|---|
| outer zone | 80 | 0.4 |
| inner zone | 19.9 | 1.6 |
| near bed zone | 0.1 | 23 |
| viscous sublayer | - | about 150 *) |

*) situation $H=10$ m; $u=1.0$ m s$^{-1}$; Chezy$=60$ m$^{1/2}$ s$^{-1}$

**Table 2.** Distribution of turbulent zones in an open channel flow.

It should be remarked that although in principle the processes in both the field and laboratory are quite similar, differences in time scales, mentioned in the previous section, can result in significant scale effects.

According to formula (1) the Kolmogorov microscales are determined by the turbulent energy dissipation, which varies strongly over the water depth, especially in the vicinity of the bed, and is strongest in a zone close to the bed. This is illustrated in Fig.6, which shows some results of Nakagawa and Nezu (1975) concerning the structure of turbulence in open channel flow. Their experiments were conducted in a tilting flume 0.50m wide, 0.30m deep and 15m long. The turbulent velocities were measured by hot-film anemometers. The figure shows that in the fully developed turbulent flow the turbulent energy production and turbulent energy dissipation are nearly in balance with each other. However near the wall the production is larger than the dissipation, whereas near the water surface the turbulent energy dissipation is larger than the turbulent energy production. These differences are mainly compensated by transfer of turbulence through diffusion (cf. Hinze 1975).

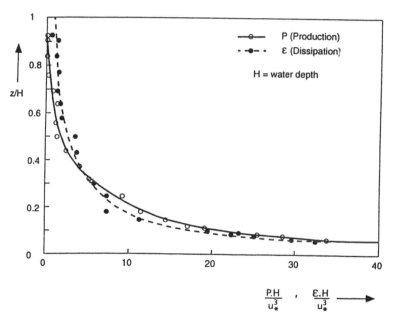

**Figure 6.** Distribution of turbulent energy production and dissipation in flowing water along a smooth wall. (data from Nakagawa and Nezu, 1975).

Starting from these results and applying equation (4), an estimate can be made for the rms velocity gradient G at a number of levels in the water column (Van Leussen, 1994):

$z/H \approx 0$      $: G = u_*^2/\nu$   (viscous sublayer)                         (8)

$z/H = 0.1H$     $: G = 4.34 \cdot (u_*^3/\nu.H)^{1/2}$                              (9a)

$z/H = 0.5H$     $: G = 1.57 \cdot (u_*^3/\nu.H)^{1/2}$                              (9b)

$z/H = 0.8H$     $: G = 1.04 \cdot (u_*^3/\nu.H)^{1/2}$                              (9c)

Corresponding values, and their variation over the tidal cycle, are presented in Fig.7 for measurements in the Ranselgat and in the Turbidity Maximum. The very high values of

G in the near vicinity of the bed, according to formula (8), are omitted in these figures. In that relatively very narrow zone near the bed, G-values of several thousands s$^{-1}$ were obtained.

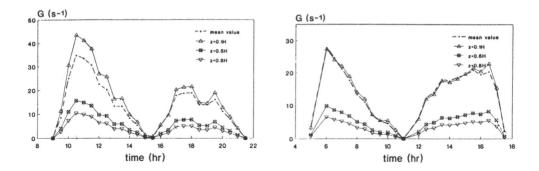

**Figure 7.** Variation of the rms velocity gradient G during a tidal cycle. (a) Ranselgat (27/06/90). (b) Turbidity Maximum (20/06/90).

The figures show somewhat higher G-values during flood than during ebb. Both figures illustrate the asymmetric path over the tidal cycle. The mean G-values are shown to vary in the range of 0 to 30 s$^{-1}$.

Starting from these G-values, estimates of the Kolmogorov microscale were made by application of the formulas (1) and (4). Fig.8 shows these vales both for the Ranselgat and Turbidity Maximum location. Assessments of the length scales in the near vicinity of the bed are also included in these figures. Under the influence of flowing water the Kolmogorov length scales in the water column are shown to vary in the range of 200 $\mu$m to 1 mm. In the upper half of the water column they are larger than in the lower part. In the near vicinity of the bed the smallest eddies have sizes in the range of 10 to 50 $\mu$m.

In Fig.8 these microscales are compared with the floc sizes, obtained from the VIS-measurements in the locations Ranselgat and Turbidity Maximum. Both figures show that during high current velocities the floc sizes are in the range of the Kolmogorov microscale. In the turbidity maximum area the floc sizes were somewhat smaller, possibly due to a reduced mixing as a result of some stratification, so that less resuspended fine sediment is transported to the upper part of the water column, as well as possibly due to a lower cohesivity and therefore slower floc growth rates. At slack tide the Kolmogorov microscale is strongly increasing as a result of decreased turbulence intensity. However, the large aggregates settle down in this situation owing to the reduced vertical mixing. This corresponds to the fact that at slack tides the macroflocs were scarcely observed, while during high current velocities there was an abundance of macroflocs.

With sufficient long residence times for the suspended flocs during the periods of relatively low turbulence, larger floc sizes are expected. This is confirmed by the results of *in situ* floc size measurements in the Zaire Estuary by Eisma (1993). In that case, in

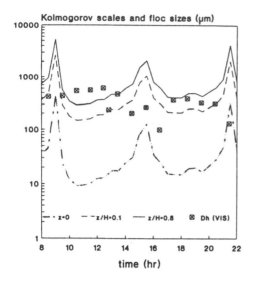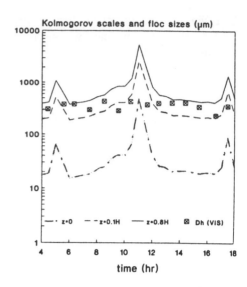

**Figure 8.** Comparison of the sizes of macroflocs, observed by VIS, with the various Kolmogorov microscales of turbulence. a) Ranselgat. b) Turbidity maximum.

the surface water the flocs did not grow above a size of about 800 $\mu$m. Only in the deeper canyon, with low turbulence level and relatively long residence times due to the large water depth (130m), did the flocs continue to grow to sizes of about 2000 $\mu$m.

The Kolmogorov microscales in the vicinity of the bed (Fig.8) show a good correspondence with the sizes of the more firmly bound microflocs. One could imagine that such flocs are cycled a large number of times through such zones of high turbulent energy dissipation. It could be hypothesized that biological processes in particular are responsible for their cohesivity and firmness, whereas the turbulence structure with its floc breakup properties is responsible for their ultimate size.

It may be concluded that turbulence is an important factor in aggregation dynamics in estuaries. In particular, their disruptive properties govern significantly the size distributions of the macroflocs. They are expected to be in a sort of quasi-equilibrium, passing continuously through zones of relatively high and low turbulence intensity. In nature the bottom boundary layer plays an important role as a zone of relatively high turbulence intensity. Both the laboratory experiments and the field measurements showed that the Kolomogorov microscale of turbulence is a good reference for the largest sizes of the fragile macroflocs.

## THEORETICAL ASPECTS

An overview of the behaviour of fine-grained sediment particles in a turbulent flow was given in Section 2. From the point of our interest in the Kolmogorov microscales, we have a special interest in the flow of the turbulent energy through the various scales of motion. The energy of the turbulence is obtained from the main flow field by the largest eddies and, by vortex stretching, transferred to smaller eddies. The small scale motions are statistically independent of the relatively slow large-scale turbulence of the mean flow. In this way the behaviour of the intermediate scales is governed by the transfer of energy from the large scales to the smallest eddies, where the energy will be dissipated due to the viscosity of the flow. This "cascade model" of Kolmogorov (1941) has proven to be a valuable model for describing the phenomena in a homogeneous isotropic turbulence field.

Because of conservation of angular momentum, the angular velocity in the direction of the stretching increases. So longer eddies, or vortices, result with reduced radii and increased velocity gradients. Discontinuities of velocity gradients would develop if there was no smoothing effect of viscosity. However when the turbulent energy is transferred to the smallest eddies, it will be dissipated by viscosity. From dimensional reasoning one can show that the size of these eddies is of the order of the Kolmogorov microscale $\eta$: see formula (1).

More recently it was found that the strongest vorticity is organized in very elongated thin tubes (Siggia, 1981; Vincent and Meneguzzi, 1993, 1994; Jiménez et al., 1993). In particular numerical experiments, solving the Navier Stokes equations with the most powerful computers, have contributed more insight into the properties of turbulence. From numerical simulations strong coherent elongated vortices (called "worms") were found to have thicknesses between the Kolmogorov microscale $\eta$ for dissipation and the Taylor microscale $\lambda$. Their lengths seem to be of the order of the integral scale for turbulence. Generally these strong vortices fill only a small percentage of the space.

The vorticity tubes seem to be produced by shear instabilities. These flow instabilities lead to the appearance of sheets, which tend to be bent and rolled-up, producing incipient vorticity tubes. During this process the vorticity is observed to increase (Vincent and Meneguzzi, 1994).

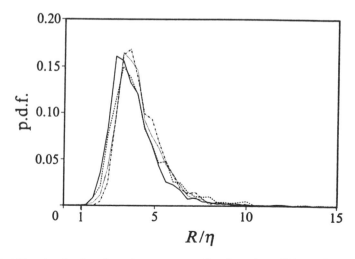

**Figure 9.** Probability density function of the worm radius R at four different Reynolds numbers $Re_\lambda = u'\lambda/\nu = 35, 61, 94$ and $168$. (Jiménez et al. 1993).

It may be expected that during this process suspended flocs are enclosed and rotate with the rotating eddy. The fragile flocs will be broken up in regions of maximum velocity gradient, say at the boundary between the rotating eddy and the surrounding fluid. In fact a continuous process is going on : entrapment of suspended flocs, rolling up within the developing vortex and ultimately floc breakup at the boundaries of the rotating vortex when it reaches its smallest size.

Fig.9 shows the probability density function of the worm radius for different Reynolds numbers (Jiménez et al., 1993). It shows the average radius of these intense vorticity structures is approximately $3\eta$-$5\eta$. Fig.9 also shows that the regions of higher vorticity reach minimum dimensions in the order of the Kolmogorov microscale $\eta$. At that level the highest velocity gradients occur, so in particular, at these dimensions, the extremely fragile aggregates are expected to be broken up. This corresponds to the finding that the Kolmogorov microscale could be seen as a good approximation for the limiting value of the sizes of the fragile aggregates.


## CONCLUSIONS

Turbulence is an important factor in aggregate dynamics in estuaries. It stimulates both the floc growth (number of particle collisions) and the floc breakup (turbulent shear stresses). Results from both laboratory experiments and field measurements have shown that, due to the very fragile nature of the aggregates, the final size distribution of the aggregates is governed by the disruptive properties of the tubulence field. Only when the floc growth is very slow, will this ultimate limit not be reached, such as in the case of low sediment concentrations, or a sediment with a low cohesivity (low effective collision efficiency factor).

To understand the effect of turbulence, a number of turbulent zones should be discerned of relatively high and low turbulence intensity. The zones and the local velocity gradients are summarized in the Tables 1 and 2. The fine sediment particles are continuously moving through zones of varying turbulence intensity, which means that they participate in a continuous process of aggregation and breakup. The ultimate sizes of the flocs, as mentioned before, result from an equilibrium between floc growth in zones of relatively low turbulence intensity and floc breakage in the areas of high turbulence.

The Kolmogorov microscales, being a measure of the smallest eddies in the turbulent flow, proved to be a good reference for the maxium sizes of the fragile macroflocs. During the field measurements over a full tidal cycle the Kolmogorov microscales varied in the range of $200\mu m$ to 1 mm. In the upper half of the water column they are larger than in the lower part. In the near vicinity of the bed the smallest eddies had sizes in the range of 10 to 50 $\mu m$ .

The Kolmogorov microscales in the near vicinity of the bed showed a good correspondence with the sizes of these more firmly bound microflocs. One could imagine that such flocs are modified through cycling a large number of times through such zones of high turbulent energy dissipation. It could be hypothesized that biological processes in particular are responsible for their cohesivity and firmness, whereas the turbulence structure with its floc breakup properties has an important effect on their ultimate size.

The correspondence between the Kolmogorov microscales for turbulence and the size of the fragile macroflocs can be explained from results of recent turbulence research using numerical modelling techniques, which have shown that the strongest vorticity is organized in very elongated thin tubes, so-called "worms", with radii down to the Kolmogorov microscale of turbulence.

It is hypothesized that the flocs are trapped by these "worms" during the process of vorticity production in tubes resulting from shear flow instabilities. Their ultimate size is determined by the dimensions of these rotating vortices when they reach their minimum diameter at the point of dissipation of their energy by viscosity.

# REFERENCES

Batchelor GK (1976) Developments in microhydrodynamics. In: W.T.Koiter (Ed.) Theoretical and Applied mechanics. North Holland Publishing Company. 33-55.

Camp TR, Stein PC (1943) Veloctiy garadients and internal work in fluid motion. Journal of the Boston Society of Civil Engineers, XXX: 219-237

Cutter LA (1966) Flow and turbulance in a stirred tank. Am. Inst. Chem. Eng. Journal, Vol. 12, No 1, 35-45.

De Boer GBJ (1987) Coagulation in Stirred Tanks. Dissertaion. University of Eindhoven. 162 pp.

Dobbins WE (1944) Effect of turbulence on sedimentation. Transactions ASCE, Vol. 109, 629-656.

Eisma D (1986) Flocculation and de-flocculation of suspended matter in estuaries. Neth. J. Sea Res., 20 (2/3): 183-199

Eisma D (1993) Flocculation and de-flocculation of suspended matter in estuaries. Arch. Hydrobiol./Suppl. Bd. 75 (Untersuch. Elbe-Aestuar 6):1-14.

Francois, RJ (1985) Studie van de uitvlokking van kaolinietsuspensies met behulp van aluminiumsulfaat. Ph D Thesis, University of Leuven, Belgium (K.U.L.) (in Dutch; "Study of the coagulation and flocculation of kaolinite suspensions with aluminium sulfate"), 339 pp.

Friedlander SK (1977) Smoke, Dust and Haze. Fundamentals of aerosol behavior. Wiley, New York, 317 pp.

Hinze JO (1975) Turbulence, 2nd Ed. McGraw-Hill, New York, 790 pp.

Hunt JR (1986) particle aggregate breakup by fluid shear. In: Mehta A.J. (ed) Estuarine Cohesive Sediment Dynamics. Springer Verlag, Berlin Heidelberg New York: 85-109

Jiménez J, Wray AA, Saffman PG, Rogallo RS (1993) The structure of intense vorticity in isotropic turbulence. J. Fluid Mech. 255: 65-90.

Kolmogorov AN (1941) The local structure of turbulence in incompressible viscous fluid for very large Reynolds numbers. C.R. Acad Sci URSS, 30: 301; Dissipation of energy in locally isotropic turbulence. C.R. Acad Sci URSS 32: 16

Levich VH (1962) Physicochemical Hydrodynamics. Prentice-Hall, Engelwood Cliffs, NY, 700 pp.

Mehta AJ (1973) Depositional behavior of cohesive sediments. Ph D Thesis, University of Florida, Gainesville, Florida.

Mehta AJ, Partheniades E (1975) an investigation of the depositional properties of flocculated fine sediments. J Hydr Res, 13(4):361-381

Nakagawa H, Nezu I (1975) Turbulence of open channel flow over smooth and rough beds. Proc. Japan Soc. Civ. Eng, 241: 155-168

Pandya JP, Spielman CA (1983) Floc breakage in agitated suspensions, Effect of agitation rate. Chemical Engineering Science, 38(12): 1983-1992

Parker DS, Kaufman WJ, Jenkins D (1972) Floc break-up in turbulent flocculation process. J.San Eng Div, Proc Am Soc Civ, Engrs, 98: 79-99

Rouse H (1938) Experimentation on the mechanics of sediment suspension. Fifth International Congress for Applied Mechanics. 550-554.

Siggia ED (1981) Numerical study of small scale intermittency in three dimensional turbulence. J. Fluid Mech. 107: 375-406.

Tambo N, Hozumi H (1979) Physical characteristics of flocs - II. Strength of flocs. Wat Res, 13: 409-419

Tennekes H, Lumley JL (1972) A first course in turbulence. MIT Press, Cambridge, 300 pp.

Thomas DG (1964) Turbulent disruption of lfocs in small particle size suspensions. American Institute of Chemical Engineers Journal, 10(4): 517-523

Tomi DT, Bagster DF (1978) The behaviour of aggregates in stirred vessels. Part I and II. Transactions of the Institution of Chemical Engineers, London, Vol.56, 1-18.

Van Leussen W (1986) Laboratory experiments on the settling velocity of mud flocs. In: S.Y.Wang, H.W.Shen and L.Z.Ding (Eds.) River Sedimentation, Vol III, School of Engineering, University of Mississippi, 1803-1812·

Van Leussen W (1988) Aggregation of particles, settling velocity of mud flocs - a review. In: J.Dronkers and W.van Leussen (Eds.) Physical Processes in Estuaries. Springer-Verlag, Berlin, New York, 348-403.

Van Leussen W (1994) Estuarine macroflocs and their role in fine-grained sediment transport. Ph D Thesis, University Utrecht, 488 pp.

Van Leussen W, Cornelisse JM (1993) The determination of the sizes and settling velocities of estuarine flocs by an underwater video system. Neth. J. Sea Res. 31(3): 231-241.

Vincent A, Meneguzzi M (1993) The spatial structure and statistical properties of homogeneous turbulence. J.Fluid Mech. 255: 1-20.

Vincent A, Meneguzzi M (1994) The dynamics of vorticity tubes in homogeneous turbulence. J.Fluid Mech. 258: 245-254.

# 4     Temporal variability in aggregate size and settling velocity in the Oosterschelde (The Netherlands)

**W. B. M. TEN BRINKE**

*Institute for Marine and Atmospheric Research Utrecht (IMAU), Department of Physical Geography, Utrecht University, The Netherlands*
*Present address: Rijkswaterstaat, Institute for Inland Water Management and Waste Water Treatment (RIZA), Arnhem, The Netherlands*

## INTRODUCTION

Many researchers have found that aggregate settling velocity increases exponentially with sediment concentration up to the point of hindered settling (Owen, 1971; Puls & Kuehl, 1986; Pejrup, 1988; Van Leussen, 1994) due to increased flocculation at higher concentrations. This dependence is described by $W = kC^m$, where $W$ = settling velocity, $C$ = suspended sediment concentration and $k$ and $m$ are constant. For different estuaries different values for $k$ and $m$ have been derived (Fig. 1).

The large scatter in Fig. 1 may be due to (1) disturbance of aggregates during sampling, (2) variations in hydrodynamics and sediment properties of different tidal systems, and (3) temporal variations in the process of flocculation and settling. In the absence of information on the temporal variability of aggregate properties, the intercomparison of results of different researchers is of little value. Aggregate properties may vary in time because aggregate size and settling velocity depend on the shear in the water column, which varies during ebb and flood, during the neap-spring tidal cycle and because of waves. Also these aggregate properties including aggregate strength vary in time because the organic part of the suspended matter varies in time. It is well-known that suspended aggregates are made up of organic and inorganic matter and that especially organic polymers released by algae and bacteria are sticky and therefore significantly affect the process of aggregation (Avnimelech *et al.*, 1982; Degens & Ittekot, 1984; Alldredge & Silver, 1988). The variability in composition, density, strength and adherence properties of aggregates also results in a variability in the erodibility, the critical shear stress for erosion ($\tau_e$), of recently settled fine-grained sediment.

The temporal variability in the settling velocity and the critical shear stresses for deposition and erosion of aggregates has important consequences for fine-grained sediment transport since this transport is largely based on scour and settling lag effects. Field measurements should therefore be carried out which focus on this temporal variability in settling and erosion properties.

The temporal variability in aggregate size and settling velocity has been studied in the Dutch Oosterschelde. Aggregates were studied *in situ* from video registrations by using a

*Cohesive Sediments.*
Edited by Neville Burt, Reg Parker and Jacqueline Watts
©1997 John Wiley & Sons Ltd.

specially designed underwater camera system (Van Leussen & Cornelisse, 1993) and from microscopic observations of diver-sampled aggregates. The results refer to calm weather conditions, low current velocities and a rather constant suspended sediment concentration < 10 mg l⁻¹. The observed variability is therefore mainly due to biological processes (Ten Brinke, 1993). The impact of variations in shear and suspended sediment concentration, however, will also be discussed.

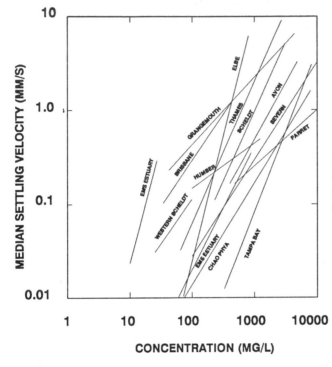

**Fig. 1.** Median aggregate settling velocity versus fine-grained suspended sediment concentration according to several investigators.

## AREA OF RESEARCH

The Oosterschelde (Eastern Scheldt) is a tidal basin in the southwestern part of the Netherlands (Fig. 2). The aggregates have been studied at 2 locations in the basin: (A) in a small channel in the western part with an average water depth of 9 m, and (B) in a main channel in the northeastern branch of the basin with an average depth of 25 m.

The construction of a storm-surge barrier in the mouth and two dams in the landward branches of the basin, completed in 1987, have reduced tidal volumes in the channels by about 30% in most of the basin and by up to 80% in the northeastern branch. Depth-averaged current velocity is now 0.8 m s⁻¹. At the locations where the aggregates have been studied, however, depth-averaged current velocity is always less than 0.3 m s⁻¹. The significant wave height is highest in the western part: 1.2 m. The tidal range has been reduced by about 12% and is now 3.25 m at Yerseke. Fresh water input into the system, already low before the construction of the works (70 m³ s⁻¹), is now insignificant (25 m³ s⁻¹). Salinity is fairly constant and high throughout the Oosterschelde (>30‰). Fine-grained suspended matter concentration at both locations at the times of aggregate observations and samplings was less than 15 mg l⁻¹ (Table 1).

The areas of research act as sediment traps showing fine-grained sediment accumulation rates of 5-10 cm yr$^{-1}$ since April 1987 (Ten Brinke *et al.*, 1995).

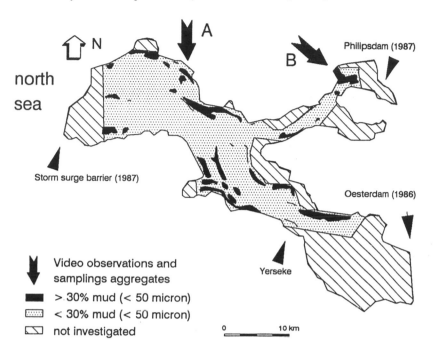

**Fig. 2.** Map of the Oosterschelde tidal basin, showing the civil-engineering works, the areas with muddy deposits, and the locations where the suspended aggregates were filmed and sampled.

**Table 1.** Fine-grained suspended sediment concentration during the measurements with the Video In Situ (VIS) camera system (( ) = number of samples).

Fine-grained suspended sediment concentration (mg l$^{-1}$):

| Date | Depth | LWS | Flood | HWS | Ebb |
|---|---|---|---|---|---|
| August 30 1989 | surface (-1m) | 9.0 (2) | 8.8 ± 1.8 (5) | 13.5 (2) | 11.0 ± 1.8 (4) |
| | bottom (+1m) | 11.0 (2) | 11.6 ± 3.7 (5) | 13.5 (2) | 11.1 ± 2.8 (7) |
| September 11 1991 | surface (-1m) | 4.0 (2) | 2.8 ± 1.0 (4) | 3.5 (2) | 4.3 ± 1.3 (4) |
| | bottom (+1m) | 5.5 (2) | 7.0 ± 2.2 (4) | 8.0 (2) | 7.3 ± 1.3 (4) |
| November 6 1991 | surface (-1m) | 3.3 ± 0.6 (3) | 4.8 ± 1.0 (4) | 4.0 (2) | 4.0 (2) |
| | bottom (+1m) | 6.7 ± 0.6 (3) | 9.3 ± 1.5 (4) | 6.5 (2) | 6.0 (2) |

## METHODS AND ANALYSES

### Video camera system

August 30 1989 and September 11 and November 6 1991 settling aggregates were filmed at the locations A (1989) and B (1991) respectively (Fig. 2). Video observations were made at different stages during ebb and flood by a submersible video system developed by Van Leussen & Cornelisse (1993). This video system consists of a vertical tube through which the aggregates are settling in still water. Before entering this tube, the aggregates are captured in a funnel-type capture/stilling room without being destroyed. The settling process in the tube is recorded by means of a CCD-camera and a 0.4 mm light sheet, both housed in a water tight underwater housing. More details can be found in Van Leussen & Cornelisse. At location A the video camera system was operated at 1 'm above the bottom. At location B the video camera system was operated at the maximum depth for which the instrument was constructed: 19 m below water surface (6 m above the bottom).

Part of the video registrations was affected by water movements within the sensing area. These movements were probably due to pressure differences between the inside and the outside of the settling column resulting from currents. Since these movements affect the settling velocities of the aggregates, the video registrations showing these movements were omitted. Only the video registrations showing aggregates settling vertically through a stagnant water volume were used. Movements of the water volume within the sensing area could easily be observed from numerous not-settling particles.

The size and settling velocity of the aggregates were measured from a television screen using a ruler and a stopwatch. The smallest particles that could be measured from the screen quite accurately were 50 $\mu$m. All visible particles > 50 $\mu$m were measured.

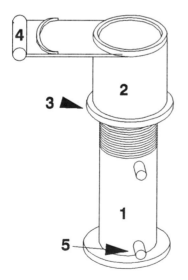

**Fig. 3.** Sampling device for collecting undisturbed aggregates.

### Sampling tubes

During a period of six weeks, from May 6 to June 11 1991, aggregates were sampled at location B (Fig. 2) with specially designed sampling tubes of the Netherlands Institute for Sea Research (Eisma et al., 1983) (Fig. 3). These tubes consist of a perspex tube (1), which is filled with water, and a part on top of the tube (2) which can be screwed up and down and which contains a slide-valve. In between these two parts a Nuclepore filter is mounted on top of the tube by means of a small perspex ring (3) that can be loosened or tightened. When the sampling tube is taken down by a diver, the upper part is screwed downwards and the valve (4) is closed so that the valve rests on the filter.

When the sampling depth is reached, the diver opens the slide-valve and gently screws the top part upwards. Then the valve is closed again and a water volume of 50 ml has been trapped above the filter. The sampling tube is brought on board of the ship and the water is sucked through the filter very carefully with a handpump. This pump can be connected to openings in the lower part of the sampling tube (5). A set of 5 tubes was used which were fastened in a frame. The frame with tubes was lowered from a ship into the water and thus did not move free with the current. Considering the very weak currents at this location, disturbance of aggregates due to the shear of water flowing over the tubes was assumed to be negligible. Sampling was performed at water slack and maximum current at 1 m below water surface and at 2 m above the bottom. 0.4 $\mu$m irgalan black Nuclepore filters were used, which were stored cool after use in plastic petridiscs for SEM examination.

Along with the sampling of aggregates by means of tubes, water samples were taken from which the phaeophytin-*a* concentration was determined. Seawater was filtered through Whatman GF/C glass fiber filters. After extraction in 90% acetone, the samples were analysed by HPLC (slightly modified after Mantoura & Llewellyn, 1983).

## RESULTS

Average aggregate size and phaeophytin-*a* concentration during the phytoplankton spring bloom of 1991 are shown in Fig. 4. Average aggregate size near the bottom was over twice the aggregate size in the upper water column. Average aggregate size at flood and HWS increased during the bloom. An increase during ebb and LWS is less apparent. Both near-bottom aggregate size and phaeophytin-*a* concentration at flood and HWS peaked in week 4. Fine-grained sediment concentration was fairly constant throughout the bloom, except for an anomalous high concentration at week 1.

**Table 2.** Aggregate settling velocities according to the Video In Situ (VIS) camera system.

Aggregate settling velocity (mm s$^{-1}$):

| Date | Depth | LWS | Flood | HWS | Ebb |
|---|---|---|---|---|---|
| August 30 1989 | median | 0.55 | 1.09 | 1.74 | 1.20 |
| | mean | 0.72 ± | 1.17 ± | 1.96 ± | 1.39 ± |
| | | 0.61 | 0.61 | 1.20 | 0.75 |
| September 11 1991 | median | 1.14 | * | 2.39 | 0.81 |
| | mean | 1.31 ± | * | 3.08 ± | 1.33 ± |
| | | 0.93 | | 2.74 | 1.85 |
| November 6 1991 | median | * | 0.72 | 0.60 | 0.63 |
| | mean | * | 0.92 ± | 0.99 ± | 0.83 ± |
| | | | 0.78 | 1.03 | 0.77 |

The video results on aggregate size and settling velocity are summarized in Tables 2 and 3. Median and mean aggregate settling velocity were smaller during November 6 than during August 30 and September 11. Median and mean settling velocities at HWS for the summer observations were three to four times as large as the median and mean settling velocities for the November aggregates. The differences between the summer and autumn aggregate settling velocities were significant according to the $t$-test ($p < 0.005$). The differences in aggregate size were much less pronounced. The differences between the maximum sizes of the summer and autumn aggregates were significant for the HWS aggregates only ($P < 0.05$).

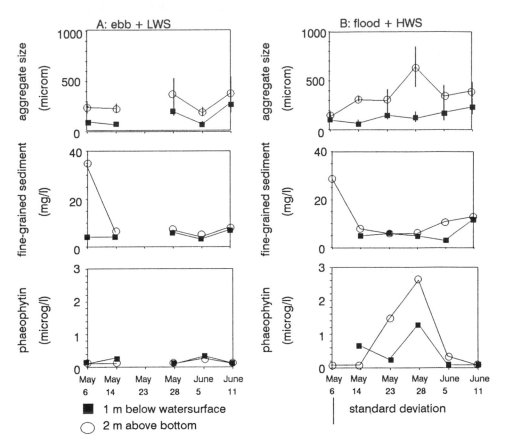

**Fig. 4.** Average aggregate size, the phaeophytin-a and suspended sediment concentration in the northeastern branch of the Oosterschelde during May-June 1991.

## DISCUSSION

Clearly, aggregates at concentrations $< 10$ mg l$^{-1}$ can be much larger and may settle much faster than relationships between aggregate properties and sediment concentration in the literature suggest. This is due to the fact that aggregation not only depends on suspended sediment concentration but also on the shear in the water column. A reduction of the shear increases the likelihood of aggregate growth after collision, the collision efficiency. Apparently the shear at the areas of research was sufficiently small so that breakup of aggregates was limited. The combined effect of turbulent shear and sediment concentration on aggregate properties is still unclear. Generally, at relatively low shear, aggregate size and settling velocity are considered to increase with sediment concentration because the number of collisions increases. At higher shear, however, the effect of sediment concentration is dominated by the effect of floc breakup. This was schematically presented by Dyer (1989) (Fig. 5). In fact, reality is even more complicated than the conceptual diagram suggests since the availability of sticky organic polymers also affects the collision efficiency, aggregate strength and density. Probably the numbers along the axes of Fig. 6 vary throughout the year.

**Table 3.** Maximum aggregate diameter according to the Video In Situ (VIS) camera system.

Aggregate size ($\mu$m):

| Date | Depth | LWS | Flood | HWS | Ebb |
|------|-------|-----|-------|-----|-----|
| August 30 1989 | median | 200 | 250 | 300 | 200 |
| | mean | 255 ± | 305 ± | 364 ± | 249 ± |
| | | 159 | 200 | 259 | 169 |
| September 11 1991 | median | 183 | * | 488 | 305 |
| | mean | 201 ± | * | 508 ± | 369 ± |
| | | 123 | | 296 | 257 |
| November 6 1991 | median | * | 279 | 248 | 245 |
| | mean | * | 307 ± | 323 ± | 334 ± |
| | | | 198 | 260 | 230 |

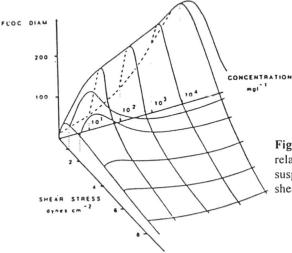

**Fig. 5.** Conceptual diagram showing the relationship between aggregate size, suspended sediment concentration and shear stress (from Dyer, 1989).

Aggregation of cohesive particles depends on collisions due to Brownian motion, fluid shear, and differential settling. In coastal waters Brownian motion is unimportant since the suspended sediment is already present as aggregates. Lick *et al.* (1993) showed that the steady state median aggregate diameter is much larger when differential settling is the dominant mechanism for aggregation than when fluid shear is the dominant mechanism. From laboratory experiments they quantified the relationship between aggregate diameter ($D_a$), concentration (C), and shear (G): $D_a = 10.5(CG)^{-0.4}$. Fig. 6A shows this relationship for a range of C and G that has been determined for the area of research. Fig. 6B shows the relationship for settling velocity ($W_s$) assuming that $W_s = 5.44D_a^{0.92}$ (Ten Brinke, 1994[a]). The sizes and settling velocities at low shear and low sediment concentration agree with the Oosterschelde results. The shape of the diagram, however, is much different from the conceptual diagram of Dyer (1989). A reduction of aggregate size when concentration increases and shear is low, seems unrealistic.

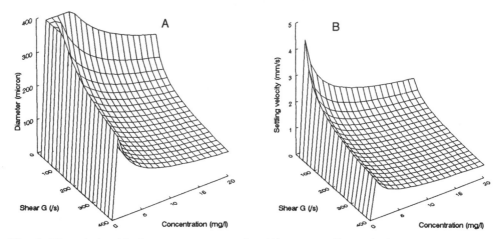

Fig. 6. The relationship between aggregate size (A) resp. settling velocity (B) and suspended sediment concentration and shear according to Lick et al. (1993).

At the area of research aggregates strongly increase in size as they settle through the water column. At the low current velocities as observed at the locations where the aggregates were sampled, differential settling is important for aggregation (Eisma, 1991). According to Van Leussen (1994) differential settling is important when the shear in the water is less than 10 s⁻¹. Van Leussen discussed the combined effect of turbulent shear and differential settling on the sizes of aggregates. He derived a formulation relating the increase of aggregate size in time to sediment concentration, shear and the collision efficiency:

$$\frac{D}{D_0} = e^{\frac{4}{3\pi} \times \alpha \times \phi \times (10+G) \times t} \qquad (1)$$

where

| | | |
|---|---|---|
| D | = aggregate diameter at t=t (m) |
| $D_0$ | = aggregate diameter at t=0 (m) |
| $\alpha$ | = collision efficiency factor |
| $\phi$ | = sediment volume per watervolume |
| G | = shear (s⁻¹) |
| t | = time (s) |

This formula can be used to estimate the collision efficiency factor $\alpha$ during the phytoplankton spring bloom of 1991 from the aggregate sizes near the water surface ($D_1$) and near the bottom ($D_2$):

$$\frac{D_2}{D_1} = e^{\frac{4}{3\pi} \times \alpha \times \phi \times (10+G) \times (t_2 - t_1)} \qquad (2)$$

Assume

$$\phi = \frac{C}{\Delta\rho} = \frac{C}{a \times D_a^p} \qquad (3)$$

and

$$(t_2 - t_1) = \frac{h}{W_s} \qquad (4)$$

where
C = concentration suspension (kg m$^{-3}$)
$\Delta\rho$ – excess aggregate density (kg m$^{-3}$)
$D_a$ = aggregate diameter (mm)
a,p = coefficient
h = depth surface aggregates (2 m below water surface) minus depth near-bottom aggregates (23 m below water surface)
$W_s$ = settling velocity (mm s$^{-1}$)

and assume $\quad W_s \quad$ = 2.39 mm s$^{-1}$ (= median at HWS September 1991)
so $\quad (t_2 - t_1)$ = 21,000/2.39 = 8800 s
then formula (2) becomes

$$\frac{D_2}{D_1} = e^{\frac{4}{3\pi} \times \alpha \times \frac{C}{a \times D_a^p} \times (10+G) \times 8800} \qquad (5)$$

from (5) follows

$$\alpha = \frac{3 \times \pi \times a \times D_a^p}{4 \times C \times (10+G) \times 8800} \times \ln(\frac{D_2}{D_1}) \qquad (6)$$

assume $\quad \Delta\rho \quad$ = 15.19$D_a^{-1.16}$ at flood + HWS
$\qquad\qquad \Delta\rho \quad$ = 15.64$D_a^{-0.81}$ at ebb + LWS (Ten Brinke, 1994[b)])
and $\quad\;\; G \quad$ = 10 s$^{-1}$, upper shear limit at which differential settling is important for aggregation,

then
for flood + HWS

$$\alpha = \frac{3 \times \pi \times 15.19 \times D_a^{-1.16}}{4 \times C \times 20 \times 8800} \times \ln(\frac{D_2}{D_1}) \quad (7)$$

and
for ebb + LWS

$$\alpha = \frac{3 \times \pi \times 15.64 \times D_a^{-0.81}}{4 \times C \times 20 \times 8800} \times \ln(\frac{D_2}{D_1}) \quad (8)$$

The values for $\alpha$ during the spring bloom of 1991 are shown in Fig. 7. These values agree with results in the literature (Edzwald, 1972). One might expect the collision efficiency to depend on the availability of sticky polymers. Neither an increase during the bloom nor a peak value at week 4, however, was calculated for $\alpha$. This may be due to the rather crude assumptions underlying the results.

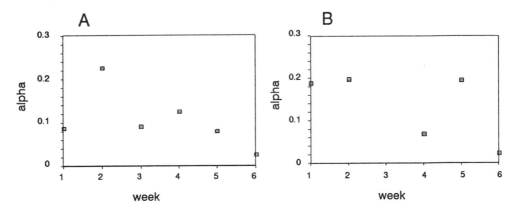

**Fig. 7.** The collision efficiency factor $\alpha$ during the phytoplankton spring bloom of 1991.

ACKNOWLEDGEMENTS.-This research was financially supported by Rijkswaterstaat. The author wishes to thank J.H.J. Terwindt and D. Eisma for the stimulating discussions on the aggregation of suspended sediment, E. van de Ketterij and P. van de Wege for diving for suspended aggregates, J.M. Cornelisse, K. Kuyper and H. Meier for their assistance with the submersible video camera, the Netherlands Institute for Sea Research for placing the SEM and a set of sampling devices at my disposal, J. Beks for his assistance with the SEM, and Rijkswaterstaat for the use of the submersible video camera.

# REFERENCES:

Alldredge, A.L. & M.W. Silver, 1988. Characteristics, dynamics and significance of marine snow. Progr. Oceanogr. 20: 41-82.

Avnimelech, Y., Troeger, B.W. & L.W. Reed, 1982. Mutual flocculation of algae and clay: evidence and implications. Science 216: 63-65.

Brinke, W.B.M. ten, 1993. The impact of biological factors on the deposition of fine-grained sediment in the Oosterschelde (The Netherlands). PhD-thesis Utrecht University, 252 pp.

Brinke, W.B.M. ten, 1994[a]. Settling velocities of mud aggregates in the Oosterschelde tidal basin, determined by a submersible video system. Est. Coast. Shelf Sci. 39: 549-564.

Brinke, W.B.M ten, 1994[b]. In situ aggregate size and settling velocity in the Oosterschelde tidal basin (The Netherlands). Neth. J. Sea Res. 32: 23-35.

Brinke, W.B.M. ten, Augustinus, P.G.E.F. & G.W. Berger, 1995. Fine-grained sediment deposition on musselbeds in the Oosterschelde determined from echosoundings, radio-isotopes, and biodeposition field experiments. Est. Coast. Shelf Sci. 40: 195-217.

Degens, E.T. & V. Ittekot, 1984. A new look at clay-organic interactions. Mitt. Geol.-Palaeont. Institut Univ. Hamburg, Heft 56: 229-248.

Dyer, K.R., 1989. Sediment processes in estuaries: future research requirements. J. Geoph. Res. 94: 14,327-14,339.

Edzwald, J.K., 1972. Coagulation in estuaries. PhD-thesis University of North Carolina, 204 pp.

Eisma, D., 1991. Particle size of suspended matter in estuaries. Geo-Marine Letters 11: 147-153.

Eisma, D., Boon, J., Groenewegen, R., Ittekot, V., Kalf, J. & W.G. Mook, 1983. Observations on macro-aggregates, particle-size and organic composition of suspended matter in the Ems estuary. Mitt. Geol.-Palaeont. Inst. Univ. Hamburg, SCOPE/UNEP Sonderband, Heft 55: 295-314.

Leussen, W. van, 1994. Estuarine macroflocs & their role in fine-grained sediment transport. PhD-thesis Utrecht University, 488 pp.

Leussen, W. van & J.M. Cornelisse, 1993. The determination of the sizes and settling velocities of estuarine flocs by an under water video system. Neth. J. Sea Res. 31: 231-241.

Lick, W., Huang, H. & R. Jepsen, 1993. Flocculation of fine-grained sediments due to differential settling. J. Geoph. Res. 98: 10,279-10288.

Mantoura, R.F.C. & C.A. Llewellyn, 1983. The rapid determination of algal chlorophyll and carotenoid pigments and their breakdown products in natural waters by reverse-phase high performance liquid chromatography. Anal. Chim. Acta 151: 297-314

Owen, M.W., 1971. The effect of turbulence on the settling velocities of silt flocs. Proc. Int. Ass. Hydr. Res.:27-32.

Puls, W. & H. Kuehl, 1986. Field measurements of the settling velocities of estuarine flocs. In: Wang, Sheng and Ding (eds). Proc. Thrd. Int. Symp. Riv. Sed., Mississippi: 525-536.

Pejrup, M., 1988. Flocculated suspended sediment in a micro-tidal environment. Sediment. Geol. 57: 1-8.

# 5    Direct measurements of settling velocities in the Owen tube: a comparison with gravimetric analysis

**M. P. DEARNALEY**
*HR Wallingford Ltd, UK*

## ABSTRACT

Over the last two decades the Owen Tube or derivatives of the original design have been used extensively to determine the settling velocities of suspensions of cohesive material in the field. These gravimetric measurements have been used as elements of basic research programmes and also as important input parameters for predictions of siltation rates.

During a two year research programme at HR Wallingford undertaken between 1989 and 1991 consideration was given to an alternative technique for analysing the settling behaviour of cohesive material. This technique is based upon obtaining high magnification video images of the processes occurring within the settling column. The video images are then analysed using a PC based image analysis system which enables settling velocities and floc dimensions to be determined. The technique has been applied on a number of occasions over the past four years and was included in the European Intercomparison Exercise that took place on the River Elbe in June 1993.

Analysis of the processed field data indicates floc break-up, reflocculation and the development of significant circulations within the settling column during the withdrawal period required for the gravimetric analysis (typically a duration of about an hour). The interpretation of the visual data obtained by the video image analysis technique has indicated median settling velocities significantly greater than those obtained from the gravimetric technique.

The paper describes the field and analysis methods and results from the recent Intercomparison Exercise on the Elbe. During this exercise an experiment was undertaken whereby video image analysis was undertaken on the results of filming a settling column whilst the standard gravimetric technique was applied. The paper also discusses the engineering applicability of the results obtained and raises the question as to which technique provides the most representative rate of settlement of flocs in the natural environment?

## INTRODUCTION

When in-situ settling columns were first introduced an increase in the observed median settling velocity of an order of magnitude was found compared to previous laboratory based techniques for determining settling velocities (Owen 1976). The inference was that mixing and resuspension of the

*Cohesive Sediments.*
Edited by Neville Burt, Reg Parker and Jacqueline Watts
©1997 John Wiley & Sons Ltd.

material in the laboratory could significantly affect the size and therefore the settling velocity distribution of the material in suspension. The use of the Owen Tube was considered a major advance.

Individual flocs can readily be observed by eye within an Owen Tube. It can also be observed that the fiocs do not always appear to move vertically downwards. In fact observations will lead to the conclusion that the environment within an Owen Tube is, at certain stages, highly turbulent. How this turbulence may affect the material in suspension in the column and whether there is significant reflocculation within the Owen Tube has been a subject of much interest (Burt, 1984).

When particles settle through a fluid of finite extent there are two effects, the fluid pulled along by the particle must produce a return flow since it cannot pass through the walls of the containing column, and since the fluid is stationary at a finite distance from the particle, there is distortion of the flow pattern which reacts back on the particle (Alien, 1968). The larger particles within the column being more affected by the finite extent of the column.

If two particles separated by only a few particle diameters move through a viscous fluid, the fluid flows around the particles in such a way that the resulting viscous force is greater than that acting on a single combined particle. Hence the terminal settling velocity is smaller than would otherwise be experienced. It is suggested that volume concentrations below which this effect is negligible are in the range 0.05% to 3% (Allen, 1968).

Most measurements of fall velocity in the field and laboratory are made indirectly and generally do not measure both the settling velocity and size of a given particle (which can lead to the determination of an effective floc density). If required the particle size distribution is inferred through the settling velocity distribution, an assumed particle density and Stoke's equation.

**GRAVIMETRIC DETERMINATION OF SETTLING VELOCITY**

The Owen Tube designed at HR Wallingford (Owen, 1976) or a derivative of this design such as that marketed by Valeport, has been the main instrument for determining settling velocities of estuarine suspensions. The tube is essentially a water sampling device which doubles as a bottom withdrawal column. The tube has an internal diameter of 50mm and is about 1 m in length. The tube is lowered into the water and held at the required depth, aligning itself with the flow. A trigger is operated which causes the tube to trap a sample of the estuarine water. It is assumed that since the tube aligns itself with the flow, and because there is flow through the tube, the sample obtained in this manner is representative of the floc distribution within the estuary.

The tube is retrieved and as it is brought above the surface the column is returned to the vertical. A clock is started at this time and samples are then withdrawn at given times. The chosen time intervals are dependent upon assumed concentrations and settling velocities. Typically for estuarine waters the sampling is undertaken in about one hour. Subsequent gravimetric analysis of the individual samples and processing of the results leads to a settling velocity distribution from which it is then possible to infer a median particle size from Stoke's Law.

There are a number of limitations associated with the Owen Tube technique. Most of these limitations also apply to other derivatives of the instrument;

i)      In low concentrations (less than about 50mgli) the mass of material for analysis is very small and standard gravimetric analysis techniques are less able to determine these small masses accurately.

ii)     As the column is being retrieved from the sampling depth to the surface some settling will occur. When the column returns to the vertical a density current can form within the column causing substantial mixing and possibly effecting the floc size distribution in the suspension. This is a particular problem at higher concentrations. Delft Hydraulics have developed a modified field instrument based on a side withdrawal system which returns to the vertical immediately the sample is obtained to overcome this particular problem (van Leussen, 1988).

iii)    The sampling by bottom withdrawal typically takes about an hour to complete. This means that reflocculation can occur within the column after the sample has been retrieved. Puls (1988) concluded that floc formation as a consequence of differential settling within the column increases the measured settling velocities appreciably and that a reduction factor needs to be applied. However, since differential settling is a phenomena that may be important in some engineering problems it is possible that in some cases the settling velocity determined from the settling column is appropriate for use without any reduction factor. Similarly, the important field process of hindered settling can occur in the Owen Tube resulting in a reduction in measured settling velocity.

iv)     It is also possible that during the sampling period external influences such as air temperature may become important, causing changes in viscosity and setting up convection currents within the column.

v)      The Owen Tube itself and the method of capturing the sample will affect the floc size distribution of the sample which is obtained. The method of trapping the sample is likely to cause significant turbulence within the column. The raising of the column above the water surface into position for withdrawal of samples will also induce turbulence and initiate circulations within the column. It is thus possible, that to an extent, the flocs in the samples retrieved are not fully representative of those in the estuarine water body. However, this problem is inherent in all sampling devices. Some of these problems are aggravated by sampling in strong currents.

## VIDEO IMAGE ANALYSIS TECHNIQUES

### OPERATION

As a result of an investigation into available instrumentation packages for determining the size and settling velocity of cohesive sediments a video image analysis technique was developed at HR (Dearnaley, 1991). Images are obtained using a standard CCD video camera with c-mount to Nikon mount adaptor, 200mm bellows and standard 135mm Nikon lens. The camera and light source are set up perpendicular to one another and focused on the settling column. Typically images of about 3mm by 4mm can be obtained with this set up. This gives an approximate resolution of 20 microns. Using a reversed 35mm lens gave an image size of approximately 1 mm square and correspondingly higher resolution. The depth of focus in the image is estimated to be about 0.1 mm and consequently very good images can be obtained even with a simple light source such as a standard spot lamp. The narrow depth of field is the means by which the calibration of the image is undertaken with varying bellows extensions providing different magnifications.

The output from the camera is then recorded on video tape. The output from a time date generator can also be added to the recording. The format of the recording can either be PAL or RGB compatible. At the recording stage it is also possible to send the video output to the PC based image analysis system where real time filtering and image enhancement effects are available so that the quality of images for subsequent processing can be appraised. The settings of the camera and light source can be modified if required.

The system is operated in the same manner whether filming is being carried out in the field or in the laboratory. In the field the Valeport derivative of the Owen Tube makes a suitable sampling device and also has the added advantage of enabling a direct comparison with the gravimetric technique to be made. Once a sample has been obtained filming should commence almost immediately, in order that the flocs are analysed in as close to their in-situ state as possible. In practice filming can typically start within one to two minutes of the sample being taken.

## ANALYSIS

There are many different commercially available image processing systems for determining object size, shape, position etc. The PC based system used here was obtained from Digithurst and operated satisfactorily from a recorded video source. It was enhanced to include a routine for grabbing two successive images from a video tape at a prescribed time interval.

The image is calibrated using the recorded bellows extension and then the system is set up to capture two frames. Video frames comprise two interlaced fields with a time interval of 0.02 seconds, thus successive frames are separated by 0.04 seconds. The image processing card has a minimum capture interval of 0.25 seconds. The required interval between the two frames is selected (typically 0.3 to 0.5 seconds) and then, with the video tape playing, the images are grabbed.

Once a pair of images are grabbed some of the analysis is automatic. A grey scale level (0-255) is chosen at which to form a cut off, this thresholding results in the monochrome image being converted to a binary image which can then be subjected to various analysis routines. Parameters such as detected object size, intensity, position (x-y co-ordinates in pixels) and orientation can be automatically determined. From these records it is possible to produce a size distribution for detected objects in the image. Providing that an appropriate thresholding level has been chosen this can be assumed to be representative of the floc size distribution. By overlaying binary images from two successive images with a known time interval it is possible to determine the relative movement between the two frames. This process can be either automated or manual (Dearnaley, 1991). For the Elbe data the manual analysis technique was adopted.

It is thus possible to determine the size and settling velocities of individual flocs in a pair of images. As noted above, the motion of the flocs within the settling column is often dominated by circulation and turbulence within the column. As a result, in order to satisfactorily determine the gravitational settling velocity of the flocs it is necessary to make assumptions about the fluid motions through the section of the column that is in focus. The first assumption is to consider that the fluid motion is uniform across the image area. This is not always the case and may make certain parts of recordings unusable for settling velocity analysis. The second assumption is to assume that the fastest upwards travelling floc, or slowest downwards travelling floc, depending upon the natural fluid motion in a pair of images, is actually moving at the same speed as the local flow. This speed can then be used as a corrector to give a settling velocity/size distribution. The results of this process are shown in Figures 1a and 1b. Figure 1a shows the raw data from ten pairs of image with both upward (-ve) and downward (+ve) motion. Figure 1b shows the same data corrected for the assumed fluid speed. There appears to be very little correlation between the data in Figure 1b indicating a non-homogeneous floc distribution.

Typically ten pairs of images are analysed from a 1-2 minute period of the tape recording and the results of this analysis are used to determine a settling velocity distribution for the suspension a that time. The settling velocity distribution can be processed into a similar format as the traditional output from Owen Tube data by using Stoke's Law to determine an effective floc density and by making assumptions about the volume of the flocs.


## RESULTS

The results presented here are from a series of experiments that were made where standard Owen Tube analysis techniques were being used whilst video image analysis was simultaneously undertaken. It was thus possible to determine the settling velocity and size distribution of material in suspension at a number of stages through the gravimetric analysis period. It was also possible to determine the average velocity of the flocs in suspension whilst a sub sample for the gravimetric analysis was being withdrawn and to qualitatively examine the patterns of acceleration and deceleration that the flocs underwent as a result of withdrawal of a sample.

Figure 2 shows a set of typical results from the image analysis of data from six time intervals corresponding to periods prior to the withdrawal of the first sample from the tube (less than one minute after the tube was placed in the vertical) and then after the first to fifth withdrawals have been made (after 1, 3, 6, 10 and 15 minutes have elapsed). The curve derived from the gravimetric analysis is also shown. It can be seen that there is a gradual decrease in the median settling velocity derived from the image analysis with time. The other important observation is that the median settling velocity based on the gravimetric analysis is an order of magnitude less than that derived from the image analysis. At times of withdrawal the fluid motion within the column was found to be in the range 20-30mm/s. Two orders of magnitude greater than the gravimetrically determined median settling velocity. To some extent the withdrawal acts to damp out some of the initial circulations set up within the column when it is first obtained.

A further interesting observation concerning reflocculation and hindered settling within the tube can be made. For the experiment the filming took place approximately 300mm above the base of the settling column. Withdrawal of each of the eight samples lowers the water level in the column by about 125mm. A conceptual maximum settling velocity for the flocs in the field of view at any time can be derived assuming that flocs settle continually from the moment the column is placed in the vertical and that no floc growth occurs. The table below shows these conceptual maximum velocities at the times of sampling and also shows the median and maximum settling velocity derived from the video image analysis at the given times. Figure 3 shows the six processed settling velocity against size distributions for the data tabulated below (also shown in Figure 2).

| Sample | Time after column placed in vertical | Conceptual maximum particle settling velocity (mms) | Measured Particle settling velocities (mm/s) | |
|---|---|---|---|---|
| | | | Median | Maximum |
| Pre first | <60 seconds | >9.6 | 2.0 | 4.2 |
| Post first | >60 seconds | 9.6 | 1.5 | 4.7 |
| Post second | >180 seconds | 2.5 | 1.2 | 4.5 |
| Post third | >360 seconds | 0.9 | 0.8 | 3.4 |
| Post fourth | >600 seconds | 0.3 | 0.7 | 2.4 |
| Post fifth | >900 seconds | 0.1 | 0.6 | 0.7 |

The table demonstrates that the observed maximum floc settling velocities after three minutes of settling are greater than the conceptual maximum. This indicates that circulations and reflocculation within the column are significant processes. The median and maximum settling velocities for the post fifth sample are similar because the sample is dominated by one large floc which represents almost 50% of the mass flux (see Figure 30). This last point demonstrates another important aspect of the image analysis technique. The technique only samples a very small proportion of the settling flux, for the case presented in Figures 2 and 3 this fraction is estimated to be about 0.025%. The field of view is considered to be equivalent to about 0.1% of the cross section of the column.

Consequently, the representativeness of the section where the filming took place must be questioned. It is also noted that the image analysis can not resolve the smallest flocs. However, it is considered that for typical estuarine suspensions the technique is probably able to resolve about 80-90% of the settling mass. The most important point, however, is that by analysing images obtained shortly after the sample is obtained the results should be more representative of the size and settling velocity distribution of the flocs in the estuarine body than that inferred from the results of the standard Owen Tube gravimetric analysis which lasts about an hour.

## SUMMARY

The results of this comparison raises an interesting question over the applicability of the results derived from the two techniques for engineering studies. The Owen Tube approach provides information concerning the net settling of material within a column of water taking into account the associated processes of reflocculation, hindered settling, circulations set up in the column and the effects of the withdrawal of material from the column. The video image analysis technique provides more detailed information on how individual flocs behave. The question is which is the most representative for engineering application of the rate of settlement of flocs in the natural environment?

In an estuarine body of water there are many scales of motion but at most times there is a dominant ebb or flood motion which is at a velocity two or three orders of magnitude greater than the settling velocity of individual flocs. At those times when material can settle to, and adhere to, the bed the horizontal velocities are usually at a minimum. They are however still likely to be one or two orders of magnitude greater than the settling velocities of the flocs. In the settling column vertical motions are greater than horizontal motions because of the restricted dimensions. Because of this it would appear that the processes occurring in the settling column over the period of sampling from which a siltation rate is often derived (approximately 1 hour) are likely to be unrepresentative of the natural processes occurring in an estuary when settlement to the bed occurs. There are consequently two ways forward in examining natural settling processes. The first is to continue to examine the settling process at micro scales where individual flocs can be observed. The second is to continue development of techniques which can be considered non-intrusive in terms of the processes occurring as natural settling occurs. The work of many of the participants in the Elbe Intercomparison Exercise has made significant advances along both these avenues.

Settling velocity is only one input parameter to a siltation calculation. Usually there are other parameters which can also be used for purposes of calibration. These may include critical bed shear stress evaluation, consolidation parameters and associated shear strengths. Further consideration of the settling velocity of those flocs that settle to the bed which are responsible for observed rates of siltation may lead to an overall improvement of the accuracy of siltation predictions.

## ACKNOWLEDGEMENTS

The author would like to express his gratitude to Mr J Spearman of Oxford Brookes University and HR Wallingford for his help with the field work during the Elbe Intercomparison Exercise and to Mr N G Feates of HR Wallingford for subsequent processing of the video image analysis data. The gravimetric analysis was carried out under the supervision of Mr C E Johnson of the Sedimentology Laboratory at HR. The author would also like to acknowledge the fruitful discussions held with colleagues during the Elbe Intercomparison Exercise and subsequent Workshop.

## REFERENCES

**Allen T (1 968)** Particle size measurement. Chapman and Hall Ltd., London

**Burt T N (1 984)** Field settling velocities of estuary muds. In, Estuarine Cohesive Sediment Dynamics. (Ed. Mehta A.J.): Proceedings of a workshop on Cohesive Sediment Dynamics with special reference to physical processes in Estuaries, Tampa, Florida, November 12-14 1984.

**Dearnaley M P (1991)** Flocculation and settling of cohesive sediments. HR Wallingford, Report No. SR 272.

**Van Leussen W (1988)** Aggregation of particles, settling velocity of mud flocs: A review. Proc. Int. Symp. Physical Processes in Estuaries. Springer, New York. pp347-403.

**Owen M W (1976)** Determination of the settling velocities of cohesive muds.   HR Wallingford, Report No. IT 161.

**Puls W, Kuehi H and Heywonn K (1988)** Settling velocity of mud fiocs: Results of field measurements in the Elbe and Weser estuary.  Proc.  Ant.  Symp.  Physical Processes in Estuaries. Springer, New York. pp404-424.

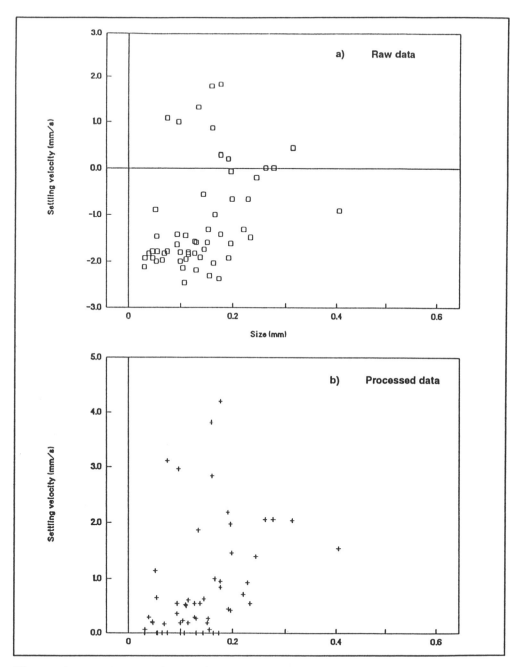

**Figure 1**     Comparison of raw and processed data

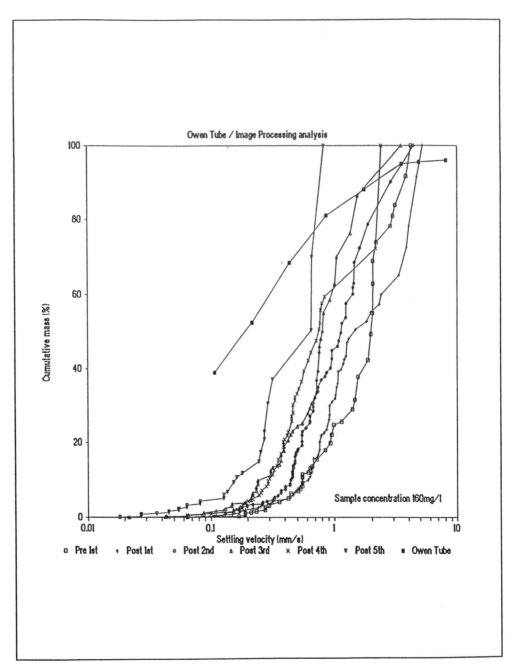

**Figure 2    Cumulative mass versus settling velocity:
Comparison of analysis techniques**

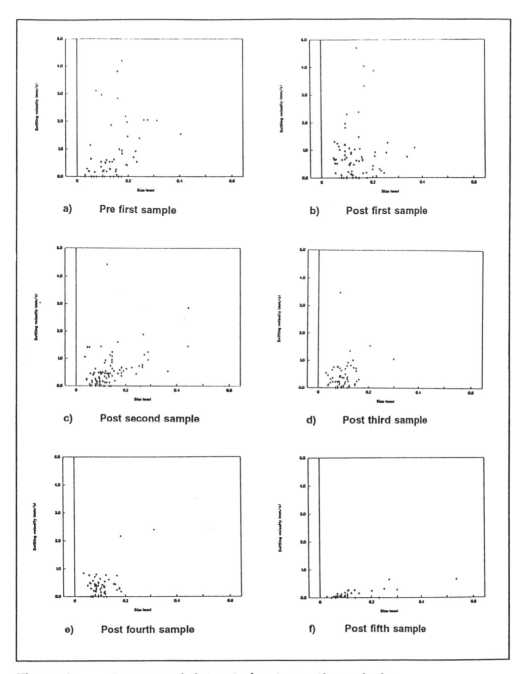

Figure 3        Processed data at six stages through the
              gravimetric analysis period

# 6 Estimation of settling flux spectra in estuaries using INSSEV

**M. J. FENNESSY[1], K. R. DYER[1], D. A. HUNTLEY[1] AND A. J. BALE[2],**
[1]*Institute of Marine Studies, University of Plymouth, Drake Circus, Plymouth PL4 8AA, UK*
[2]*NERC Plymouth Marine Laboratory, Prospect Place, Plymouth PL1 3DH, UK*

## 1 INTRODUCTION

Cohesive sediment fluxes in estuaries, particularly vertical flux, are difficult to measure directly. Numerical models that simulate estuarine fluxes are dependent on reliable field data. Many simulations have made use of single terms for floc settling velocity, often influenced by the Suspended Particulate Matter (SPM) concentration. Such terms are limited in their ability to reflect spectral changes that operate over tidal and seasonal scales. Recent investigations have shown that floc settling characteristics are influenced by seasonal changes in the organic particulate matter and local turbulence (ten Brinke, 1993; van Leussen, 1994; Fennessy *et al.*, 1994b). These influences are considered to alter floc density spectra, flocculation efficiency and resistance to break-up. The data processing method described in this paper is able to produce spectral detail of floc populations in a variety of formats, and is therefore able to provide unique field data.

Techniques are presented which transform observed floc populations into estimates of concentration and settling flux spectra. The floc population raw data is obtained with INSSEV (IN Situ SEttling Velocity), an estuarine bed-mounted instrument that video records settling flocs (Fennessy *et al.*, 1994a). INSSEV collects the sample of suspended particulate matter with a decelerator chamber equipped with flap doors at either end which close at a rate proportional to the ambient current velocity (Fig 1). After a suitable delay, to allow residual turbulence to decay, a slide door, in the floor of the decelerator, is opened to allow flocs to enter the stilled settling column where they are recorded using a high magnification video system. For each recorded floc the volume is measured assuming spherical shape and the density is calculated using the Stokes equation in order to obtain the floc dry mass.

Standardised sampling techniques ensure that the observed floc population in a sample is derived from a known volume of water. This allows concentration and settling flux spectra with respect to size, settling velocity or effective density bands, to be assembled from the individual floc dry mass values. This paper presents mass spectra with respect to size bands, specifically those generated by a *Malvern* Laser Particle Sizer, fitted with a 300 mm lens and modified for *in situ* deployment (Bale and Morris, 1987). A comparison is made between an INSSEV volume distribution and one produced with the *Malvern* sizer.

*Cohesive Sediments.*
Edited by Neville Burt, Reg Parker and Jacqueline Watts
©1997 John Wiley & Sons Ltd.

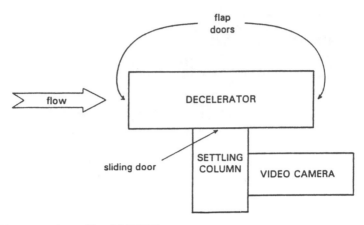

**Figure 1** Diagrammatic profile of INSSEV.

Several correction and calibration techniques have been developed for the INSSEV raw floc data to achieve the spectral formats, and these are described and discussed in the following sections, together with characteristics of the instrument system.

## 2   AMBIENT SETTLING VELOCITY

INSSEV obtains raw floc data, size (horizontal diameter) and settling velocity, by making observations of individual flocs in a stilled settling column charged with filtered water of a slightly higher salinity than the ambient estuarine water. The salinity and temperature of the ambient and column water are measured during deployment so that the density difference can be calculated. This is achieved through the use of the standard Stokes settling velocity equation:

$$w_s = \frac{D^2(\rho_f - \rho_w)g}{18\mu} \tag{1}$$

where $D$ represents the observed horizontal floc diameter, and the bracketed term represents effective density, the difference between the floc bulk density and the density of the water through which the floc is falling. Effective density is also referred to as excess density, differential density, or density contrast.

The Oseen modification (Schlichting, 1968; Graf, 1971) to the standard Stokes equation may be used throughout the following procedures although its effect is very small on all but the largest flocs.

$$w_{s_{oseen}} = \frac{D^2(\rho_f - \rho_w)g}{18\mu} \cdot \frac{1}{1 + 0.1875Re} \tag{2}$$

The effect is to increase the effective density, and hence the dry mass calculation, by the product of 0.1875 times the floc Reynolds Number, $Re$. The techniques presented in this paper use only the unmodified Stokes Equation.

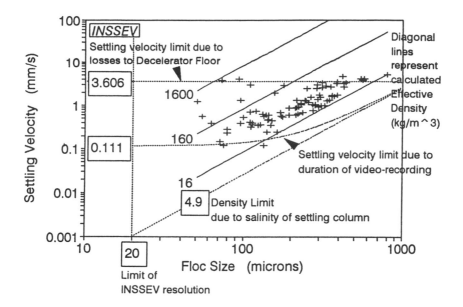

**Figure 2** Floc Settling Velocity and Size scattergraph for one sample, showing calculated Effective Density and instrumentation limit lines. The three solid diagonal lines represent constant calculated effective density relative to the ambient water.

Two corrections have to be applied to the observed individual settling velocity values. First, they are all multiplied by 1.03 to compensate for drag induced by the inner walls of the settling column (Allen, 1975). Second, effective density in relation to the column water, $\rho_{e_{col}}$, is calculated using the rearranged Stokes equation:

$$\rho_{e_{col}} = \left(\rho_f - \rho_{w_{col}}\right) = \frac{18\mu\left(w_{s_{col}} \times 1.03\right)}{D^2 g} \tag{3}$$

so that the difference in density, $\rho_{w_{diff}}$, due to higher salinity in the settling column than the ambient salinity, can be added.

$$\rho_{e_{amb}} = \rho_{e_{col}} + \rho_{w_{diff}} \tag{4}$$

The settling velocity is then recalculated using the Stokes equation:

$$w_{s_{amb}} = \frac{D^2 \rho_{e_{amb}} g}{18\mu} \tag{5}$$

to obtain the Ambient Salinity Settling Velocity, which is always slightly higher than the observed column settling velocity. It is the Ambient Salinity Settling Velocity value which is used exclusively for the settling velocity term, $w_s$, in the remainder of this paper. An example of a scattergraph showing settling velocity against size is shown in Fig 2.

## 3   INSTRUMENTATION LIMITS

Four lines are generated from the instrument operating parameters which enclose the data on the example scattergraph in Fig 2. The scattergraph provides a visual check on the validity of the sample. Strictly, data points should not occur outside these boundaries. The fact that some of them do is due to residual turbulence, particularly in the decelerator. The four limit lines are produced as follows:

### Floc Size Resolution

This is always 20 μm. The detection level of the video system is in the region of 7 - 10 μm but the arbitrary lower limit of 20 μm was chosen to maintain the integrity of size measurements for smaller particles.

### Density Limit

This is due to the settling column being charged with a higher salinity water than the ambient salinity of the estuary. It is plotted as a diagonal line on the scattergraph (Fig 2), in the same way as the calculated constant effective density lines, using the density difference, $\rho_{e_{diff}}$ , in the standard Stokes settling velocity equation.

$$w_{s-\rho l.lM} = \frac{D^2 \rho_{e_{diff}} g}{18\mu} \tag{6}$$

### Maximum Settling Velocity

The Maximum Ambient Salinity Settling Velocity of each sample, $w_{s-MAX}$ is due to the faster falling particles settling to the floor of the decelerator before the slide door to the settling column is opened. It is obtained by dividing the height of the decelerator (100 mm) by the elapsed time, $e_{so}$ , from the start of the flap doors closing to when the slide door is open above the video camera object plane. The value $e_{so}$ is made up of the duration of the flap door closure, the turbulence decay time, half the slide travel time and a few seconds of built-in delays for system checks (see Section 5.3, Equation 17).

$$w_{s-MAX} = \frac{100}{e_{so}} \tag{7}$$

The limit $w_{s-MAX}$ for any sample can be increased at the time of deployment by reducing the turbulence decay time, $t_{td}$ , the programmed interval between flap doors closing and slide opening. It is typically set to between 10-20 seconds. However, a reduction in $t_{td}$ may increase the likelihood of residual turbulence in the decelerator transferring to the top of the settling column. If a large proportion of flocs are above the calculated limit line then turbulence was still present in the decelerator at the time that the slide door to the settling column was opened.

**Minimum Settling Velocity**

This is due to the duration, $t_{rec}$, of the video recording observation after the slide door to the settling column is open above the video camera object plane. First a value is calculated for the minimum column settling velocity, $w_{s-columnMIN}$, which is the velocity of the slowest particle that could fall the 109 mm through the column to appear on the video image in the time of the video recording.

$$w_{s-columnMIN} = \frac{109}{t_{rec}} \tag{8}$$

This minimum column settling velocity cannot be plotted as a horizontal line on the $y$-axis (Fig 2), because the scattergraph plots the ambient salinity settling velocity of the observed flocs. In all the scattergraphs with a density difference the limit line is plotted as an ascending curve using the following equation.

$$w_{s-MIN} = w_{s-columnMIN} + \frac{D^2 \rho_{e_{diff}} g}{18\mu} \tag{9}$$

By definition, this line will always be above the Density Limit Line.

## 4 BACKTRACKING

Backtracking is another technique used to check the validity of the individual floc data. It traces each floc back up the column into the decelerator using individual settling velocity and elapsed time data. Both the observed column settling velocity and the calculated ambient salinity settling velocity are used in the backtracking calculations. The Camera Control Unit frame counter, recorded onto the video tapes, provides a time base that gives every floc in the sample a unique time reference as it passes the camera axis (a horizontal plane in the middle of the video image). Figure 3 shows a graph of the results of the technique for the sample shown in Fig 2.

If all the flocs are shown to be within the vertical height of the decelerator at the time of sampling then an absence of turbulence in the decelerator can be assumed. The flocs on the left of the graph are the faster falling flocs, those towards the right hand side are slower falling. In the example, the floc positions at sampling, shown above the decelerator ceiling on the left, are those that were probably kept in suspension by residual turbulence, for a short while, during the time that the slide door was open. The flocs well below the line of the decelerator floor are those which were probably given an assisted entry into the settling column, by residual turbulence, during the time the slide door was open.

Backtracking results for each sample are statistically analysed. Like the limit lines on Fig 2 the "backtrack" scattergraph (Fig 3) provides a visual indication of the extent of turbulence. Residual turbulence in the decelerator can arise from two sources: the ambient

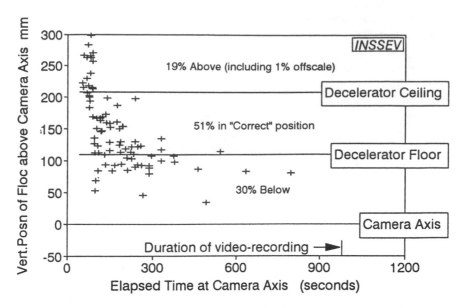

**Figure 3** Backtracking plots showing vertical position of each floc in Fig 2 at time of sampling.

turbulence of the estuary; and, if the sample was obtained when INSSEV was not correctly aligned with the current flow, turbulence generated by eddies shed from the leading edge of the decelerator side wall. Data from early field trials has been used to improve the operation of the instrument during deployments and this has enhanced the validity of later samples.

## 5  MASS SPECTRA FROM INDIVIDUAL FLOC DATA

A unique feature of the INSSEV instrument system is the opportunity to present the observed floc data in several spectral formats: number, volume or mass against floc size, settling velocity or effective density. This is made possible because the data are known to have come from a specific volume of water. The following processing stages are required to achieve the spectral formats:

1. Dry Mass calculation for each floc
2. Volume referencing of the observed sample
3. Sample volume corrections
4. INSSEV SPM concentration calculation
5. Independent Calibration
6. Band selection
7. INSSEV observed reduction of calibration
8. Modifications for comparison with other instruments
9. Settling Flux

The remainder of this Section details the processing stages used to obtain mass spectra for two parameters: SPM Concentration and Settling Flux.

## 5.1 Dry Mass calculation for each floc

Mass calculation, the product of volume and density, is made more difficult for porous structures like flocs. Effective density, $\rho_e$ , is calculated using the Stokes Equation in Section 2 (Equations 1, 3 and 4). This equation assumes that the particle (floc) is spherical and therefore the volume is also taken as spherical, derived from the horizontal diameter, $D$. An approximate value for Floc Dry Mass may be obtained from the following equation:

$$M_{dry_f} \cong \frac{\pi D^3 \rho_e}{6} \tag{10}$$

Unfortunately this volume times density equation is too simplistic. Two areas of error have to be addressed:

1. Density composition of the particulate matter
2. Interstitial water

The Stokes Equation density notation is appropriate when dealing with single grains composed of single density matter and which have no interstitial water, eg. quartz spheres. For aggregates, in this case estuarine flocs, which may be composed of mineral and organic particulates as well as interstitial water, the definitions of density and effective density need to be more specific. For the remainder of this section, the definitions listed in Table 1 will be used.

**Table 1** Density subscript symbols used in equations.

| | |
|---|---|
| $\rho_e$ | Floc Effective Density ($\rho_f - \rho_w$) |
| $\rho_w$ | Density of Water through which Floc is falling |
| $\rho_f$ | Floc Bulk Density |
| $\rho_m$ | Dry Density of Mineral Component of SPM |
| $\rho_o$ | Dry Density of Organic Component of SPM |
| $\rho_{mo}$ | Mean Dry Density of SPM - see Equation 13 |
| $\rho_{iw}$ | Density of Floc Interstitial Water, $\rho_{iw} = \rho_w$ unless otherwise stated |
| $\rho_{e_{np}}$ | Mean Effective Density for solid (non-porous) aggregate ($\rho_{mo} - \rho_w$) |

### 5.1.1 Density composition of the particulate matter

McCave (1975) provides a definition of the various components of Floc Density, which are used to obtain dry mass (the first two terms) of aggregates in the ocean:

$$V_m \rho_m + V_o \rho_o + V_w \rho_w = V_f \rho_f \tag{11}$$

where the total volume of the floc, $V_f$, is defined as the volume of both the mineral, $V_m$, and organic, $V_o$, components of the SPM and the volume of the interstitial water, $V_w$. Equation 11 assumes that the density of floc interstitial water is the same as the density of the surrounding, or ambient, water. If this is not the case then, using $V_{iw}$ to represent floc interstitial water, Equation 11 becomes:

$$V_m \rho_m + V_o \rho_o + V_{iw} \rho_{iw} = V_f \rho_f \qquad (12)$$

The mean dry density of the floc, $\rho_{mo}$, is obtained from:

$$V_{mo} \rho_{mo} = V_m \rho_m + V_o \rho_o \qquad (13)$$

McCave (1975) made the generalisation that deep ocean aggregates had a 60:40 mineral:organic ratio, producing a mean dry density of 1591 kg·m$^{-3}$. This was based upon the assumption that the mineral dry density was 2500 kg·m$^{-3}$ and the organic dry density was 1030 kg·m$^{-3}$. Calculation reveals that the 60:40 ratio was in fact a mass ratio, which translates into a volume ratio of 38:62, mineral to organic. Estuarine flocs have typical mass ratios of 90:10, mineral:organic, which translates into a volume ratio of 78:22. This typical example is based upon a mineral dry density of 2600 kg·m$^{-3}$ and an organic dry density of 1030 kg·m$^{-3}$, and results in a mean dry density, $\rho_{mo}$, of 2256 kg·m$^{-3}$.

When mineral to organic ratios are known, such as those derived from loss on ashing during gravimetric analysis, the constituent volumes and densities can be calculated.


### 5.1.2 Interstitial Water

Although Equation 10 is a reasonable approximation for low density flocs where the volume of floc interstitial water, $V_{iw}$, is approaching the total floc volume, $V_f$, it is not correct and becomes less accurate with increasing effective density. This is because Equation 10 uses $\rho_e$, which is $(\rho_f - \rho_w)$, in calculating dry mass, ie. $(V_f \rho_f - V_f \rho_w)$, whereas the equation required is one that calculates the floc bulk mass minus the mass of interstitial water only, $(V_f \rho_f - V_{iw} \rho_{iw})$ ; see Table 1 for subscript definitions.

The error in Equation 10 is due to the fact that $V_{iw}$ never equals $V_f$ and *decreases* with increasing effective density. A more accurate calculation of floc dry mass can therefore be made by reducing the volume, and hence mass, of the water that is subtracted. This can only be done with INSSEV data by estimating the volume ratio of particulate matter to interstitial water from the floc effective density. If flocs are highly consolidated aggregates, with hardly any interstitial water, then $V_{iw}$ approaches zero. If they have an extremely low effective density, then $V_{iw}$ approaches $V_f$. Thus there is an inverse linear relationship between $\rho_e$ over $\rho_{e_{np}}$ and $V_{iw}$ over $V_f$, which can be used to define $V_{iw}$. A value is required for $\rho_{e_{np}}$, which is the difference between the mean dry density of mineral and organic components of the SPM, $\rho_{mo}$, and the ambient water density, $\rho_w$. In estuaries $\rho_{mo}$ will range between about 2000 - 2600 kg·m$^{-3}$; 2256 kg·m$^{-3}$ has been used for the examples in this paper. An accurate value for $\rho_{mo}$ can be obtained from gravimetric analysis but the following equations are not very sensitive to the typical variation found in most estuaries.

$$V_{iw} = V_f\left(1 - \frac{\rho_e}{(\rho_{mo} - \rho_w)}\right) \tag{14}$$

Equation 14 can be incorporated into a more accurate version of Equation 10.

$$
\begin{aligned}
M_{dry_f} &= V_f\rho_f - V_{iw}\rho_{iw} \\
&= V_f\left((\rho_e + \rho_{iw}) - \left(1 - \frac{\rho_e}{\rho_{mo} - \rho_w}\right)\rho_{iw}\right)
\end{aligned}
\tag{15}
$$

Taking $\rho_{iw}$ to be the same as $\rho_w$, Equation 15 is then reduced to the following, which is soluble with INSSEV and Master Variable data.

$$M_{dry_f} = \frac{\pi D^3}{6} \frac{\rho_e \rho_{mo}}{(\rho_{mo} - \rho_w)} \tag{16}$$

This calculation of individual floc dry mass from *in situ* observations offers many possibilities. The method for obtaining SPM concentration and settling flux spectra are detailed in the following sub-sections. The dry mass data may also be used for exploratory simulations of the many processes involved in what is loosely termed flocculation or aggregation. INSSEV data have already shown that floc populations change significantly during the tidal cycle (Fennessy et al., 1994b). *In situ* data on collision frequency, flocculation efficiency and break-up are not yet available. Equation 16 provides an important variable for the investigation of these population properties.

## 5.2 Volume referencing of the observed sample

Before dry mass values can be summed for conversion into a Concentration value, the nature of the INSSEV sampling procedure has to be evaluated so that the volume of water from which the flocs have been collected can be defined and appropriate individual correction coefficients applied to each floc. This Sub-Section deals with the Sample Volume definition.

The representative volume of water from which the individual floc data have been obtained is termed the **video object plane sample volume**, $V_{vs}$. This notional volume is illustrated in Fig 4. It is derived from the practical (when measuring flocs from video images) object field width of 4 mm, the nominal 1 mm depth of field, and the 100 mm height of the decelerator, to give 400 mm$^3$. The INSSEV sampling operation is a dynamic procedure lasting up to several hundred seconds, followed by an observation period,

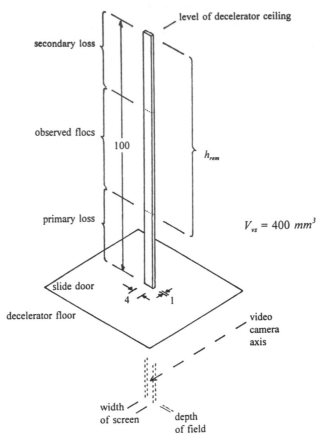

**Figure 4** Illustration of Video Object Plane Sample Volume, $V_{vs}$ , and the Primary and Secondary Losses.

typically 30 minutes. Due to the wide range of settling velocities in each sample and the nature of the Instrument, which sets the flap and slide door operating sequences according to ambient current velocity and SPM concentration, the system is not able to observe, and therefore measure, all flocs contained in the 400 mm$^3$ of the video object plane sample volume. The following sub-section explains the construction of correction coefficients which are applied individual floc dry masses to compensate for the unobserved flocs.

### 5.3 Sample volume corrections

This technique makes the assumption that the sampled floc population is evenly distributed in the decelerator at the start of sampling. It may be applied to all the flocs within a certain size, settling velocity or effective density band. This would then require the use of a mean settling velocity, calculated for all the flocs in the band selected. If the chosen band width is large then the mean settling velocity will not be particularly representative and the technique would lose accuracy. Accuracy improves with decreasing bandwidth until there is only "one floc in the band" and the settling velocity of the floc is used to calculate the correction coefficient, $b_f$, which is applied to the individual floc dry mass. It is this individual floc correction that is described in this sub-section.

The technique considers two operational periods that cause part of the notional 100 mm column (Fig 4), defining the Sample Volume, not to deliver all the flocs, contained within its height, into the settling column. These are termed the primary and secondary losses and result in the cumulative loss which generates the correction coefficient, $b_f$. The "loss" equations have units of length (notional column height) in mm, and are the product of the individual floc settling velocity and time intervals when the slide door to the column is closed or open above the video object plane.

The time interval responsible for the primary loss, $t_{1loss}$, is the elapsed time from the start of sampling to opening of the slide door, $e_{so}$,

$$t_{1loss} = e_{so} = t_{fc} + t_{td} + \frac{t_{st}}{2} + 3.4 \tag{17}$$

where $t_{fc}$ is the duration of flap closure, $t_{td}$ the turbulence decay time interval, and $t_{st}$ the slide door travel time. These first three time intervals are operationally defined by the instrument system in response to ambient master variables. They have been determined empirically during the field trials programme. The constant represents the total fixed delays due to processing time and relay stabilisation. The primary loss, $h_{1loss}$, for any observed individual floc is the product of its settling velocity and the primary loss time interval:

$$h_{1loss_f} = w_{s_f} \cdot t_{1loss} \tag{18}$$

The "loss" refers to that part of the 100 mm column (Fig 4) which contains further flocs of the same settling velocity, but which settle to the floor of the decelerator and are not observed.

In order to evaluate the secondary loss a value is required for each floc to represent the remaining height of the notional 100 mm column. This remaining height value, $h_{rem}$, is obtained using the result from Equation 18.

$$h_{rem_f} = 100 - h_{1loss_f} \tag{19}$$

The primary loss calculation for faster falling flocs will result in a value for $h_{1loss_f}$ greater than 100, which makes $h_{rem_f}$ negative. A condition is required in the data processing that the largest value accepted for $h_{1loss_f}$ is 70, which means that $h_{rem_f}$ never falls below 30.

$$IF \quad h_{1loss_f} > 70 \quad THEN \quad h_{rem_f} = 30 \quad ELSE \quad h_{rem_f} = 100 - h_{1loss_f} \tag{20}$$

The time interval that determines the secondary loss, $t_{2pass}$, is the total time available for flocs to pass into the settling column. It is the duration of the slide door being open above the video object plane, $t_{so-sc}$. It is always too short a time interval for the slowest falling flocs, which means that a proportion of those falling down through the decelerator when the slide door opens will not reach the top of the Settling Column before the slide door closes.

$$t_{2\,pass} = t_{so-sc} = t_{sd} + t_{sl} + 2 \tag{21}$$

The delay between the slide door opening and closing, $t_{sd}$, is set by the estimated SPM concentratiqn at time of sampling. Secondary loss is obtained by deducting the product of individual floc settling velocity and the time available for flocs to pass into the column, $t_{2pass}$, from the height remaining after the primary loss, $h_{rem}$.

$$h_{2loss_f} = h_{rem_f} - w_{s_f} \cdot t_{2\,pass} \tag{22}$$

The primary and secondary losses are incorporated into a single equation for cumulative loss, $h_{cumuloss}$.

$$h_{cumuloss_f} = \left(100 - h_{rem_f}\right) + \left(h_{rem_f} - w_{s_f} \cdot t_{2\,pass}\right)$$
$$= 100 - w_{s_f} \cdot t_{2\,pass} \tag{23}$$

The cumulative loss equation cancels the influence of the primary loss. It is an effective equation for the slower falling flocs but does not address the effect of residual turbulence on the faster settling flocs during the primary loss time. More complex equations have been used to attempt to handle these problems but they are theoretical and haven't been tested experimentally. Instead, limiting conditions are applied during processing and attention has been paid to reducing turbulence effects at the time sampling by improved operating protocols. Equation 23 is transformed into the correction coefficient, $b_f$,

$$b_f = \frac{100}{100 - \left(100 - w_{s_f} \cdot t_{2\,pass}\right)}$$
$$= \frac{100}{w_{s_f} \cdot t_{2\,pass}} \tag{24}$$

The condition required in the application of Equation 24 is:

$$IF \quad w_{s_f} \cdot t_{2\,pass} > h_{rem_f} \quad THEN \quad b_f = \frac{100}{h_{rem_f}} \quad ELSE \quad b_f = \frac{100}{w_{s_f} \cdot t_{2\,pass}} \tag{25}$$

Coefficient $b_f$ is numerically small for the faster falling flocs but large for the slower falling flocs. However, because the slower falling flocs typically have a dry mass three to four orders of magnitude less than the faster falling flocs, the correction causes only a small change in spectral shape between the Floc Dry Mass and the Volume Corrected Floc Dry Mass (Fig 5, histogram rows 1 and 2).

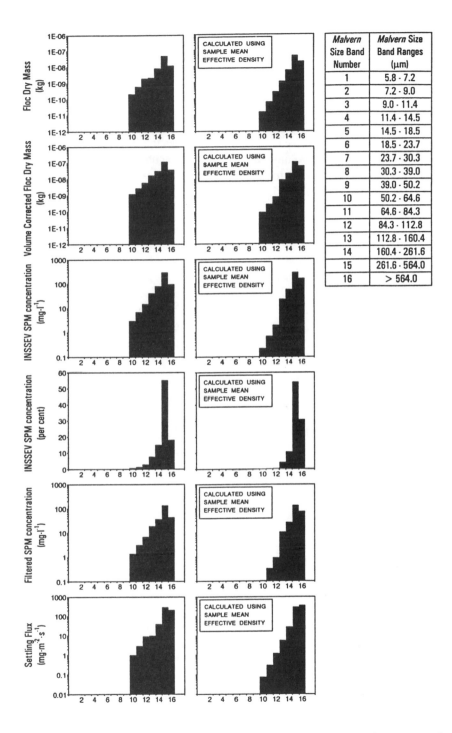

| Malvern Size Band Number | Malvern Size Band Ranges (μm) |
|---|---|
| 1 | 5.8 - 7.2 |
| 2 | 7.2 - 9.0 |
| 3 | 9.0 - 11.4 |
| 4 | 11.4 - 14.5 |
| 5 | 14.5 - 18.5 |
| 6 | 18.5 - 23.7 |
| 7 | 23.7 - 30.3 |
| 8 | 30.3 - 39.0 |
| 9 | 39.0 - 50.2 |
| 10 | 50.2 - 64.6 |
| 11 | 64.6 - 84.3 |
| 12 | 84.3 - 112.8 |
| 13 | 112.8 - 160.4 |
| 14 | 160.4 - 261.6 |
| 15 | 261.6 - 564.0 |
| 16 | > 564.0 |

**Figure 5** Mass Histograms of INSSEV (Fig 2) data by Size Band (left). Same data processed with Sample Mean Effective Density (right).

The correction coefficient, $b_f$, is applied to the floc dry mass value to obtain the Volume Corrected Dry Mass value.

$$M_{dry_{fc}} = M_{dry_f} \cdot b_f \qquad (26)$$

## 5.4 INSSEV SPM concentration calculation

Summing all the corrected floc dry mass values produces the SPM mass in the notional 100 mm column, which has the volume of 400 mm$^3$ (Section 5.2 and Fig 4). This total, corrected floc mass is converted to SPM concentration, with units of mg·l$^{-1}$, using the volume referencing coefficient of $2.5 \times 10^9$.

$$C_{inssev} = \Sigma M_{dry_{fc}} \times 2.5 \times 10^9 \qquad (27)$$

## 5.5 Independent Calibration

The INSSEV Concentration, $C_{inssev}$, can be compared with an independent measurement of SPM concentration. This will usually be a filtered sample taken at the same time. The ratio of these gives another measure of INSSEV performance, $Q_c$.

$$Q_c = \frac{C_{inssev}}{C_{filtered}} \qquad (28)$$

These ratios correlate well with results from backtracking (Section 4) and with measured turbulence characteristics.

The reciprocal of $Q_c$ is used a calibration coefficient and applied to the chosen banded values of corrected floc dry mass.

$$C_{filtered_{band}} = \frac{C_{inssev_{band}}}{Q_c} \qquad (29)$$

Histograms of Concentration by Size Band are presented in Fig 5, histogram rows 3-5.

## 5.6 Band selection

Depending on which type of mass spectra are required the corrected individual floc masses are banded. An advantage of the INSSEV system is the opportunity to select the band intervals of other instrumentation, to allow comparison. In this paper the data is shown in size bands; but this method can be modified in the final stages for settling velocity or effective density banding. Because all three of the usual banding variables (size, settling velocity and effective density) range over several orders of magnitude, it is convenient to use log scales for the banding. The scale used for the size banding used in Figs 5 and 6 is that of the *Malvern Instruments'* laser particle sizer (see Section 6). The bands are related to the optical configuration of the instrument when fitted with a 300 mm lens.

In addition to the specific banding intervals used in this paper for comparison with the *Malvern* sizer, the following standard banding intervals (Table 2) are used for all INSSEV deployments to allow intersite and seasonal comparison of INSSEV data. Single term ratios are formed from these spectral outputs for analysis. For example, in the case of size bands a boundary is created at 126 μm ($10^{1.8} \times 2$), which is convenient for the 125 μm division, proposed by Eisma (1986), as the partition between microflocs and macroflocs. A useful ratio is formed by placing the mass of macroflocs over the mass of microflocs.

Table 2 INSSEV Standard Histogram Bands

| Variable | units | Interval Description | Range | No. of Bands |
|---|---|---|---|---|
| Size | microns | fifth decade × 2 | 20 - 2,000 | 10 |
| Settling Velocity | mm·s$^{-1}$ | quarter decade | 0.01 - 100 | 16 |
| Effective Density | kg·m$^{-3}$ | third decade | 1 - 100,000 | 15 |

The band intervals have also been selected to provide a set of available bands for all anticipated estuarine and fluvial environments. Normally there should be not less than six columns with floc data.

The apparently high upper value for the effective density range is to accommodate an artefact of the use of the Stokes equation. Small needle-like mineral grains have been observed in the INSSEV recordings settling in a vertical orientation. Although their mineral density is of the order of 3000 kg·m$^{-3}$ (which should result in an effective density of ~ 2000 kg·m$^{-3}$) the Stokes calculations (Equations 3 and 4) produce effective densities considerably in excess of 10,000 kg·m$^{-3}$. This is because the observed horizontal diameter is used in the Stokes equations.

## 5.7 INSSEV observed reduction of calibration

The calibration technique (described in Sub-Section 5.5) assumes that INSSEV data includes the very small size fractions, but the lower limit of size resolution is 20 μm. Strictly, the total SPM concentration, by an independent sample, $C_{filtered}$, should be reduced to take account of the mass of material not observed by INSSEV. This is practical when other instrumentation, such as a settling tube or a particle sizer, is deployed at the same time and location as INSSEV and therefore able to give an estimate of the mass of material below 20 μm. The mass of the flocs in a sample which are below 20 μm in size is relatively small but these very small aggregates and primary particles represent a significant proportion of the floc number in most estuaries. Calculation of the amount of reduction needs to take account of the mode of operation of the other instrument. In the case of settling tubes, van Leussen (1988), provides a review of their operational weakness.

## 5.8 Modifications for comparison with other instruments

INSSEV estimates of mass, volume or number spectra are derived from individual floc data measurements of two of the three Stokes variables (size, settling velocity and effective density). Thus, there is considerable scope to mimic the data output format of many other instruments (acoustic, optical and gravimetric) which may only measure one of the three variables. The stages of processing described above place the floc mass data into bands after all corrections have been made. Spectral formats are selected as required. Future work on estuarine sediment transport will benefit from field experiments where several different instrument systems are deployed together such that independent measurements can be integrated. INSSEV is well adapted for such initiatives.

Particle sizers generally measure particle (floc) volume. Mass estimates from these instruments make use of a mean value for particle density (or effective density of flocs). INSSEV data may be processed with a mean effective density calculated for the sample (applied in Section 5.1, Equation 16). In this way the spectral output can be made to mimic a volume measuring instrument. This is illustrated in Fig 5 (right hand column of histograms). The effect is to increase the proportion of the sample mass in the larger size bands. INSSEV data is able to produce mean effective density for a sample. Fennessy et al. (1994b) have shown that this mean variable undergoes considerable variation over a tidal cycle. This suggests that although volume distributions with respect to floc size are valid from particle sizers, mass distribution should not be inferred.

## 5.9 Settling Flux

Total settling flux is the sum of all the volume corrected floc dry mass values multiplied by their individual settling velocities. The following equation applies both the volume referencing coefficient and the independent calibration coefficient.

$$F = \Sigma\left(M_{dry_{fc}} \cdot w_{s_f}\right) \times \frac{2.5 \times 10^9}{Q_c} \tag{30}$$

The units are $mg \cdot m^{-2} \cdot s^{-1}$, provided Dry Mass is entered in kg and Settling Velocity entered in $mm \cdot s^{-1}$. Flux within any band is the sum of the flocs that fall within the selected band. An example of settling flux mass spectra is shown in the bottom pair of histograms in Fig 5. The values represent settling flux in still water. Turbulence parameters need to be applied for ambient estuarine conditions.

The importance of the larger flocs in settling flux is illustrated in Fig 5, rows 5 and 6 (left hand histograms). In this sample 73% of the SPM concentration is in the two largest size bands, but this increases to 89% for settling flux. When turbulence is taken into account, smaller flocs are more likely to remain in suspension, so the larger floc sizes become increasingly important.

## 6   INTERCOMPARISON WITH LASER PARTICLE SIZER

During many of the field trials of INSSEV, samples were taken at the same time as the *In Situ* Laser Particle Sizer developed at Plymouth Marine Laboratory (Bale and Morris, 1987). This is an adaptation of the *Malvern Instruments'* laser diffraction instrument which, for estuarine use, employs a 300 mm lens. This generates the size band intervals that have been used for the histograms of INSSEV data in Fig 5 and 6.

INSSEV Tamar Sample 2 (HW-1.23) taken on 17 March 1993 is compared with a sample taken with the *Malvern* deployed alongside, with 0.6 m between sampling points. Both

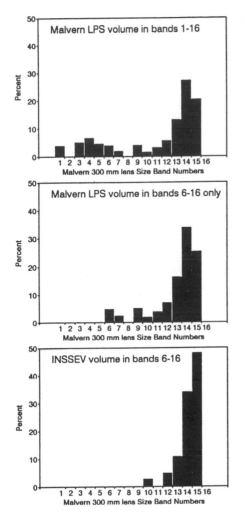

| Malvern Size Band Number | Malvern Size Band Ranges (μm) |
|---|---|
| 1 | 5.8 - 7.2 |
| 2 | 7.2 - 9.0 |
| 3 | 9.0 - 11.4 |
| 4 | 11.4 - 14.5 |
| 5 | 14.5 - 18.5 |
| 6 | 18.5 - 23.7 |
| 7 | 23.7 - 30.3 |
| 8 | 30.3 - 39.0 |
| 9 | 39.0 - 50.2 |
| 10 | 50.2 - 64.6 |
| 11 | 64.6 - 84.3 |
| 12 | 84.3 - 112.8 |
| 13 | 112.8 - 160.4 |
| 14 | 160.4 - 261.6 |
| 15 | 261.6 - 564.0 |
| 16 | > 564.0 |

**Figure 6** Comparison of floc volume distribution from the Tamar Estuary obtained with *in situ* Malvern Laser Particle Sizer and similar output from INSSEV.

instruments sampled at 0.5 m above the bed at the same time. These samples have been used for comparison because they were both considered to be valid in terms of their respective instrumentation checks. Fig 6 shows the size band distributions of the measured floc volumes.

The top *Malvern* histogram has been truncated below the INSSEV size threshold of 20 μm to produce the middle histogram; percentages are proportionately increased but the distribution in bands 6 - 16 is the same. The INSSEV histogram (bottom) does not show any SPM in bands 6 - 9 but this is probably due to the small size of the observed population, or the minimum settling velocity limit. The shape characteristics are similar, but with INSSEV showing a slightly larger proportion of larger floc sizes. The absence of material in size band 11 of the INSSEV histogram is reflected in the distribution shape between bands 9 - 12 of the Particle Sizer.

# 7  CONCLUSIONS

The derivation, within this paper, of concentration and settling flux spectra from size and settling velocity observations of floc samples illustrates the versatility of using individual floc data.

Assembling floc data into many sample mean or spectral formats allows intercomparison with a wide range of existing instrumentation, important for data validation exercises.

The ability to calculate the dry mass of individual flocs within a known volume offers considerable potential for numerical simulations of floc aggregation and disaggregation.

Instruments which only measure floc size, or volume, distribution and then apply a uniform floc density for the sample to obtain floc mass distribution will overestimate the mass contained in the larger floc size bands.

# REFERENCES

Allen T. (1975) *Particle Size Measurement*, Chapman and Hall, London, 454pp.

Bale A. J. and A. W. Morris (1987) In situ measurement of particle size in estuarine waters. *Est. Coast. Shelf Sci.*, **24**: 253-263.

Eisma D. (1986) Flocculation and de-flocculation of suspended matter in estuaries. *Neth. J. Sea Res.*, **20**: 183-199.

Fennessy M. J., K. R. Dyer and D. A. Huntley (1994a) INSSEV: an instrument to measure the size and settling velocity of flocs in situ. *Mar. Geol.*, **117**: 107-117.

Fennessy M. J., K. R. Dyer and D. A. Huntley (1994b) Size and settling velocity distributions of flocs in the Tamar Estuary during a tidal cycle. *Neth. J. Aq. Ecol.*, **28**: 275-282.

Graf W. H. (1971) *Hydraulics of Sediment Transport*, McGraw-Hill, New York, 513pp.

McCave, I. N. (1975) Vertical flux of particles in the ocean. *Deep-Sea Res.*, **22**: 491-502.

Schlichting H. (1968) *Boundary Layer Theory*, McGraw-Hill, New York, 747pp.

ten Brinke W. B. M. (1993) *The impact of biological factors on the deposition of fine-grained sediment in the Oosterschelde (The Netherlands)*. PhD Thesis, University of Utrecht, 252 pp.

van Leussen W. (1988) Aggregation of particles, settling velocity of mud flocs. A review. In: J. Dronkers and W. van Leussen, *Physical Processes in estuaries*, Springer-Verlag, Berlin Heidelberg, pp 347-403.

van Leussen W. (1994) *Estuarine macroflocs and their role in fine-grained sediment transport*. PhD Thesis, University of Utrecht, 488 pp.

# SETTLING AND CONSOLIDATION

# 7 Consolidation of cohesive sediments in settling columns

G. C. SILLS

*Department of Engineering Science, Oxford University, UK*

## Background

An understanding of the behaviour of cohesive sediment, either on its own or mixed with non-cohesive sediment, is required in a wide range of civil engineering applications. This includes the design of new structures such as barrages and breakwaters, the management of ports and harbours where navigation depths must be maintained, the management of waste disposal and the operation of sewers.

All these applications require the use of system models to provide predictions of overall behaviour, where such models incorporate process models of the appropriate components. The success of the system model depends crucially on the accuracy of the process models, since only if these have been appropriately set up will the system model be able to predict events under conditions that have not previously occurred. In general, the system models for the behaviour of cohesive sediment must incorporate the processes of settling, consolidation, erosion and transport.

Laboratory experiments play an important role in developing process models, since it is generally easier to control the boundary conditions, and to make more precise measurements of the relevant factors and variables. These advantages have been recognised in the use of settling columns to examine the process of settling and consolidation, providing insight into fundamental behaviour as well as a means of determination of soil parameters. Examples include Been (1980), Been and Sills (1981), Berlamont et al (1992), Burt and Parker (1984), LCHF (1978), Michaels and Bolger (1962), Migniot and Hamm (1990), Ockenden and Delo (1990),Tan et al (1988), Tan et al (1990), Toorman and Berlamont (1989).

The purpose of this paper is to examine the constraints and accuracies of settling column experiments, along with the interpretation of the results and their subsequent use.

## Effective stress

Typical sediment concentrations in rivers, estuaries and coastal regions range from a few milligrams per litre to a few hundred, with higher concentrations in dredged spoil and in dumped waste slurries. Most settling column experiments therefore start from sediment slurries of initial bulk density in the range up to about 1.2 Mg/m$^3$, typical of the range of applications. After being introduced into the column with uniform density, the sediment surface settles, leaving clear water above. Measurements usually include the rate of surface settlement, bulk density by gamma or X-ray techniques and pore water pressure on the side of the column.

*Cohesive Sediments.*
Edited by Neville Burt, Reg Parker and Jacqueline Watts
©1997 John Wiley & Sons Ltd.

At low initial densities, the sediment slurry behaves as a dense fluid, with the pore water pressures equal to the total vertical stress throughout the bed. This condition is typically referred to as one of hindered settling, in which the rate of surface settlement is proportional to the slurry density, and has been described by Kynch (1952). As settlement proceeds, a denser layer builds up from the bottom of the bed in which the pore water pressures are less than the total vertical stress, so that the sediment weight is partly supported by the soil structure developing in the bed (Been and Sills 1981). The transition density is referred to as the "structural density". Its significance lies in the fact that, for densities above the structural density, the concept of effective stress is viable. This concept was proposed by Terzaghi (1936). Paraphrasing his definition, the first part defined the effective stress $\sigma'$ as the amount by which the total stress $\sigma$ is greater than the pore water pressure $u$; and the second part claimed that all changes in the state of a soil (that is, strains or changes in stiffness or strength) are uniquely related to changes in the effective stress. Despite the fact that experiments have since shown that soils exhibit time and strain rate dependent changes without corresponding changes having occurred in the effective stress, (discussed in the context of settling column experiments by Sills (1995)) the concept has remained a fundamental one for the analysis of soil behaviour. It may be noted that the effective stress has no physical existence, cannot be measured directly, and can only be obtained as the difference between total stress and pore pressure. The traditional soil parameter associated with the effective stress to represent the strain condition of the soil is the void ratio, $e$, defined as the volume of voids as a ratio of the volume of solids. It is appropriate that this is a volumetric measure rather than a gravimetric one such as concentration or density, since it is expected that such phenomena as fluid flow through soil will be determined by characteristics of interparticle spacing and electrochemical activity rather than the mass of the particles.

Settling column experiments, therefore, provide the facility to monitor the behaviour of sediment slurries over a complete range of densities, from values below the structural density in conditions of hindered settling to values above the structural density, when soil conditions prevail. Pore water pressure measurements allow the slurry condition to be described accurately as being either a dense fluid, with pore pressures equal to total vertical stress, or a soft soil in which effective stresses exist.

**Settling column experiments**

The results presented in the paper have been obtained at Oxford, using a standard methodology that has been developed over a number of years. This consists of a single input of soil slurry into an acrylic settling column of internal diameter 102mm, with pore pressure measurements using a transducer calibrated immediately before use and density measurement using an X-ray system moving at 100 mm per minute. Specific soil information is presented later.

**Density measurement and accuracy**

Gamma rays or X-rays are passed through the settling column, and collected in a detector assembly consisting of a sodium iodide crystal and photo-multiplier. The count rate generated can be related exponentially to density through a minimum of two independent calibration points. Since the relationship between count rate and density is exponential, the accuracy of the density calculation will decrease as the density increases. Gamma ray systems have higher energy levels than the X-ray system (for example, the gamma energies of [133]Barium are higher than 660 keV, (MAST G6M report 1993), by comparison with the Oxford X-ray operating energy of 160 keV), so the measurement is less likely to be atomic number dependent, a consideration that can be important in slurries of waste containing heavy metals. On the other hand, the overall intensity available from the X-ray system is much higher, so that the radiation beam can be highly collimated to give high spatial resolution, and a given count rate accuracy can be obtained in a much shorter interval of time. It is therefore feasible to traverse the settling column from top to bottom in a few minutes, thus obtaining continuous profiles. In contrast, to compensate the longer time needed for a reading, the gamma-ray system is often used only at discrete levels. The X-ray system has the safety advantage that it can be switched off, but it is much more expensive to install than the gamma-ray system.

The accuracy of the X-ray system with automatic data logging can be determined by considering the repeatability of a profile, and the method and accuracy of calibration. A count rate profile is obtained by traversing the X-ray system up and down the settling column. Random variations occur due to fluctuations in the X-ray output and the photomultiplier system, while systematic variations are due to changes in the column material and density of the sediment with height. For an experiment in which the sediment mass remains constant, the calibration parameters are obtained from the measured density of the overlying water and from the condition that the area obtained by integrating over the density profile is equal to the total stress at the bottom of the bed. Experience with this calibration process for typical bed depths of 500 mm indicates that the total stress condition at the base can always be satisfied to within ±0.002 kPa. Overall, the density accuracy obtained with the X-ray system and calibration by integration is better than ±0.02 $kN/m^3$, (±0.002 $Mg/m^3$), for densities around 12 $kN/m^3$. The corresponding accuracy for a density of 13 $kN/m^3$ is better than ±0.04 $kN/m^3$, (±0.004 $Mg/m^3$).

The accuracy of the density and the total stress determination may be altered if the calibration relies on separate calibration samples. Experiments have shown that there can be differences in absorption characteristics between similar acrylic tubes, and it is known that acrylic absorbs water over a period of time, so that there is a risk of the calibration density changing during the course of an experiment. An inaccuracy of 0.05 $kN/m^3$, (0.005 $Mg/m^3$), in a calibration sample of density around 12 $kN/m^3$, produces a similar inaccuracy in the density profile at the same density, and a higher one at higher densities.

Accuracies quoted for the gamma-ray systems in the MAST G6M report (1993) are of the order of ±0.1 $kN/m^3$, (±0.01 $Mg/m^3$), although there is no indication of the density at which this accuracy applies. At best, for a continuous profile, this would correspond to an accuracy of total stress of the order of ±0.05 kPa at the bottom of a 500 mm sediment bed. If the density profile had been measured at discrete points, then the total stress accuracy would depend on the interval between readings, and could be very much less.

The results used to illustrate typical density results are taken from experiment KC2, on mud dredged from the Ketelmeer, a lake in the Netherlands. The Liquid and Plastic Limits lie in the ranges 82-88% and 43-48% respectively. The slurry was poured at an initial density of 11.62 $kN/m^3$, (1.184 $Mg/m^3$), into the settling column through a tube whose outlet was kept just below the surface of the rising mud level, to reach an initial height of 600 mm. Typical density profiles for consolidation due to self-weight are shown in Figure 1, (Sills and Yuan 1993). The profile for 14 days is a superposition of the upward and downward profiles, showing typical repeatability. Where features occur in both profiles, they must be real density or column variations, so that this example illustrates the sensitivity of the measurement system.

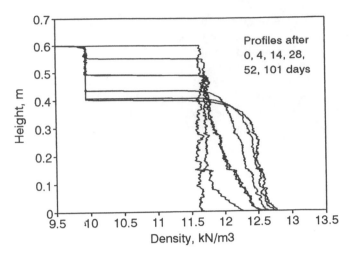

**Figure 1**   Density profiles for self-weight consolidation of Ketelmeer mud, experiment KC2, initial density 11.37 kN/m$^3$, (1.159 Mg/m$^3$), initial height 600 mm

It also illustrates the magnitude of the effect of the small random differences due to fluctuation in the X-ray output. The apparent density variations at 0.15 and 0.27 m from the base were visible in the profile of the water filled column, and are therefore due to non-uniformities in the acrylic column rather than to real density changes. Their influence on the effective stress calculation is very small, so it is not worth using the water profile to adjust subsequent count rate profiles, although this could be done if desired.   It may be noted that the effect reduces at higher sediment densities.  It can be seen that there is a steady increase of density upwards from the bottom of the bed, a pattern which is characteristic of self-weight consolidation of the mud, where the pore pressures dissipate first from the bottom of the bed.

**Pore pressures**

Pore pressures are measured at ports let into a settling column wall, separated from the sediment by a porous element, and connected to a pressure transducer or a standpipe system. The limiting factor in accuracy is then that of the pressure transducer or the accuracy of reading the water level in a standpipe system, bearing in mind capillary effects in the latter case.  The MAST G6M report (1993) quotes accuracies at different research centres of ±1 mm head of water and ±2 mm head of water, depending on the arrangement, corresponding to ±0.01 kPa and ±0.02 kPa.

The system at Oxford uses a multiplexing unit designed by Bowden (1988), to connect a single pressure transducer first to a calibration unit and then in turn to each pore pressure location. A second calibration can be carried out immediately afterwards if desired.  This system ensures that zero drift of the transducer is not a problem.  Great care is taken to ensure that no air is trapped in any of the connecting lines, and the unit can be flushed during an experiment.  The accuracy of repeatability of the calibration is better than 1 mm head of water, or 0.01 kPa.  In measurements in the overlying water where, by definition, the excess pore pressure must be zero, values of excess pore pressure (i.e., the amount by which the measured pore pressure exceeds the equilibrium hydrostatic value) in the range ±0.01 kPa are consistent with this

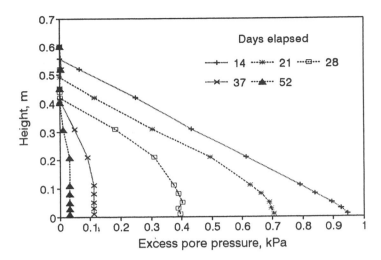

**Figure 2** Excess pore pressure profiles during self-weight consolidation of Ketelmeer mud, experiment KC2

accuracy. In most cases in the consolidating bed, the excess pore pressure distribution follows a smooth curve over the depth of the bed, and an accuracy of better than ±0.02 kPa is a very reasonable claim. Occasionally, however, an isolated reading seems to be inconsistent with the general trend. Possible explanations are partial clogging of the porous filter or the presence of gas either in the filter or the line to the transducer, both of which could occur during the course of an experiment. Whatever the reason, the provision of a number of pore pressure ports - at least ten - over a 500 mm height, coupled with more than one port at the same height, make it possible to assess individual results in the context of the overall pattern.

Figure 2 shows a series of excess pore pressure profiles from the Ketelmeer mud experiment KC2 already introduced (Sills and Yuan 1993). In this example, the results show the typical pattern associated with self-weight consolidation, in which the initial excess pore pressure distribution is triangular, and water drains upward from the base of the bed, so that the excess pore pressures dissipate first from the bottom.

**Effective stress and its correlations with void ratio and permeability**

Overall, given that the accuracy of effective stress calculation depends on total stress calculations and pore water pressure measurements, it is clearly important to recognise the limitations that may exist on interpretations about soil behaviour at low effective stress levels. Results using the Oxford X-ray system suggest that uncertainty in the effective stress calculation is typically less than ±0.03 kPa.

Once the effective stresses have been calculated, a stress-strain relationship for the soil can be produced by correlation with the void ratios at the same height in the bed. These can be obtained from the density profiles by using an appropriate specific gravity for the sediment. This effective stress void ratio correlation is shown in Figure 3 for the Ketelmeer mud experiment KC2. There is a close similarity between the individual results, with a slight indication that the later curves lie below the earlier ones. Other experiments on different

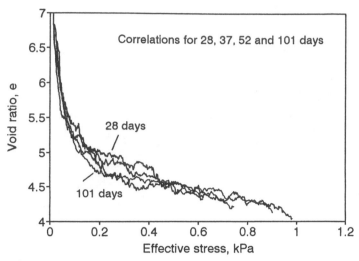

**Figure 3** Void ratio effective stress relationship for Ketelmeer mud, experiment KC2

experiments have shown much more pronounced downward separation of the void ratio effective stress curves with time, and it has been suggested (Sills 1995) that this provides evidence of creep compression, where void ratio changes occur at constant effective stress. In the present case, the compression is too low to be reliably attributed to creep, particularly in view of some of the later discussion.

**Influence of the conditions of the experiment**

In an ideal settling column experiment, there would be no influence of the method of input of the sediment or of the settling column walls. In practice, however, the sediment behaviour will be influenced by both these effects. Elder (1985) reported the results of experiments by Thomas using Combwich mud, an estuarine silty clay, containing approximately 40% clay sizes, with Liquid and Plastic Limits in the ranges 62-63% and 29-32% respectively. The initial densities lay in a range from 11.0 kN/m$^3$ (1.12 Mg/m$^3$) to 11.6 kN/m$^3$ (1.18 Mg/m$^3$) with initial depths of slurry from 500 mm to 950 mm. Some of the slurries were mixed with salt water. The experiments demonstrated that the rate of surface settlement was apparently unaffected by the column diameter for a range of diameters between 100 mm and 180 mm, provided that the sediment slurry was introduced either through a tube whose outlet was raised to keep it always at the level of the mud surface, or by using a spreader to disperse the slurry. Following this analysis, the former method of introduction of the slurry has been adopted as the standard method for tests at Oxford.

The normal assumption in analysing settling column results is that settlement and flow are one-dimensional, so that uniform conditions exist across any horizontal plane. To check this, Elder (1985) made direct measurements of moisture content from the centre line to the side wall at the end of a number of column tests using Combwich mud. Most of his measurements were made at the end of experiments in which the pore pressure dissipation had been close to complete before the column was dismantled, and these showed very little lateral variation. In one of the tests, however, the pore pressures were still comparatively high when the experiment was finished after 518 hours. This was an experiment of initial slurry height 1748 mm, and

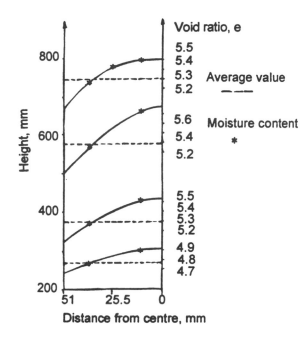

**Figure 4**    Variation of void ratio across column at the end of experiment DME5 on Combwich mud

density 11.9 kN/m³ (1.21 Mg/m³), which is above the structural density for Combwich mud under the conditions of a single input. Effective stresses therefore existed from the start of the experiment. By the end of the experiment, the bed height had reduced to less than 1400 mm. Figure 4 shows the void ratios calculated from the moisture content measurements, with a speculative variation across the radius, assuming symmetry about the centreline. The value obtained from the X-ray density profile is marked for comparison. It can be seen that the void ratio is lower near the column wall than near the centre, a result that implies that higher effective stresses existed near the wall than at the centre. This could be explained by preferential drainage up the column walls, leading to an excess pore pressure gradient across the horizontal plane. The effective stresses would then indeed have been highest at the walls for as long as the pore pressures were above the final equilibrium hydrostatic value. There must therefore be considerable doubt about whether true one-dimensional conditions exist throughout a typical settling column experiment.

Another way in which non-uniform conditions could develop in the horizontal plane is through friction between the sediment and the column walls. This would lead to a reduction in the total stress transmitted through the consolidating bed, as well as to non-uniformity of the total stress across the base. Measurements have been made of the vertical total stress in the centre of the column base for different sediments at different initial densities. Figure 5 shows results from KC2, in which the vertical total stress decrease is plotted against time, compared with the calculated vertical effective stress increase at the position of the lowest pore pressure measurement, 10 mm up from the base of the column. The bed height is also plotted. It can be seen that the pattern of loss in total stress on the centreline is very similar to that of the effective stress increase and the bed surface settlement. Thus, the total stress loss increases with the increasing stiffness and yield stress that occurs as the soil consolidates.

113

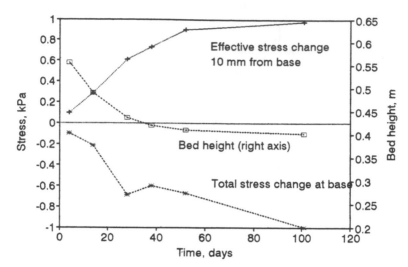

**Figure 5** Variation of measured vertical total stress at the centre of the column base and calculated vertical effective stress near the base in experiment KC2, Ketelmeer mud

This reduction in the vertical total stress can be explained by two different mechanisms. In the first case, a preferential drainage path up the side of the column would cause higher vertical effective stresses here than in the centre, so that the consolidating bed would develop as an annulus of stiffer soil around a softer centre. This stiffer annulus could then support an increasingly large proportion of the buoyant weight of the soil, causing a redistribution of the vertical total stress away from the centre. In the second mechanism, friction develops between the sediment bed and the column walls, with the soil arching between the walls. In this case, some or all of the buoyant weight would be carried by wall friction that increased as the soil consolidated. There would be a rotation of principal stresses across the horizontal plane, from vertical on the centreline to a direction other than vertical at the walls, where vertical shear stresses existed. It may be noted that the effect of the first mechanism would be enhanced by a smooth column surface, since this would be likely to offer less resistance to water flow, and therefore a greater possibility of preferential drainage, while the effect of the second mechanism would be stronger in the case of a rough boundary. Some horizontal non-uniformity therefore seems inevitable.

There is no evidence that the measured total stress at the centre of the base can be used to determine the magnitude of the effective stress acting on the sediment. The effective stress change shown in Figure 5 has been calculated from the total stress obtained by integrating the KC2 density profiles. If the measured total stress at the bottom of the bed had been used instead, there would be practically no development of effective stress, and this would be inconsistent with the observed settlement and density increases. The total stress reduction is therefore an indication of some horizontal variability, but the driving process for the consolidation is still the self-weight stress. It is, however, clear that the experimental results provide mean values rather than local values.

The sediment bed in experiment KC2 was loaded beyond the self-weight stress condition by applying a surcharge through a piston on top of the bed. Figure 6 shows two self-weight density profiles, for 52 and 101 days, along with three profiles produced by a surcharge

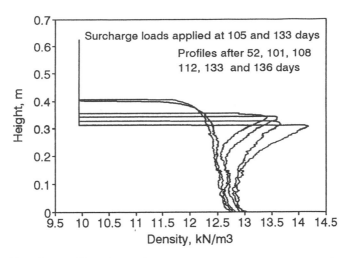

**Figure 6**    Density profiles for self-weight and surcharge loading of Ketelmeer mud, experiment KC2

increment of around 2 kPa applied after 105 days, and a sixth profile following a second surcharge increment after 133 days.    The effect of the surcharge load is to increase the total stress level throughout the bed.  The pore water drains first from the top of the consolidating bed, so that the effective stresses  increase here earlier than they do deeper in the bed, thus producing a higher density surface layer.    Corresponding excess pore pressure profiles are shown in Figure 7.  For consistency with the concept of excess pore pressure as the difference between the pore pressure and the hydrostatic pressure, (and thus a measure of the pressure head causing water flow) the total stress is marked on this Figure as the excess total stress, or the amount by which the total stress is greater than the hydrostatic pressure.  It can be seen that the stress recorded in the piston through which the surcharge is applied varies as the bed consolidates, possibly due to variable friction between the piston seal and the surrounding column walls.  The actual effective stress in the bed is a combination of the applied surcharge effective stress and the self-weight effective stress already considered.  For a comparatively small surcharge load, the effective stress is higher at the bottom of the bed, but as drainage occurs under higher surcharge levels, higher effective stresses work their way down from the top of the bed.    Figure 8 shows the effective stress void ratio relationships obtained for these results.  The first two data sets identify the self-weight correlation.  The next data set shows the effect of the first surcharge loading, which increases the effective stresses in the top part of the bed to values only a little below those already existing in the lower part of the bed.  It can be seen, however, that the void ratios associated with the upper part are lower than those of the lower part of the bed.  This is, of course, consistent with the density profiles for the same period.  Later data sets for the same load increment show similar correlations, while the final data set, obtained after the next load increment, shows that the even higher densities of the surface layer are now associated with higher effective stresses also.  The pattern is of two correlations, almost parallel to each other, one representing and extending the self-weight correlation, and the other describing a structure which exists in the upper part of the bed where a collapse has occurred to a lower void ratio under the steep effective stress gradient.  A reasonable explanation is that the initial deposition process produced a very open framework, which was susceptible to collapse under the high stress gradients subsequently imposed.

115

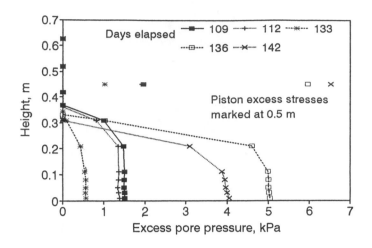

**Figure 7** Excess pore water pressure profiles for surcharge loading of Ketelmeer mud, experiment KC2. The current surcharge load is marked for each profile

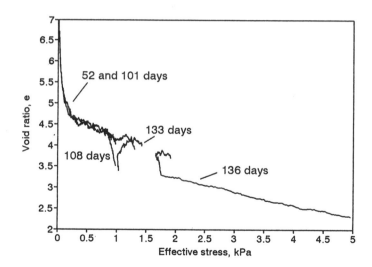

**Figure 8** Void ratio effective stress relationship due to self-weight and surcharge loading for Ketelmeer mud, experiment KC2

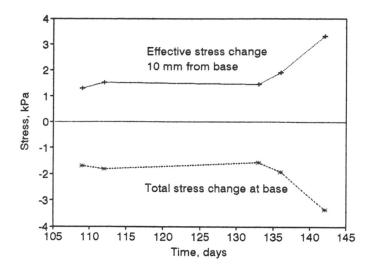

**Figure 9** Variation of measured vertical total stress at the centre of the column and calculated vertical effective stress near the base for surcharge loading in experiment KC2, Ketelmeer mud

Figure 9 shows the development of effective stress calculated for a level near the bottom of the sediment bed compared with the loss of total stress recorded at the centre of the base of the column. As for the self-weight stages, there is a strong correlation between the two. During the first load increment, applied after 105 days, there is little change in the vertical effective stress, and the total stress loss also remains steady. The effect of the second increment, applied after 133 days, was to increase the effective stresses throughout the bed, and the total stress loss increased also. Both the mechanisms described earlier would explain this observed behaviour, and wall friction, at least is likely to exist.

Total stress losses of similar magnitudes have been observed in other experiments on Ketelmeer mud, (Sills and Yuan 1993) and in experiments on other muds, including Bothkennar mud, from the EPSRC site near Grangemouth, Scotland. Been and Sills (1981) noted that much smaller losses, of an order of magnitude less than those reported here, occurred in experiments on Combwich mud, an estuarine silty clay, in which the initial densities were much lower. However, the effective stress levels that developed were also lower. Other experiments on these and other sediments have shown a range of losses, up to a maximum for KC2. The actual amount would be expected to be related both to the sediment itself, to the conditions of the experiment, such as the initial density and initial height, and also to the sediment/column friction and the occurrence of preferential drainage.

**Discussion**

It has been shown that horizontal non-uniformity can occur across the settling column., as indicated by differences in void ratio at the centre and at the sides of the column at the same elevation in one particular experiment. These differences were lower near the bottom of the column, where the effective stresses reach their highest values under self-weight consolidation. However, in other experiments, very little horizontal void ratio variation occurred, so that it is not an inevitable phenomenon. In other experiments, the vertical total stress measurement at the centre of the base of the column suggests that there was a non-uniformity of total stress and effective stress. Two mechanisms have been proposed, one relating to preferential drainage paths and one to the development of shear stresses at the column wall. It is likely that these two processes occur together. The variation across the column cannot be detected by the X-ray measurement system, so that a range of densities will be represented by a mean value, with the corresponding effective stress variation also being reduced to a single value. Such averaging of parameters occurs in many experiments, and is not necessarily a problem, as it is unlikely to introduce apparent causative relationships that do not exist. It will therefore still be expected to provide a qualitative understanding of sediment behaviour. However, there is a need for caution in the application of quantitative relationships obtained from laboratory tests to the prediction of field conditions, although there may well be no better method of obtaining the required parameters.

The results presented in this paper have come from experiments in which the initial density of the sediment slurry was higher than the structural density, where effective stresses exist and the behaviour can be discussed in terms of traditional soil concepts. Similar conclusions may also be drawn from other experiments in which the initial density was lower than the structural density, and the initial stage of the experiment included a period of hindered settling.

## Conclusion

This paper has dealt with two aspects of the use of settling columns in laboratory experiments. The first conclusion is that a high resolution of effective stress and void ratio calculations can be obtained using the X-ray system to measure densities, coupled with pore pressure measurement with a single transducer calibrated immediately before use. The second conclusion is that these measured and calculated values represent mean values across a horizontal diameter of the column, and that the measurements available do not allow an estimation of the variability that exist across this section. Measurement of the water content across horizontal sections should perhaps become a standard conclusion to a settling column test, and an additional vertical total stress transducer near the wall of a column could provide useful information. At present, it is not possible to do more than recognise that the accuracy of an effective stress void ratio correlation may be considerably less than the accuracies obtained from calibrations of the measuring systems used.

## Acknowledgments

I am very grateful to members of the soil mechanics group at Oxford for their considerable assistance over the years with the settling column research. Mr. R. Morton has maintained the X-ray stability, and Mr. C. Waddup and Mr. F. Yuan carried out the experiments on Ketelmeer mud. I have had fruitful discussions with many members of the soil mechanics group, including Dr R. Gonzalez, Dr M. Rutherford, Mr B. Slujitoru and Mr Yuan, and with Ms. B. Wichman of the Rijkswaterstaat. Some of the research quoted was supported by the Rijkswaterstaat in the Netherlands, and I am grateful to Ms Wichman for her agreement to quote the Ketelmeer results used in this paper.

# References

Been, K. (1980) Stress-strain behaviour of a cohesive soil deposited under water. D.Phil. thesis, Oxford University.

Been, K. and Sills, G.C. (1981) Self-weight consolidation of soft soils: an experimental and theoretical study. Géotechnique XXXI 4, pp. 519-535.

Berlamont, J., van den Bosch, L. and Toorman, E. (1992) Effective stresses and permeability in consolidating mud. 23rd Int. Conf. on Coastal Engineering, Venice. ICCE, pp. 135-136.

Bowden, R.K. (1988) Compression behaviour and shear strength characteristics of a natural silty clay sedimented in the laboratory. D.Phil. thesis, Oxford University.

Burt, T.N. and Parker, W.R. 1984 Settlement and density in beds of natural mud during successive sedimentation. Hydraulics Research, Wallingford, Report IT 262, 15pp.

Elder, D. McG. (1985) Stress-strain and strength behaviour of very soft soil sediment. D.Phil. Thesis, Oxford University.

Kynch, G.J. (1952) A theory of sedimentation. Trans. of the Faraday Society, Vol. 48, pp. 166-176.

LCHF (1978) Rheological study on mud from Zeebrugge. General Report, Lab. Central d'Hydraulique de France, 66pp. (in French).

MAST G6M (1993) Coastal morphodynamics: on the methodology and accuracy of measuring physico-chemical properties to characterize cohesive sediments. Co-sponsored by Commission of the European Communities Directorate General XII. Prepared as part of the EC MAST-I Research Programme.

Michaels, A.S. and Bolger, J.C. (1962) Settling rates and sediment volume of flocculated kaolin suspensions. Industrial and Engineering Chemistry Fundamentals, Vol. 1, pp. 24-33.

Migniot, C. and Hamm, L. (1990) Consolidation and rheological properties of mud deposits. Proc. 22nd Int. Conf. Coastal Engineering, ASCE, pp. 2975-2983.

Ockenden, M.C. and Delo, E.A. (1990) Laboratory testing of muds and the application of results. Abstract Volume of the Int. Workshop on Cohesive Sediments: "Towards a definition of `mud'", Royal Belgian Institute for Natural Sciences, Belgium, Brussels, Ed. S. Wartel, pp. 60-69.

Sills, G.C. (1995) Time dependent processes in soil consolidation. International Symposium on Compression and Consolidation of Clayey Soils, Hiroshima, Japan, May 1995. *SM*

Sills, G.C. and Yuan, F. (1993) Research into the consolidation behaviour of gassy slurry. Report to Rijkswaterstaat, Netherlands, on contracts DWW 555 and DWW 555A.

Tan, S.A., Tan, T.S., Ting, L.C., Yong, K.Y., Karunaratne, G.P and Lee, S.L. (1988) Determination of consolidation properties for very soft clay. Geotechnical Testing Journal, ASTM, Vol 11, No. 4, pp. 233-240.

Tan, T.S., Yong, K.Y., Leong, E.C. and Lee, S.L. (1990) Sedimentation of clayey slurry. J. Geotechnical Engineering, ASCE, Vol 116, No. 6, pp. 885-898.

Terzaghi, K. (1936) The shearing resistance of saturated soils and the angles between the planes of shear. Proc. 1st Int. Conf. Soil Mech., Vol. 1, pp. 54-56.

Toorman, E.A. and Berlamont, J.E. (1989) Report on the consolidation experiments on Doel Dock mud. Report HYD 117, Hydraulics Laboratory, K.U. Leuven (in Dutch).

# 8     Towards a new constitutive equation for effective stress in self-weight consolidation

E. A. TOORMAN AND H. HUYSENTRUYT,
*Hydraulics Laboratory, Katholieke Universiteit Leuven, Belgium*

## INTRODUCTION

When in a cohesive sediment suspension the suspended particles (flocs and aggregates) deposit on the bottom of the water column they form a loosely packed layer, which has a continuous, open network structure. Under the weight of the accumulating particles above, this structure slowly deforms due to unrigid bonds and sliding of particles, during which pore water is expelled. The understanding and modelling of this compaction process, known as consolidation, is of importance for cohesive sediment transport modelling (e.g. bed level variations, erosion strength), the management of dredged material disposal sites (e.g. storage capacity, re-use frequency) and for applications in other fields (e.g. thickening of colloidal suspensions).

A sediment suspension is an incompressible two-phase continuum of solid particles in water. Incompressibility does not exclude the possibility of local changes in bulk density of the mixture (e.g. a water column with on its bottom a saturated sediment bed). Therefore, consolidation models need to solve the mass balance equation of only one phase (see Toorman, 1996). The solid mass balance equation requires additional information on the consolidation rate (the pore water flow rate if the fluid mass balance is used), which is obtained from the semi-empirical force balance equation, which describes the momentum exchange between the fluid and the solid phase. It neglects inertia forces. This stress balance equation introduces the parameters permeability ($k$), or alternatively the effective settling rate (Toorman, 1996), and effective stress ($\sigma'$), for which empirical relationships must be found.

The search for proper constitutive relations has been continuing for several decades. Traditionally the consolidation parameters $\sigma'$ and $k$ have been assumed to be a function of the concentration alone (in geotechnical models they are expressed in terms of void ratio $e$). The earliest and most widely used empirical law for $\sigma'$ is the semi-logarithmic relationship $e = \alpha + \beta \log \sigma'$, proposed by Terzaghi (1942). Other simple polynomial relationships of order 1 or 2 have been proposed by others (see Alexis *et al.*, 1993). These elegant forms seem to describe well the behaviour of certain materials in very specific tests. But a general consensus on the form of the constitutive equation has never been achieved.

However, there are serious problems with the assumption that $\sigma'$ is only a function

*Cohesive Sediments.*
Edited by Neville Burt, Reg Parker and Jacqueline Watts
©1997 John Wiley & Sons Ltd.

of the concentration. Different types of experiments clearly show a time dependent behaviour. Elder & Sills (1984) seem to be the first to bring this point to the attention. In this paper an overview will be presented of the different observations and arguments that suggest a more complicated dependence. Several possible forms of the constitutive law for effective stress are critically evaluated by means of experimental data and theoretical considerations.

## STRESS BALANCE FOR A CONSOLIDATING SATURATED SOIL

When there is relative motion between the particles and fluid, there is momentum exchange between the phases because of friction at the contact surfaces. During sedimentation this leads to drag and lift forces on the particles. In a quiescent suspension the fluid pressure $p$ equals the weight or total stress $\sigma$ of the fluid column, which is computed by integration of the density profile:

$$\sigma(z) = \sigma_i + \int_z^{z_i} \rho g \, dz \qquad (1)$$

with: $z$ = vertical coordinate (positive upwards); $\sigma_i$ = overburden pressure or load at $z_i$. During consolidation the collapsing soil skeleton forces pore water to move out of the squeezed pores. However, there is much flow resistance due to friction in the narrow drainage paths, formed by the connecting pores. Consequently there is an increase of the pore fluid pressure due to the weight of the solid particles. The pore pressure $p$ can become significantly higher than the hydrostatic pressure $p_0$. The difference, the excess pore pressure $p_e$, increases with depth according to an increase in weight, but does not outbalance the buoyant weight $\sigma_0 = \sigma - p_0$, because the skeleton has the capacity to carry part of its own weight thanks to its continuous network structure. On the other hand the deformation of the skeleton is slowed down by the retarded pore water movement. The fraction of the load, carried by the soil structure is the effective stress, which is defined as:

$$\sigma' = \sigma - p = \sigma - (p_0 + p_e) = \sigma_0 - p_e \qquad (2)$$

The presence of an excess pore pressure is an indication that the consolidation process is still continuing. The driving force for the pore water flow is the excess pore pressure gradient. Generally the flow is upward since the excess pore pressure is zero in the supernatant fluid above the mud layer. The balancing of the excess pressure gradient by the flow resistance is expressed by the semi-empirical Darcy-Gersevanov law (Gersevanov, 1934). It can be written as (Schiffman et al., 1985):

$$-\frac{1}{\rho_w g} \frac{\partial p_e}{\partial z} = \frac{1-\phi}{k}(u-v) \qquad (3)$$

where: $k$ = permeability, $g$ = gravity constant, $\rho_w$ = water density, $\phi$ = sediment volume fraction $(1-\phi$ = porosity$)$, $u$ = fluid velocity, $v$ = averaged sediment particle velocity. In general the permeability is not a constant, but depends on the concentration and possibly other factors (even direction, which is a result of orientation of flocs of the plate-like clay particles). By substitution of eq.(2), eq.(3) can be rewritten as:

$$\frac{1}{g}\frac{\partial\sigma'}{\partial z} = -\Delta\rho_s\phi + \rho_w(1-\phi)\frac{u-v}{k} \tag{4}$$

where: $\Delta\rho_s = \rho_s - \rho_w$, with $\rho_s$ = sediment density. This equation, known as the *saturated soil stress balance*, describes how the effective stress on each layer of particles is the result of the downward force of the buoyant weight of the particles minus the upward friction force due to Darcian flow of the ambient fluid.

In order to eliminate the fluid velocity, the average velocity of the two-phase system will be defined as $U = \phi v + (1-\phi)u$. The stress balance equation can then be written as:

$$\frac{1}{g}\frac{\partial\sigma'}{\partial z} = -\Delta\rho_s\phi + \rho_w\frac{U-v}{k} \tag{5}$$

In the case of sedimentation without bottom (or external) drainage $U = 0$. The solids sedimentation rate $v$ can now be eliminated by combining eq.(5) with the mass balance equation, giving an equation of the form (Toorman, 1996):

$$\frac{\partial\phi}{\partial t} + \frac{\partial(v\phi)}{\partial z} = \frac{\partial\phi}{\partial t} - \frac{\partial}{\partial z}\left[w_0\phi + \frac{w_0}{\Delta\rho_s g}\frac{\partial\sigma'}{\partial z}\right] = 0 \tag{6}$$

where: $w_0 = k\Delta\rho/\rho_w$ = stress-free settling rate. In the case that $\sigma'$ is assumed to be a function of $\phi$, the chain rule can be applied to the effective stress gradient, transforming this term into a Fick-law type term with a diffusivity coefficient $D$ proportional to $\partial\sigma'/\partial\phi$ (Toorman, 1996). Solution of the obtained equation allows the computation of the time evolution of the density profile. This requires the knowledge of the constitutive laws for permeability and effective stress.

Pore water flow and network deformation stop where the hydraulic gradient becomes zero ($\partial p_e/\partial z = 0$). This pore pressure equilibrium can also be described by the equation:

$$\frac{\partial p_e}{\partial z} = \frac{\partial p_e}{\partial\sigma_0}\frac{\partial\sigma_0}{\partial z} = \frac{\partial p_e}{\partial\sigma_0}g\Delta\rho \quad\Rightarrow\quad \frac{\partial p_e}{\partial\sigma_0} = 1 - \frac{\partial\sigma'}{\partial\sigma_0} = 0 \tag{7}$$

Notice that local pore pressure equilibrium is not necessarily equal to process equilibrium ($\partial\sigma'/\partial t = 0$). Indeed, the effective stress still increases as long as pore pressure equilibrium is not attained over the whole column, i.e. as long as consolidation continues.

## EFFECTIVE STRESS MEASUREMENT

### Self-weight consolidation

In self-weight consolidation tests the local effective stress is obtained as the difference of the total stress, computed with eq.(1) using measured density profiles, and the pore pressure, measured with capillary tube pore pressure probes (e.g. Bowden, 1988).

A major problem is the accuracy of effective stress measurements, which depends on the accuracy of the measured densities (1%) and pore pressures (error $\leq 10$ Pa). Although the theoretical error is only of the order of 10 Pa, in practice it is about one order higher. The accuracy of effective stress data depends very much on its value relative to the total stress. In the beginning of consolidation, effective stress values are small ($<1\%$ of $\sigma$), i.e. the error on $\sigma'$ can reach $100\%$. Since the absolute error remains the same, the relative error decreases as consolidation proceeds and effective stresses grow.

Just below the surface, the effective stress cannot be determined accurately, because there total stress and pore pressure are of the order of their accuracy. According to the experience at the K.U.Leuven it seems to be impossible to obtain a higher accuracy than an absolute error on effective stress of 50-100 Pa, which seems to confirmed by data from elsewhere. This implies that no really meaningful measurements can be made in the top layer, which thickness can be roughly estimated as follows. Assume a homogeneous density distribution in the top layer and equilibrium. In that case there are no excess pore pressures and the effective stress is then equal to the buoyant weight, i.e. $\sigma' = \Delta\sigma = \Delta\rho g \Delta z$. Hence, when $\Delta\rho = 100$ kg/m$^3$, $\Delta z$ must be larger then 0.1 m to be meaningful, and even larger when there is no equilibrium.

## Geotechnical measuring techniques

The most commonly used geotechnical device is the oedometer or consolidometer, in which the deformation of a soil sample under different loads is recorded (e.g. Cargill, 1983). This method produces data for high concentrations and loading, much higher than those obtained in self-weight consolidation and therefore is not useful for sediment beds in nature. The constant rate of strain (CRS) method (Cargill, 1986; Znidarčič et al., 1986) is designed to measure at lower concentrations (down to 240 g/l) and effective stresses as low as 100 Pa. Comparison between oedometer and CRS data, presented by Cargill (1983), show dependence of measured $\sigma'$ values at a certain sample density on the technique (up to 60% difference) and on the initial concentration.

## ANALYSIS OF EFFECTIVE STRESS DATA

### Effective stress profiles and density dependence

The time evolutions of the measured density (using a gamma-transmission probe) and excess pore pressure profiles and the corresponding effective stress profile for one consolidation column experiment with River Scheldt mud are shown in figures 1.A-1.C. The experimental procedure is described by Berlamont et al. (1993). On the bottom the gradient of the excess pore pressure is zero (fig.1.B). This corresponds with what logically follows from Darcy's law when there is no pore water flux (see eq.7).

Assuming that the effective stress is a function of concentration alone, following the traditional assumption, substitution of this empirical relationship into eq.(6) and application of the differentiation in parts would transform the effective stress term in eq.(6) into a diffusive one. Figure 2 presents the effective stress data as a function of the local density. The results show significant scattering of the data. This may be partially explained by the relatively large error on the density measurement ($\pm$ 20 kg/m$^3$). But this is not the major cause. It is observed that at locations where the density hardly increases with time, the effective stress does increase significantly (particularly at the bottom). There is also clearly a trend of *gradually* increasing effective stress with depth. This is particularly evident in the lower part of the column where the density discontinuity due to the sand layer does not cause a discontinuity in the effective stress profile. These two observations explain the scatter in the vertical direction.

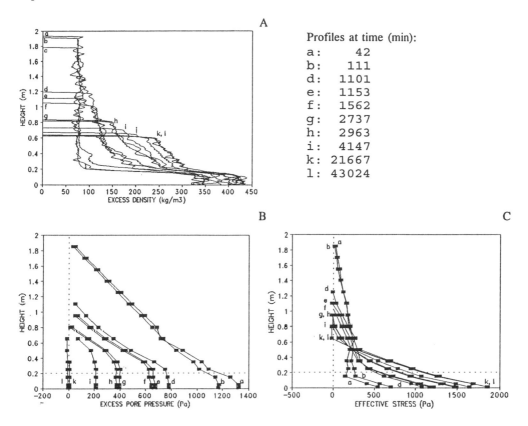

A

Profiles at time (min):

a:       42
b:      111
d:     1101
e:     1153
f:     1562
g:     2737
h:     2963
i:     4147
k:    21667
l:    43024

B

C

**Figure 1.** Measured density (A), excess pore pressure (B) and effective stress (C) profiles for a settling column test with a mixture of river Scheldt mud and fine sand (sand content 60%). Initial density: 1097 kg/m³. The density profiles (measured with a gamma-transmission probe) clearly show the segregation of sand, which occurred at the beginning of the test, resulting in a nearly 0.2 m thick sand layer on the bottom. The remaining suspension has an initial density of 1075 kg/m³, i.e. 23% of the material has segregated. Horizontal dotted line in figs. B and C indicates surface of segregated sand layer. (Irregularities in the top two $p_e$ profiles indicate malfunctioning of some piezometers).

By definition the value of the effective stress at the interface is zero, independent on the mud layer surface concentration, which increases due to consolidation. Therefore, the traditional constitutive equation where $\sigma'$ is only a function of concentration cannot be correct, because it implies an increasing discontinuity of $\sigma'$ at the interface (*Elder & Sills,* 1984). All the profiles show that the effective stress gradually decreases to zero at the interface between suspension and consolidating bed (fig.1.C).

A few proposals can be found in the literature to solve the interface problem. Pane & Schiffman (1985) assumed that there is a relatively thin transition layer between

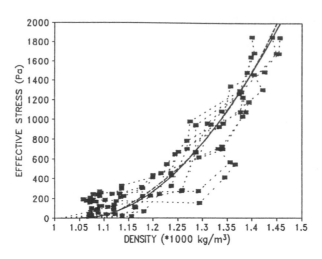

**Figure 2.** Effective stress data from fig.2b as a function of local density. Full line = exponential fit; dashed line = power law curve fit.

suspension and consolidating mud where the structure is in formation and thus the interparticle stresses are only partially developed. Consequently, they proposed to overcome the problem by multiplying the effective stress term with a factor $\beta(\phi)$ to account for the transition from settling ($\beta = 0$) to consolidation ($\beta = 1$). Apart from these conditions, the form of the function $\beta$ has never been investigated. They applied this function only over a thin layer around the interface. In the remainder of the mud layer they kept on following the traditional view, such that their function only guaranties a smoothing of the discontinuity. Been & Sills (1981) allowed the effective stress-void ratio relationship to change with time by modifying the empirical constitutive equation such that a zero effective stress at the interface is enforced, independent of the surface concentration. This yields good model results.

**Effective stress versus buoyant weight**

Considering the fact that the effective stress in a top layer of constant density increases gradually with depth from zero at the interface, it seems more evident to express the effective stress as a function of a parameter which can account for this. A natural choice for this purpose is the buoyant weight $\sigma_0$, which is by definition the difference between the total stress $\sigma$ and the hydrostatic pressure $p_0$. The buoyant weight equals the effective weight of the soil skeleton when it is completely self-supporting. Consequently, $\sigma_0$ is the upper limit of the effective stress, reached when consolidation has stopped.

Figure 3 shows the effective stress as a function of buoyant weight at different times. As expected the effective stress increases with time and reaches its equilibrium value, which is the buoyant weight. Figure 3 shows that the equilibrium condition, eq.(7), is reached much earlier at the bottom because this layer is rich of non-cohesive sand. The part of the layer in equilibrium increases with time. Figure 3 also suggests that the mud layer surprisingly already had an initial structure, because there is a measurable effective

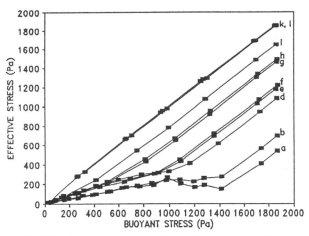

**Figure 3.** Effective stress as a function of buoyant weight and time.

stress, larger than the error, for the earliest profiles. This is anomalous and indicates serious experimental errors.

It may seem logical from the above considerations to connect the effective stress in a point to the weight of the layer above this point. If a unique relationship of the form $\sigma' = f'(\phi)\sigma_0$ existed, the above mentioned problem of the gradual decrease of $\sigma'$ would be solved in a natural way. However, the data plotted as the non-dimensional ratio of $\sigma'/\sigma_0$ versus concentration clearly show that this is not the case. Hence this formulation is not satisfactory.

## The effective stress gradient

Figure 3 confirms that the effective stress *gradually* decreases to zero at the interface (where $\sigma_0 = 0$). The slopes of the curve seem to increase with time and reach the equilibrium value 1. The slope can be written as:

$$\frac{\partial \sigma'}{\partial \sigma_0} = \frac{\partial \sigma'}{\partial z}\frac{\partial z}{\partial \sigma_0} = \frac{-1}{g\Delta\rho}\frac{\partial \sigma'}{\partial z} = f(\phi,...?) \tag{8}$$

In figure 4 the slope is plotted versus the density $\rho$ in order to verify whether there is any relationship. This graph shows less scatter than figure 2. Hence, maybe the slope $\partial\sigma'/\partial\sigma_0$ is a more useful measure of the structural development. The same idea had already been suggested independently by Le Hir (1991). Integration of eq.(8) and of the Darcy-Gersevanov equation (eq.(3)) yields:

$$\sigma' = \int_{z}^{z_i} g\Delta\rho\, f(\phi,...)dz = \int_{z_i}^{z} g\Delta\rho\, (1 - w/w_0)\, dz \tag{9}$$

where: $w = -v$. This equation expresses that from each layer a certain fraction $f = 1 - w/w_0$ contributes to the load carried by the plane at depth $z$. If it is true that the effective stress

127

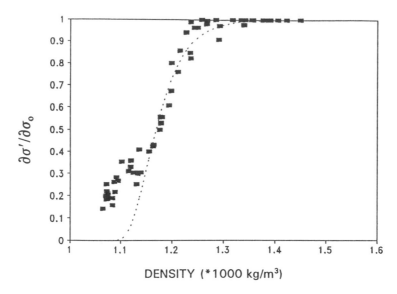

**Figure 4.** Slope of the curves from figure 3 as a function of density (selected data points) and curve fit (dotted line) of the form $\exp(b\phi)/(a+\exp(b\phi))$.

gradient is a unique function of concentration, as may be suggested by figure 4, then $w$ should be as well and the theory of Kynch (1952) would be applicable and should lead to unique relationships between settling rate and concentration. This is generally not the case however. An important mathematical reason why Kynch's hypotheses causes a problem is the need for obtaining a zero (i.e. balanced) sediment flux on the bottom. In a consolidation model usually a zero sediment flux is specified as bottom boundary condition. Since the concentration at the bottom continuously increases with time (until equilibrium), a zero sediment flux can only be obtained by outbalancing the net settling flux, which is a function of the concentration, with a diffusive flux (Toorman, 1992 & 1996). Notice that the latter is obtained with the traditional relationship $\sigma'(\phi)$ but not with $f(\phi)$. Another, maybe even more important reason for not obtaining unique relationships with Kynch's method is the flocculation history and the polydispersity of natural sediments. The latter results in different redistribution (or sorting) of the grains, depending on the conditions of bed formation (Toorman & Berlamont, 1993).

Substitution of eq.(8) into the stress balance, eq.(5) with $U = 0$, gives a relationship between $k$ and $v$:

$$\frac{-v}{k} = \frac{\Delta\rho_s}{\rho_w}\phi\left[1 - \frac{\partial\sigma'}{\partial\sigma_0}\right] \quad \Rightarrow \quad \log(k\Delta\rho/\rho_w) = \log(-v) - \log\left[1 - \frac{\partial\sigma'}{\partial\sigma_0}\right] \tag{10}$$

with: $\Delta\rho = \Delta\rho_s\phi = \rho-\rho_w$. Figure 5 compares the settling rate $w_s$ $(= -v)$, computed with Kynch's method (Toorman, 1992), with $1 - \partial\sigma'/\partial\sigma_0$. According eq.(10) the log-difference between the two curves is a measure of the permeability. The shapes of both curves confirm that the second type of behaviour observed in the settling rate plot in figure 5,

128

characterized by the curvature (Toorman, 1992), corresponds to consolidation. Indeed, the settling rate, obtained with Kynch's method, is the settling rate related to the total sediment flux, i.e. it contains the effective stress contribution. This observation thus seems to confirm that there is certain truth in the assumption that the effective stress gradient is a function of concentration.

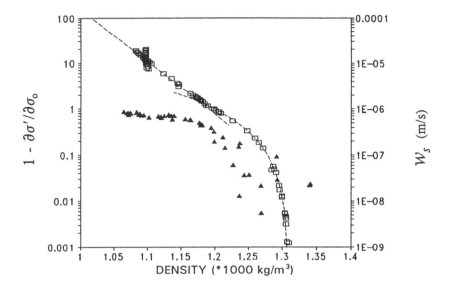

**Figure 5.** Comparison of the settling rate [□] and $1-\partial\sigma'/\partial\sigma_0$ [▲]. Dashed lines are curve fits for the two deposition modes defined by Toorman (1992).


**Proposed new formulation for effective stress**

Careful analysis of the two previously investigated ideas has revealed several inconsistencies. Yet, there seems to be truth in both types of equations. Therefore, it is suggested to combine the two to write the effective stress induced sediment flux reduction as:

$$\frac{w_0}{g\Delta\rho_s}\frac{\partial\sigma'}{\partial z} = w_0\phi\frac{\partial\sigma'}{\partial\sigma_0} = -w_e(\phi)\phi + D_e(\phi)\frac{\partial\phi}{\partial z} = (w-w_0)\phi = S - w_0\phi \qquad (11)$$

where: $w_e$ = settling rate reduction due to effective stresses; $D_e$ = effective stress induced diffusivity; $S$ = total sediment flux. This formulation can solve all the mathematical problems. The physical interpretation or justification seems more difficult. The concentration gradient term can be understood intuitively since the number of interparticle contacts increases with concentration and hence the capacity to carry more weight increases. When the concentration increases downward (negative gradient) the settling flux is reduced and vice versa.

Data of settling column experiments with non-cohesive, rigid, spherical glass beads

129

(Shannon *et al.*, 1963 & 1964) show that (for the higher concentrations, i.e. $\phi > 0.15$) the diffusive flux is negligible and the data can be predicted accurately with the theory of Kynch (1952). In this case density discontinuities occur. This observation gives a strong indication that the diffusive flux should be attributed to the deformation of the aggregate network structure. Shannon *et al.* (1964) attribute the occurrence of diffusion for lower concentrations to segregation of the larger particles and possibly to agglomeration as well. For this case, the estimated experimental values for $w_e$ versus $\phi$ show a similar trend as in figure 4 (Toorman, 1995).

Since the total flux $S$ becomes zero when consolidation has stopped, it follows from eq.(11) that the ratio of density gradient to density equals the ratio of diffusivity and net settling rate $w_0 - w_e$. The fact that equilibrium density profiles of cohesive sediment show a slope (see fig.1.A) implies that the diffusivity is not zero. Hence, there is clear evidence for the existence of the two contributions.

### Effective stress growth in time

Figure 6 shows the time evolution of the measured effective stress for the lowest three piezometers in a consolidation column. The difference between the three data sets corresponds exactly with the difference in buoyant weight between the measurement points, because in the bottom part of the mud layer the excess pore pressure is dissipated (see fig.3). The three stages in the curves correspond exactly with the stages in the settling curve, which clearly shows that the effective stress increases due to compaction above.

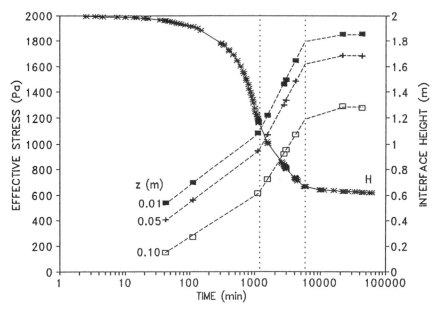

**Figure 6.** Effective stress growth, measured by the bottom three piezometers (z = distance from bottom), and consolidation curve [*].

# CONCLUSIONS

Measurements of excess pore water pressures in consolidation columns clearly indicate that the effective stress is not simply a function of concentration alone. The data for self-weight consolidation tests show that the effective stress increases gradually from zero at the interface to a certain value, which is bounded by the upper limit of the buoyant weight of the sediment layer above. Based on theoretical considerations and experimental data, a new formulation for the effective stress is proposed, which satisfies the requirements. The settling flux reduction by the effective stress consists of two terms of which the effect of one is diffusive in nature. The first term is related to the structural support reaction force at the interparticle contact points and the diffusive term to the deformation of the network structure. The formulation should be verified and further investigated for the determination of the two concentration depending empirical parameters by using it in numerical simulations and by experimental data. The influence of the sorting of the different types and sizes of grains should be investigated as well. If the new formulation still proves to be unsatisfactory, the validity of the Darcy-Gersevanov law should be questioned seriously.

**Acknowledgements**: This work has been undertaken as part of the MAST-2 G8 Coastal Morphodynamics Programme, funded partly by the Commission of European Communities, Directorate General for Science, Research and Development, under contract MAS2 CT92-0027. The first author holds the position of post-doctoral researcher granted by the Belgian National Fund for Scientific Research.

# REFERENCES

Alexis, A., P. Thomas and S. Gallois (1993). *Consolidation of cohesive sediments*. Final Report for the MAST-1 G6M Programme, Civil Eng. Dept., I.U.T. St. Nazaire (in French).

Berlamont, J., L. Van den Bosch and E. Toorman (1992). Effective stresses and permeability in consolidating mud. *ASCE 23rd Int. Conf. on Coastal Engineering*, (Venice, Oct.4-9), pp.2962-2975.

Berlamont, J., M. Ockenden, E. Toorman and J. Winterwerp (1993). The characterisation of cohesive sediment properties. *Coastal Engineering*, 21:105-128.

Been, K. & G.C. Sills (1981). Self-weight consolidation of soft clays: an experimental and theoretical study. *Géotechnique*, 31(4):519-535.

Cargill, K.W. (1983). Procedures for prediction of consolidation in soft fine-grained dredged material. DOTS Technical Report D-83-1, U.S. Army Eng. Waterways Experiment Station, Vicksburg.

Cargill, K.W. (1986). The large strain, controlled rate of strain (LSCRS) device for consolidation testing of soft fine-grained soils. Technical Report GL-86-13, U.S. Army Eng. Waterways Experiment Station, Vicksburg.

Elder, D. McG. and G.C. Sills (1984). Time and stress dependent compression in soft sediments. *ASCE Symp. on Prediction and Validation of Consolidation* (San Francisco, Oct.1984), Paper SM048.

Gersevanov, N.M. (1934). Dinamika Mekhaniki Grunkov. (*Foundations of soil mechanics*), 2, Gosstroiisdat (in Russian).

Gibson, R.E., G.L. England and M.J.L. Hussey (1967). The theory of one-dimensional consolidation of saturated clays - I. *Géotechnique*, 17:261-273.

Kynch, G.J. (1952). A theory of sedimentation. *Trans. Faraday Society*, 48:166-176.

Lee, K. and G.C. Sills (1981). The consolidation of a soil stratum, including self-weight effects and large strains. *Int. J. Numerical and Analytical Methods in Geomechanics*, Vol.5:405-428.

Le Hir, P. (1991). Presented at the MAST1/G6M Workshop (Wallingford, Sept.1991).

Pane, V. and R.L. Schiffman (1985). A note on sedimentation and consolidation. *Géotechnique*, 35:69-72.

Schiffman, R.L., V. Pane and V. Sunara (1985). Sedimentation and consolidation. *Flocculation, sedimentation and consolidation. Proc. Engineering Foundation Conference* (Moudgil & Somasundaran, eds.), pp.57-121.

Shannon, P.T., Stroupe, E. and Torey, E.M. (1963). Batch and continuous thickening. Basic Theory. Solids flux for rigid spheres. *I&EC Fundamentals*, Vol.2(3):203-211.

Shannon, P.T., Dehaas, R.B., Stroupe, E. and Torey, E.M. (1964). Batch and continuous thickening. Prediction of batch settling behaviour from initial rate data with results from rigid spheres. *I&EC Fundamentals*, Vol.3(3):250-260.

Terzaghi, K. (1942). *Theoretical soil mechanics*. J. Wiley, New York.

Toorman, E.A. (1992). Modelling of fluid mud flow and consolidation. PhD Thesis, Hydraulics Laboratory, Civil Eng. Dept., K.U.Leuven.

Toorman, E.A. & J.E. Berlamont (1993). Settling and consolidation of mixtures of cohesive and non-cohesive sediments. *Advances in Hydro-Science and Engineering*. Proc. ICHE'93 (Wang, ed.), pp. 606-613, University of Mississippi.

Toorman, E.A. (1995). Sedimentation of non-cohesive rigid spherical glass beads. Report HYD151. Hydraulics Laboratory, K.U.Leuven.

Toorman, E.A. (1996). Sedimentation and self-weight consolidation: general unifying theory. *Géotechnique*, 46(1):103-113.

Znidarčič, D., R.L. Schiffman, V. Pane, P. Croce, H.Y. Ko and H.W. Olsen (1986). The theory of one-dimensional consolidation of saturated clays: part V, constant rate of deformation testing and analysis. *Géotechnique*, 36(2):227-237.

# GENERAL DESCRIPTIVE
# DYNAMICS/CASE STUDY

# 9     Turbidity maxima formation in four estuaries

A. M. W. ARUNDALE, E. J. DARBYSHIRE, S. J. HUNT,
K. G. SCHMITZ and J. R. WEST
*School of Civil Engineering, The University of Birmingham,
Birmingham B15 2TT, UK*

## ABSTRACT

Data from two medium tidal range estuaries, the Tamar and tidal Trent, and two high tidal range estuaries, the Severn and Parrett, are used to examine the variation in the controlling influences on fine sediment transport and the uncertainties that still exist. The results suggest that tidal range, river flow, wind generated waves, water depth, channel width and inter-tidal zones can be influential. Future work should aim at field studies designed to improve the understanding of temporal and spatial variations and the relative importance of the mechanisms that control sediment and solute transport. This ought to lead to an improvement in the predictive capability of cohesive sediment transport models.

## INTRODUCTION

The presence of finely divided cohesive sediment in suspension or lying on sub-tidal and inter-tidal zones in estuaries has important ecological and economic implications. Fine sediment in the water column retards photosynthesis at concentrations of 100-300 mg/l and hence controls primary production in the water column. The deposition and erosion cycles on inter-tidal zones influence primary productivity, invertebrate populations and hence the dependent wading bird and fish species. A number of heavy metal and organic pollutants have an affinity for clay particles and thus pollution management requires a good knowledge of cohesive sediment transport. The distribution of inter-tidal sediments affects the ability of coasts to resist erosion and sub-tidal deposition leads to considerable expenditure each year on dredging for navigational requirements. Given the importance of fine sediments in estuaries there is a need to develop mathematical models so that the effects of proposed engineering and environmental management strategies can be assessed. Good model predictive capability requires a sound understanding of the underlying transport processes and the response of those processes to external forcing functions.

The transport of fine sediment involves the processes of sedimentation, consolidation and erosion. These processes may be of significance with respect to time scales of seconds to a number of years. This makes modelling a challenge and temporal and spatial averaging of the governing equations unavoidable. A further difficulty arises from the flow field in estuaries not being fully understood due to the complexity induced by tides, buoyancy from rivers, waves and irregular channel geometry. These influences often

*Cohesive Sediments.*
Edited by Neville Burt, Reg Parker and Jacqueline Watts
© 1997 John Wiley & Sons Ltd.

produce the very significant feature of estuarine fine sediment transport, a longitudinal distribution of suspended sediment concentration which exhibits a peak in the upper reaches, usually called the turbidity maximum.

In recent years a considerable effort has been made to understand estuarine sediment transport processes. Much of this effort has focused on laboratory settling and erosion studies as these are easier and cheaper than field studies. There have been relatively few detailed field studies aimed at elucidating the complex interaction of tidal, wave, fluvial and channel topographical effects on estuarine sediment transport. Those studies that have been undertaken have usually focused on a single estuary.

The objective of this paper is to consider briefly some aspects of results from four estuaries having different characteristics in order to illustrate the variations in the controlling influences on fine sediment transport and the uncertainties that still exist. The results are used as a basis for suggesting the focus for future work.

## PREVIOUS WORK

The transport of solute and particles in estuaries may be given by

$$\frac{\partial c}{\partial t} + \frac{\partial (u_i c)}{\partial x_i} + w_s \frac{\partial c}{\partial x_3} = \frac{\partial}{\partial x_i}\left(\varepsilon_{ci} \frac{\partial c}{\partial x_i}\right) + S \tag{1}$$

where $c$ = turbulent mean concentration, $u_i$ = components of turbulent mean velocity ($i = 3$ is the vertical direction), $w_s$ = fall velocity ( = 0 for solutes), $\varepsilon_{ci}$ = eddy diffusivity coefficient and S is a source/sink term (erosion and deposition at the bed for sediments). A simpler area averaged form for solute transport is

$$\frac{\partial (c_A A)}{\partial t} + \frac{\partial (c_A u_A A)}{\partial x} = \frac{\partial}{\partial x}\left(AD\left(\frac{\partial c_A}{\partial x}\right)\right) + S_A \tag{2}$$

where subscript A indicates cross sectional average, A = channel cross sectional area and D = dispersion coefficient which can be written in terms of components due to transverse shear effects ($D^t$) and vertical shear effects ($D^v$) (West and Mangat 1986).

Although solute and suspended sediment transport are strongly influenced by turbulence the above temporal and spatial averaging are essential in order to produce mathematical models that are compatible with field data and computer hardware limitations.

The concept of turbidity maximum formation has evolved from simple tidally averaged gravitational circulation and tidal pumping concepts to a more refined understanding involving intra-tidal variations (Dyer 1986, West et al. 1990). The later can be monitored by a little persistence in making salinity and suspended solids concentration measurements. The turbidity maximum is often linked with the limit of saline intrusion, the fresh-salt water interface (FSI). Departures from this concept have been reported. Multiple maxima have been observed in the upper Tay estuary (Buller 1975) due to the shallow water run-off from channels draining wide inter-tidal zones. Multiple peaks in temporal records of suspended solids concentration observed in the Thames estuary (Thorn and Burt 1978) and Severn estuary (Hydraulics Research Station 1981a) were considered to be due to erosion following preferential deposition in deeper sections of the estuary during decelerating parts of the tidal cycle.

# DATA

This paper considers data from a number of sources along with some recent observations in the upper Severn estuary undertaken by the authors (Hunt 1994, Schmitz 1996). Details of the length, river flow and tidal ranges in the lower reaches are given in Table 1.

**Table 1.** Some physical data for estuaries studied.

| Parameter | Parrett | Severn - Bristol Channel | Tamar | Trent - Humber |
|---|---|---|---|---|
| Length | 20 km | 220 km | 32 km | 144 km |
| Mean daily fluvial discharge | 1 m³/s (Chisel-borough) | 62 m³/s (Bewdley) | 23 m³/s (Gunnislake) | 85 m³/s (Colwick) |
| Mean spring tidal range | 12.3 m (Avonmouth) | 12.3 m (Avonmouth) | 4.7 m (Plymouth) | 6.4 m (Immingham) |
| Mean neap tidal range | 6.5 m (Avonmouth) | 6.5 m (Avonmouth) | 2.2 m (Plymouth) | 3.2 m (Immingham) |

The discussion of the Severn estuary refers to a time dependent depth averaged model of the lower estuary and the Bristol Channel. The model used for the simulation was TIDEWAY-2D which was calibrated against data collected by Hydraulics Research in 1980 (Hydraulics Research Station 1981b).

# TURBIDITY MAXIMUM FORMATION - MEDIUM TIDAL RANGE ESTUARIES

The tidal Trent is a sub-estuary of the Humber estuary and has a length of approximately 82 km, with its confluence with the Humber being 62 km from the North Sea. Spring tide data from June 1978 for maximum recorded suspended solids concentration (fig. 1) show a peak near to Trent Falls, the mouth of the Trent (Arundale 1996). The average salinity distribution shows a landward movement of about 10 km for the spring tide compared to the May neap tide. The daily mean flows for the survey dates were 57 and 49 m³/s for May and June respectively.

The narrowness of the concentration of suspended solids peak (~ 20 km) suggests that effects in the locality of the FSI are important. Medium tidal range estuaries with a modest river inflow may exhibit salinity induced vertical density gradients on the ebb tide. If these are sufficient to inhibit resuspension of sediment during a part of the ebb tide then they effectively cause the longitudinal dispersion that helps to counteract the fluvial seaward advection on the downstream side of the suspended solids concentration peak.

A good example of ebb tide salinity effects has been recorded recently (Darbyshire et al. 1993) in the Tamar estuary (fig. 2). The figure shows the temporal variation of depth mean values and the variation of parameters with distance above the bed. The mean time of a vertical profile reading is shown by a vertical line whose length signifies the elevation of the near surface reading. The second line at a given time shows the spatial variation

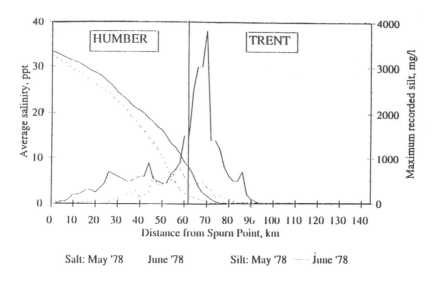

Figure 1 Longitudinal profiles of salinity and concentration of suspended solids for the Humber and tidal Trent - May (neap tide) and June (spring tide) 1978

(scales on each fig.) of a given parameter about the depth mean value given in the upper graph.

On the flood tide, the depth mean velocity and suspended solids concentrations are seen to be in phase and there are no detectable salinity induced gradients in the water column. During the early ebb tide, when salinity is present, particulate transport is very much less for a given velocity than on the flood tide. This is due to the significant solute induced vertical density gradients suppressing vertical turbulent transport processes, and hence reducing the values of the $\varepsilon_{cs}$ term in eqn. 1 and increasing the $D^v$ component of solute dispersion in eqn. 2. The suspended solids concentrations are generally less than 1 g/l and thus particulate induced density gradient effects should be minimal. The solute induced mechanism, coupled with the increase in sediment transport after the disappearance of the salinity, may help to explain the presence of suspended solids upstream of the FSI for the spring tide in fig. 1. It is very probable that the effects of solute induced density gradients are accentuated by the increase in settling velocity often caused by the enhanced flocculation induced by salinity.

The neap tide suspended solids data for the Trent show the effect of tidal range, with the peak being reduced to about 25% of the maximum spring tide value and located about 10 km downstream of the FSI, close to the confluence of the rivers Ouse and Trent. It is tempting to speculate whether the confluence or the greater channel width of the Humber estuary downstream of the confluence also exert some influence on either, or both, the transport of salinity and suspended solids for the conditions observed.

The above results emphasise the close direct, and indirect, links between salinity and suspended solids and the need for more studies to permit a deeper understanding.

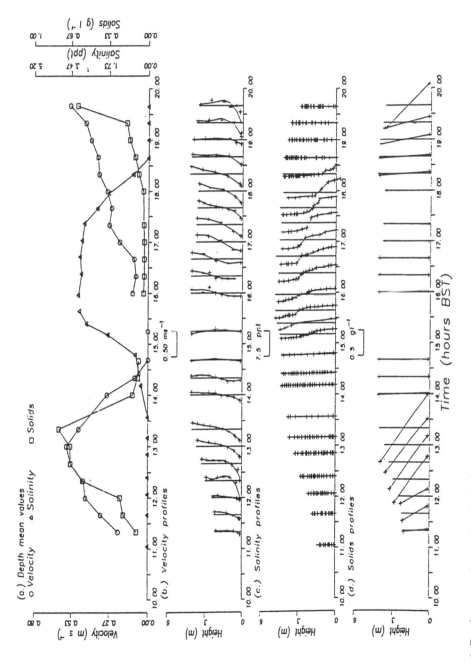

**Figure 2** Depth mean values and vertical profiles of velocity, salinity and concentrations of suspended solids, Calstock, R. Tamar, 7th August 1991

## TURBIDITY MAXIMA FORMATION - HIGH TIDAL RANGE ESTUARIES

Data from a Hydraulics Research Station study of the Severn estuary in 1980 (fig. 3) show a distinct peak in the region of 170 km from a seaward boundary defined as 20 km west of Lundy Island. The increased tidal range leads to higher suspended solids values by about a factor of two, when compared with the Trent. The maximum is located approximately 40 km downstream of the FSI. More recent measurements during spring tides in the upper estuary upstream of the FSI have shown (fig. 4) concentrations of suspended solids in excess of 10 g/l for river flows of below 50m³/s (Schmitz 1996). As both surveys have common data for fairly low summer river flow conditions (30 m³/s), there is circumstantial evidence for two turbidity maxima in the Severn.

Salinity data (Hydraulics Research Station 1981b) show that weak ebb tide gradients can occur during spring tides, but they are very much less than the examples given in fig. 2. The longitudinal salinity distribution must be maintained by some dispersion process. As the FSI is usually located in the wider reaches downstream of Minsterworth, it is possible that transverse dispersion effects are important in the lower estuary.

The mechanisms by which the longitudinal turbidity distribution is maintained are less clear. Observations in the lower Severn estuary have shown that the high concentrations of suspended solids can lead to significant vertical suspended solids gradients, particularly during tidal acceleration and deceleration in the deeper parts of the estuary (Kirby and Parker 1983). In the upper estuary the flood-ebb asymmetry of the tidal wave leads to higher flood tide velocities and stronger vertical mixing. Cumulative sediment flux data from near to Gloucester in the upper estuary upstream of the FSI show (fig. 5) some

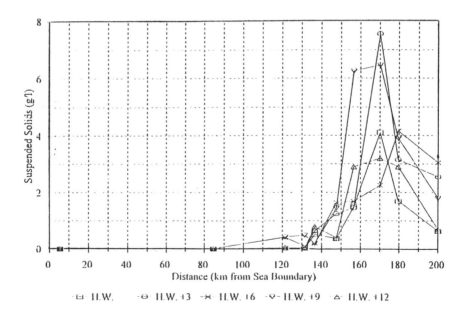

**Figure 3** Longitudinal distribution of the concentration of suspended solids, lower Severn estuary and Bristol Channel, low river flows, summer 1980.

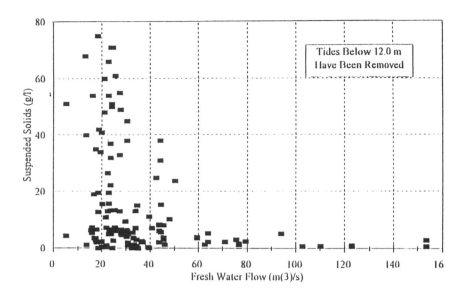

**Figure 4** Variation of concentration of suspended solids with river flow for spring tides in the upper Severn estuary

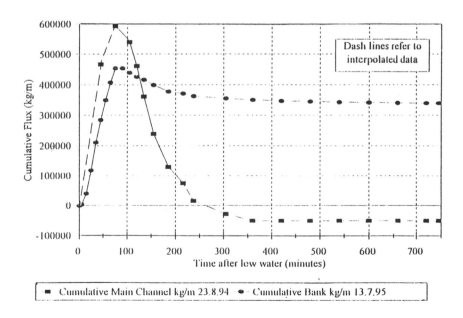

**Figure 5** Examples of main channel and near bank cumulative suspended solids fluxes near Gloucester

complexity. Near bank and main channel sites show similar cumulative flood tide fluxes of around 500,000 kg/m width. The ebb tide sediment flux per unit width is much larger in the main channel due to high near bed concentration caused by the weaker ebb tide currents. This leads to a net landward flux near to the bank and a seaward flux in the main channel for the conditions observed (Schmitz 1996). The separation of the FSI and turbidity maximum is clearly due to different mechanisms influencing the salt and suspended solids.

Observations (fig. 6) at Combwich, 5 km from the mouth of the Parrett estuary, help to elucidate further the fine sediment transport processes in high tidal range estuaries. The data show the variation of velocity or concentration of suspended solids at four points near to the bed for most of a flood and ebb tide. Most of the channel dries out over the low water period leaving a small low water channel. The suspended solids monitors have a minimum detectable concentration of 2 g/l, thus lower values appear as this value. The near bed data show sediment passing upstream with concentrations of up to 15 g/l. The concentrations are appropriate to the high tidal ranges but are inconsistent with respect to FSI dependence. The salinity values are in excess of 20 kg/m$^3$ and the high concentrations of suspended solids disappear during the ebb tide before the site dries out, suggesting that the suspended sediment is advected downstream. The existence of extensive mud flats in Bridgwater Bay at the mouth of the estuary leads to the speculation that they could act as a trap on the ebb tide or as a source on the flood tide. It could be anticipated that as the ebb flow leaves the Parrett estuary, which is heavily engineered to prevent flooding, the wider expanses of Bridgwater Bay are conducive to sedimentation.

The near bed data also show a lag in resuspension on the ebb tide, presumably due to the turbulence damping effects associated with the particulate induced density gradient effects of the near bed layer, a further influence on the $\varepsilon_{s}$ term in eqn. 1. Such effects are less likely to occur on the accelerating part of the flood tide due to shallow water, steeper longitudinal water surface gradients and hence larger bed shear stresses than occur during the ebb tide. The ebb tide sediment induced density effects provide a mechanism by which sediment can potentially undergo a tidal cycle mean landward transport under appropriate river flow conditions.

If trapping mechanisms occurs in the main Severn estuary then this would help to explain the existence of an ephemeral or permanent lower estuary turbidity maximum. The work of Kirby and Parker (1983) recorded accretion of mud in the Bristol and Newport Deeps which are located in the region of Avonmouth. These channels are the first deep water areas below the upper estuary, which only has a narrow, shallow low water channel during spring tide, low river flow conditions. It is possible that the Deeps act as ebb tide traps for fine sediment during the spring to neap part of the tidal cycle, and allow subsequent spring tides to enable the return of sediment to the water column.

In many estuaries much of the fine sediment can be found on the inter-tidal zones. The most favourable depositional conditions involve spring tide, low river flow and calm conditions. Erosion occurs when river flows increase in the upper estuary and storms occur in the wider parts of lower estuaries where there is sufficient fetch for wave generation. The Severn has areas of wide (~ 1 km) inter-tidal zones in its lower reaches which exhibit storm modulated cycles of erosion and deposition (West and West 1989).

The effects of transverse distributions of salinity and suspended solids have been investigated by using a two-dimensional time dependent depth averaged mathematical

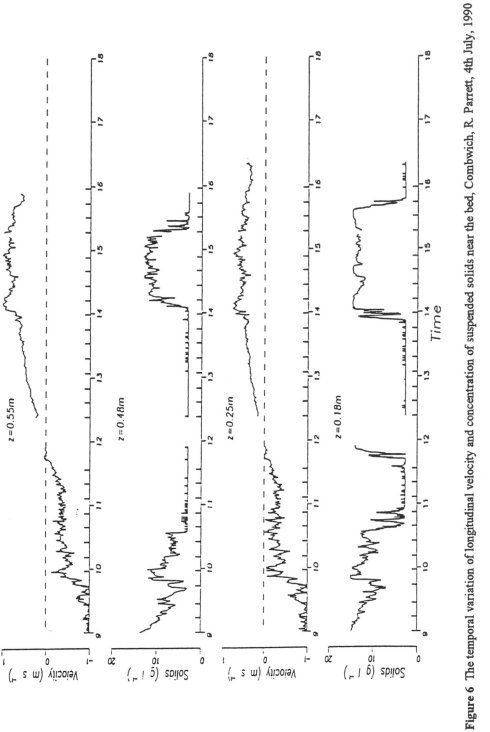

**Figure 6** The temporal variation of longitudinal velocity and concentration of suspended solids near the bed, Combwich, R. Parrett, 4th July, 1990

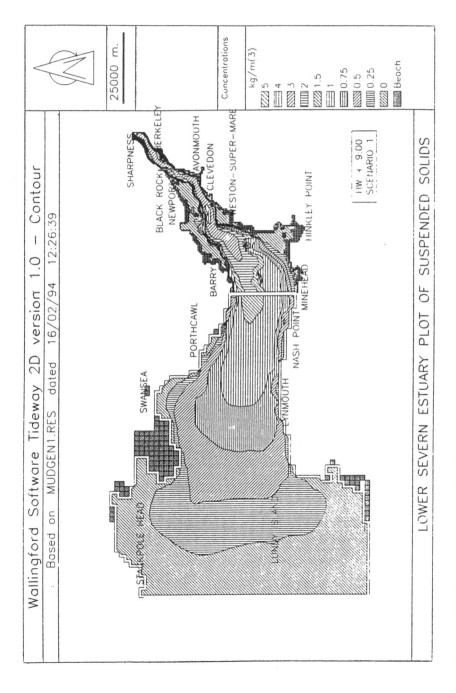

**Figure 7** Distribution of concentration of suspended solids predicted by a depth averaged model.

model of the Bristol Channel and Severn estuary. An example of the suspended solids distribution is shown in fig 7. Transverse gradients are present in both this figure and the equivalent salinity plot. Salinity shows higher concentrations in the main channel and the margins for flood and ebb tides respectively. This results in longitudinal dispersion in the one-dimensional sense, the D term in eqn. 2, and adds support to the supposition that transverse effects are influential in determining the longitudinal salinity distribution in the wider parts of estuaries.

The suspended solids concentration distribution shows the potential for a turbidity maxima to appear in the lower Severn estuary though it must be noted that in fig. 7 the results are dependent to a degree on the upstream boundary condition. Although the high concentrations occur in the main channels making mud available for accretion in the Deeps, a 50 cycle repeating spring tide showed that an equilibrium can be achieved between sediment input from the upper estuary and accretion on the inter-tidal zones such as Wentlooge Flats to the north of Avonmouth and Bridgwater Bay to the west. This provides for the possibility of a further trapping mechanism for sediments in the lower estuary.

Given the complexities of the interactions that affect solute and sediment transport in estuaries there is a need for careful studies to develop a better understanding of the controlling mechanisms. The studies should recognise the temporal changes due to changing river flow, tidal range and wave activity and exposure, as well as the spatial dependence caused by topographical influences. The greatest uncertainties are arguably in high tidal range estuaries where unfortunately conditions are often fairly difficult for field studies.

## CONCLUSIONS

1.  Cohesive sediment transport in estuaries can exhibit a complex response to changes in external forcing functions.

2.  As tidal range increases the influence of vertical solute induced density gradients decreases and the influence of vertical suspended solids gradients increases. Transverse mixing, bed topography and inter-tidal zone effects have the potential to be influential.

3.  An increase in river flow can lead to an increase in importance in vertical solute induced density gradients due to buoyancy effects and a changing influence due to transverse mixing as the FSI is moved seaward. The influence of suspended solids induced vertical gradients could increase due to mobilisation of sediments in the upper estuary or decline due to dilution effects.

4.  The influence of wind wave induced effects will tend to increase with tidal range due to the greater potential for the cycling of sediment on inter-tidal zones in high tidal range estuaries.

5.  In any estuary the individual effects dominating fine sediment transport may be expected to vary with time and space.

6.  There is usually a need for careful investigation into the temporal and spatial variation of the mechanisms controlling cohesive sediment transport in an estuary. Then perceptive decisions with respect to modelling should be made and the best achievable predictions may be obtained.

## ACKNOWLEDGEMENTS

The authors gratefully acknowledge the support of the Engineering and Physical Sciences Research Council, Hydraulics Research Wallingford and the National Rivers Authority, Severn Trent Region.

## REFERENCES

Arundale, A. M. W. (1996) Transport processes in the tidal Trent, PhD thesis, The University of Birmingham.

Buller, A. T. (1975) Sediments of the Tay estuary. II Formation of ephemeral zones of high suspended sediment concentrations, *Proc. Roy. Soc. Edin. (B)*, 75, 1-2, pp 65-89.

Darbyshire, E. J. and West, J. R. (1993) The turbidity maximum in the Tamar estuary, *Netherlands Journal of Aquatic Ecology*, 27, 2-4, pp 121-131.

Dyer, K. R. (1986) Coastal and Estuarine Sediment Dynamics, J. Wiley (Chichester).

Hunt, S. J. (1994) Transport processes in the upper Severn estuary, MPhil thesis, The University of Birmingham.

Hydraulics Research Station (1981a) The Severn Estuary - Silt Monitoring, HRS Report EX 995.

Hydraulics Research Station (1981b) Observations of tidal currents, salinities and suspended solids concentrations, HRS Report EX966.

Kirby, R and Parker, W. R. (1983) The distribution and behaviour of fine sediment in the Severn estuary and inner Bristol Channel, U.K., *Can. J. Fish. Aquat. Sci.*, 40(1), pp 83-95.

Schmitz, K. G. (1996) Turbidity Maxima in the Seven Estuary, PhD thesis, The University of Birmingham.

Thorn, M. F. C. and Burt, T. N. (1978) The silt regime of the Thames estuary, Hydraulics Research, Wallingford, Rep. No. IT 175.

West, J. R. and Mangat,, J. S. (1986) The determination and prediction of longitudinal dispersion coefficients in a narrow, shallow estuary. Estuarine, *Coastal and shelf Science*, 22, pp 161-181.

West, J. R., Uncles, R. J., Stephens, J. A. and Shiono, K. (1990) Longitudinal dispersion processes in the upper Tamar estuary. *Estuaries*, 13, (2), pp 118-124.

West, M. S. and West J. R. 1991 Spatial and temporal variations in inter-tidal zone sediment properties in the Severn estuary. Estuaries and Coasts, eds. Elliott, M and Ducrotoy, J. P., Olsen and Olsen, Fredensborg, pp 25-30.

# EQUIPMENT AND INSTRUMENTATION

# 10 Interfacial hydrodynamics and entrainment functions of currently used erosion devices

**G. GUST and V. MÜLLER**
*Department of Ocean Engineering 1,*
*Technical University Hamburg-Harburg, Germany*

## INTRODUCTION

In aquatic systems, flows due to natural and engineering processes often initiate repetitive or transient erosion-transport-deposition cycles at the interface of sediments and overlying fluid. For cohesive sediments, a complex chain of events is thus generated which affects bed stability, flow structure, material fluxes, water quality and ecosystem dynamics. To understand and quantify these processes, either individually or as network of coupled processes, both laboratory and field studies are required on the hydrodynamic driving forces and the bed responses with defined measuring protocols for modeling, verification and simulation purposes.

The current measuring protocols to quantify hydrodynamic forcing and entrainment response of cohesive sediments particularly aim at the determination of erosion thresholds and entrainment rates in unidirectional and wave-current boundary-layer flows. Typical results amenable to intercomparison are the (equilibrium) concentration curves obtained for incrementally increasing/decreasing steps of steady sub- and supercritical wall shearing stresses obtained in rotating annuli (Partheniades and Kennedy, 1966). The complex issues of secondary circulation and spatial inhomogeneity of the wall shearing stress (Sheng, 1989), bed preparation and operational range in these devices (Graham et al., 1992) spurred in recent years the development of refined rotating annuli (Krishnappan, 1993) and of alternative erosion devices of small geometry for laboratory and field use (see Tab. 1).

From a process-oriented point of view, several questions remained unanswered so far for the devices of Tab. 1, the prominent ones being 1. what parameters should be considered as representing the in-situ driving forces, 2. what best possible or reasonable reproducibility could be anticipated for the results obtained by the different devices, 3. to which field scenarios rotating annuli relate and if these are typical cases worth the operational efforts required.

Latter set of questions, extendable in general to any erosion device, arose due to several hydrodynamic and interfacial features particular to cohesive-sediment suspensions. For example, clay suspensions can switch from Newtonian to non-Newtonian flow behavior (Vanoni, 1946), in which case ensuing turbulent drag reduction by the suspensions genera-

*Cohesive Sediments.*
Edited by Neville Burt, Reg Parker and Jacqueline Watts
©1997 John Wiley & Sons Ltd.

tes bottom stresses different from those of clear-water flows typically used for calibrations (Gust, 1976). Furthermore, the way how fluid mud and consolidating suspensions with continuous vertical interfacial density variation erode is quantitatively not known so far, as are the cumulative effects of waves superimposed on currents, of biota and of chemical milieus on bottom stresses and entrainment rates.

The purpose of this paper is to provide a pool of experimental data towards the development of a universally acceptable protocol for entrainment measurements of cohesive natural sediments. Optimization criteria for such task are 1. identification of the smallest possible set of field-relevant hydrodynamical and sedimentological features to be met in simulation experiments, 2. identification of the interfacial hydrodynamics of erosion/deposition devices and 3. evaluation of these devices in view of close simulation of natural or engineering in-situ processes. To proceed towards this goal, three steps were undertaken:

1.1. Quantification of the interfacial, erosion-relevant hydrodynamic characteristics of currently available erosion devices: Intercomparison of wall shearing stresses and horizontal pressure gradients measured in replicas of erosion devices in use or under development in the lab of the author and elsewhere.

*Rationale:*

All devices aim at providing a complete set of the hydrodynamic input parameters driving the erosion processes under steady or slowly-varying (tidal) flows. Typically, a smooth bed is assumed, and from a sequence of steady-state values of wall shearing stresses exerted, combined with (equilibrium) values of mass suspended, a quantitative entrainment function is established. Design criteria of the devices, in some cases augmented by numerical simulations of the stress-generating Newtonian flow utilizing Navier-Stokes-Solvers, are currently based on the assumption that the major hydrodynamic force acting on cohesive sediments in the field is the interfacial wall shearing stress. The full stress tensor at an interface contains both tangential and normal components, though, and to provide a valid experimental protocol, it has to be shown that an entrainment device does not generate horizontal pressure gradients (localized suction or blowing) at the interface with forces different from those prevailing at field sites. Furthermore, the link between (equilibrium) concentrations of eroded material in a simulation device and local values observed in natural flows, where erosive or depositional features often experience rate changes in time, has not been demonstrated and may not exist beyond a critical (currently unknown) acceleration value.

1.2. Determination of mean steady-state entrainment functions of abiotic Kaolinite with different consolidation times.

*Rationale:*

With an erosion device whose hydrodynamic erosion/deposition features closely resemble those of natural steady boundary-layer flows (see below), a mean entrainment curve should be established under reproducible known hydrodynamic, temporal and sedimentological characteristics for a sufficiently large number of replicas permitting statistical treatment of the data set. This curve (or successors) could provide information on the variability of the erosion process. It could also serve as possible reference curve for future experiments investigating, for example, biogeochemical and flow acceleration effects which may influence the entrainment of cohesive sediments. Currently, it should provide an intercalibration standard for the various entrainment devices in use.

1.3. Comparison of these abiotic Kaolinite entrainment curves with data obtained for same sediment but different erosion time histories and durations, and with other sediments including literature data from rotating annuli and an open flow-through field system.

*Rationale:*

Deviations from the mean entrainment curve of abiotic Kaolinite should, under consideration of the individual measuring protocols and set-ups selected, permit identification of the causes underlying the results obtained. In particular, differentiation between device-specific and process-specific parameters should become possible, as well as clues for the minimum sensible parameter set required to quantify entrainment under natural conditions.

## DEVICES TESTED AND MEASUREMENT TECHNIQUES

Tab 1.: Erosion devices tested (or cited) in this study

| No. | DEVICE | REFERENCES | COMMENTS |
|---|---|---|---|
| 1. | Rotating annulus | Partheniades and Kennedy (1966) | rotating lid, flume rigid, lab device |
| | | Graham et. al. (1992) Krishnappan (1993) | rotating lid, counterrotating flume, lab device |
| 2. | Straight flume | Hüttel and Gust (1992) Gust and Li (unpublished) | smooth and rough flow, lab device smooth flow, clear water and suspension flows i. lab |
| | | Gust and Morris (1989) | seaflume, undisturbed sediments at 200m water depth |
| 3. | Propeller stirring 'Eromes' | Schünemann and Kühl (1991) | forward rotation of propeller |
| | | Müller et.al. (1995) | backward rotation of propeller |
| 4. | Vertically oscillating grid 'Plunger' | Tsai and Lick (1986) | grids of differentporosity and geometry tested |
| 5. | Small gap, central suction 'Erosion bell' | Williamson and Ockenden (1993) | bell head according to fig. 12 of report at h = 4, 6 mm tested |
| 6. | Disk stirring combined with central suction 'Microcosm' | Gust (1990) Gust et. al. (in prep.) | housing diameters 10 and 30 cm, 5 and 10 cm water depth, resp. |
| 7. | Natural flow | Gust and Weatherley (1985) Humann et. al. (in prep.) Eden et. al. (in prep.) | deep sea at 5000 m intertidal regime of the Elbe river |

*Erosion devices included in this review*

The erosion devices tested or cited in this study encompass systems which have been described in the literature and reports elsewhere either as laboratory or field devices (Tab. 1).
They were duplicated in the lab of the authors based on the information available. In those
cases where the geometric dimensions of the original devices were not fully discernible, the
data shown below are only valid for the duplicated versions. For an adequate intercomparison, bottom stresses from smooth-flume and in-situ measurements are included as well.

*Interfacial stress tensor measurements in the devices*

In all devices of Tab. 1, the magnitude of the tangential component of the stress tensor at
the bottom-fluid interface was measured by means of flush-mounted hot-film anemometry
(Bellhouse and Schultz, 1966), an unobstrusive technique to obtain the bottom stress by
means of forced heat convection (Ludwieg and Tillmann, 1949). Directions of the wall
shearing stress were determined qualitatively by the transport path of dissolving $K_2MnO_4$
crystals attached to the bottom. The horizontal gradient of the pressure was measured by a
highly sensitive wet-wet differential pressure gauge operating as part of an inductive frequency carrier bridge which converted membrane displacements into electrical signals. The
probes were placed at pre-selected radial distances into a hydrodynamically smooth PVC-
bottom. Two bottom stress probes, connected to hot-film anemometers (DANTEC model
55M10) run in constant-temperature mode, simultaneously recorded at an overheat ratio of
1.05 the skin-friction values over 5-min time series at 50 Hz per channel on a digital multi-
channel data recorder with 12 bit A/D resolution in the 0 - 5 V range. The output signals
from the differential pressure transducer (Hottinger & Baldwin, model PD1) were recorded
on the third channel. Measuring positions in the circular housing were either r = 0 or 20 mm
for the positive pressure tap and r = 40 mm for the negative tap, while the bottom stress was
recorded at r = 0, 20, 30, 40 mm for the devices with approx. 100 mm in diameter. For the
larger device (30 cm in diameter), r was selected at 0, 23, 40, 92, 110, and 130 mm. Temperature of the ambient water was monitored by a quartz thermometer with 0.02°C accuracy
(Heraeus, model Quat 100). From the time series of data, radial distributions of the unsteady
wall shearing stress and the horizontal differential pressure were obtained in the devices
after converting the raw data to physical values by means of independently obtained calibration coefficients using MATLAB routines.

*Mass entrainment measurements in the microcosm*

Mass entrainment data were obtained in a microcosm with 30 cm in diameter (device 6) by
exposing a 20-cm deep sediment column of Kaolinite, consolidated for different time intervals, to wall shearing stresses and horizontal pressure gradients of known time function and
duration at overlying water height of 10 cm. For the sediment core, commercially available
Kaolinite was mixed with oxygen-saturated fresh water both as porewater and overlying
fluid at a controlled temperature of 24(±1)°C, which then consolidated over 24 h, 48 h or
240 h. Erosion time intervals were selected at 10 min per bottom stress for the abiotic-
sediment entrainment curves established. The fluid volume was 7 l. Entrained mass was
calculated from the differences in dry mass obtained by filtering and weighing 250-ml suspension samples taken at the beginning and at the end of each 10-min time interval for
which the applied bottom stress was held constant. With each increase in bottom stress $\tau$,

the suspension generated in the prior erosion step was replaced by clear fluid and entrained mass determined anew from initial and final suspension samples. Sequences of spatially homogeneous bottom stress, each comprised of seven values in the range $0.0045 < \tau < 0.5$ N/m$^2$, were repeated five times with freshly stirred sediment consolidated in two experiment series for 24-h and 48-h intervals, each. For the 240-h consolidation time, only one sequence was executed. For the Mississippi river mud, the same methodology was chosen.

In an improved version of the microcosm, in which the generation of bottom stress at the interface was computerized to permit simulation of time-dependent patterns such as tidal stress curves, the suspension was not only sampled, filtered and weighed, but also recirculated through two density-monitoring devices. One of them was based on an optical scattering technique, and the other one on the Coriolis principle. They provided on-line registrations of the density of the suspension in the concentration range $0 < c_0 < 20$ g/l. These data were stored on a PC486 and calibrated by entrained mass values obtained by filtering and weighing 250-ml suspension samples replaced by clear water over several days. With sampling rates of 2 Hz of the on-line density measuring devices, resolution of the entrainment process became possible at time increments of 1s.

*Calibrations, error considerations and confidence limits*
The flush-mounted hot-film sensors recording the bottom stress (Gust, 1988) represent a standard tool for skin friction measurements and were calibrated in a variety of unidirectional flows. The following calibration techniques were used to date: 1. the energy-gradient technique in open channel-flow (Gust and Southard, 1983), 2. the Moody-diagram for circular (Sattel, 1994) and rectangular (Cardoso et al., 1989) duct flow, cross-checked by LDA and pressure drop measurements, 3. laminar shear flow beneath a rotating disk based on the work of Stewartson (1952), 4. intercomparison with the measured velocity gradient in the viscous sublayer of the microcosm.

The calibration range of these probes could be extended up to 40 N/m$^2$ by utilizing method 2; most intercompared erosion threshold values are in the range $< 0.5$ N/m$^2$, though, except for sediment with heavy biofilms (Grant and Gust, 1987), based on reported field data of unidirectional flows. Special experimental facilities are necessary for obtaining the high-stress calibration curves. Careful temperature control is required to maintain the selected overheat ratio in efforts to keep the overall experimental uncertainty at $\leq 5\%$ for bottom stress measurements. This condition can be relaxed when temperature-compensating, microprocessor-controlled Wheatstone bridges (Sattel, 1994) are used. Alternatively, a novel data reduction procedure can be applied which accounts for changes in the ambient temperature via a software program (Müller and Gust, 1994) utilizing an extended formulation of King's law as presented in Schlichting (1979).

The error of the pressure gauge calibrations depended on the precision at which the pressure head was established. The output range between 0 and 5 V of the transducer linearly increased with increasing water head between 0 and 10 cm. The resolution of the 12 bit A/D converter (1.22 mV/bit) was equivalent to 25 μm water height/bit or 0.25 Pa/bit. Measuring uncertainty was approximately twice as large, leading to an overall experimental uncertainty of 0.2% f.s. for the pressure measurements.

The distributions of horizontal pressure gradients and wall shearing stresses of the devices of Tab.1 are presented as survey for intercomparison. Discerning the effects of changes in experimental variables such as device geometry, frequency and amplitude of stirrer operation was considered as first-order priority. Pre- and postcalibrations of the stress probes, as well as long-term tests of their stability run independently over a duration of months, ensured that the results obtained are characteristics of the erosion devices and not measuring artefacts.

Confidence limits can be placed on the entrainment curves of abiotic Kaolinite and of Mississippi river mud at 24 h and 48 h consolidation times.

## RESULTS

*Hydrodynamic characteristics of natural flows and erosion devices - General considerations*

Erosion devices need to enact the conditions prevailing in the undisturbed natural (or engineering) hydrodynamic process during entrainment. Thus, hydrodynamic field data and those from straight open-channel flows were chosen to establish the necessary performance criteria. Field data of unsteady bottom stresses over cohesive smooth beds are rare. Some data were collected in the deep sea at 5000 m water depth and compared with smooth-bottom flume data from the laboratory by Gust and Weatherly (1985). Entrainment results for undisturbed cohesive sediments at Puget Sound/Seattle, exposed to a steadily increasing, known bottom stress by means of a seaflume, were presented by Gust and Morris (1989). Recent data on the near-bottom flow field and estimates of the bottom stress using the logarithmic-layer technique (see, e.g., Sanford et al., 1991) in the intertidal region of the Elbe estuary at Station Kollmar are shown in Fig. 1. From this station, the entrainment function for natural sediment is investigated by Humann et al. (in prep.) at $\tau < 0.5$ N/m$^2$.

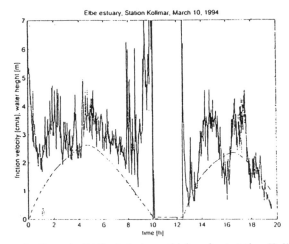

**Figure 1.** Friction velocity u∗ (solid line) during a tidal cycle at station Kollmar (Elbe Estuary). Water depth (dashed line); friction velocity u∗ was estimated by the velocity gradient technique from velocity measurements at height horizons 1 to 10 cm above ground;

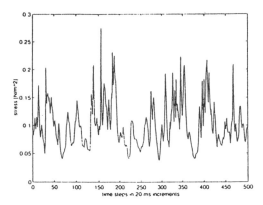 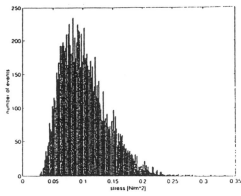

**Figure 2.**    Bottom stress τ in a straight flume (device no.2 of Tab.1) with flow obeying the universal law of the wall: a) segment of a time series at $\tau_{mean}= 0.1$ N/m$^2$, b) histogram (12000 values, 200 bins). Sampling frequency was 50 Hz (unless otherwise noted).

Turbulent flows over a smooth bottom in straight channels obey the universal law of the wall (Hinze, 1975). While turbulence characteristics and drag laws are well known in such wall-bounded shear flows due to its wide range of engineering applications in aerodynamics and hydrodynamics, high-resolution measurements and evaluations of the unsteady bottom stress in aquatic multi-phase flows (also in-situ) have only recently become more frequent due to improved skin-friction probes (Gust, 1988). Mean values of the bottom stress and its spatial distribution in steady clear-water flume flows at high flow speeds were presented, for example, by Cardoso et al. (1989). Own measurements of a smooth-flow time series of unsteady bottom stresses (Fig. 2a) and the resulting histogram (Fig. 2b) are shown for a mean value of τ = 0.1 N/m$^2$. Data on the mean horizontal pressure gradient were collected by Hüttel and Gust (1992) for smooth flume flows of clear water at τ = 0.025 N/m$^2$ with Δp/Δr at < 1 Pa/cm ($10^{-5}$ bar/cm). In hydrodynamically smooth flows, unsteady bottom stresses are always pointing in the downstream direction (Eckelmann, 1974), and localized pressure patterns always migrate downstream (Emmerling, 1973).

Flume flows with suspended sediments, both cohesive and non-cohesive, have been extensively investigated by hydraulic engineers, although mostly with abiotic sediments (Vanoni, 1946; Einstein and Chien, 1955). It has been shown that the mean-flow structure and drag laws are modified by the suspended sediments; in particular changes in von Karman's constant (see discussion by Gust, 1984), drag reducing effects (Gust, 1976) and transport characteristics in non-Newtonian suspensions (Best and Leeder, 1993) are current topics in the literature. Indications for reduced bottom stresses (affecting entrainment characteristics) by suspension-laden flows (8g/l Kaolinite in sea water) at same discharge as clear - water flows are also presented in Tab. 2, with the mean value of the bottom stress τ dropping by 22 %.

*Operational settings of devices*
The same measuring protocol was used for the devices of Tab.1 to obtain the spatial (radial) distribution of mean wall shearing stresses (in N/m$^2$) and horizontal pressure gradients (in Pa/Δr) at Δr = 40 mm (and 20 mm) for smooth, impermeable bottoms utilizing clear water

**Tab 2.:** Significant mean values and statistical moments of the hydrodynamic characteristics of bottom stress $\tau$ and pressure gradient $\Delta p/\Delta r$ for different erosion devices at $\tau \approx 0.1$ N/m$^2$ and $\tau \approx 0.3$ N/m$^2$. All devices except MKLP30 have an inner diameter of 10 cm (MKLP30 = 30cm i.d.). std = standard deviation. dp is mean pressure gradient for $\Delta r = 40$ mm in the 10-cm devices.

| | |
|---|---|
| ClearWa | : Flume, clear water |
| NotClWa | : Flume, sea water with 8g/l kaolinite |
| | |
| Bell 4 | : small gap, central suction, 'Erosion - bell', h = 4mm |
| Bell 6 | : small gap, central suction, 'Erosion - bell', h = 6mm |
| EROM 4D | : 'Eromes', forward rotation of propeller; h = 40 mm |
| EROM 4S | : 'Eromes', backward rotation of propeller; h = 40 mm |
| MKHP10 | : 'Microcosm', housing diameter = 10cm, High-Pumping modus |
| MKLP10 | : 'Microcosm', housing diameter = 10cm, Low-Pumping modus |
| MKLP10 | : 'Microcosm', housing diameter = 30cm, Low-Pumping modus |
| PlungA | : vertically oscillating grid, 'Plunger'; grid 1 (see text) |
| PlungB | : vertically oscillating grid, 'Plunger'; grid 2 (see text) |

| | frequency [rpm] (suction rate) [l/min] | $\tau$ [N/m$^2$] | std/mean | std | skewness | flatness | dp [Pa] |
|---|---|---|---|---|---|---|---|
| ClearWa | | 0.10 | 0.38 | 0.0391 | 0.8848 | 4.005 | 0.00 |
| NotClWa | | 0.08 | 0.37 | 0.0297 | 1.2063 | 5.703 | 0.00 |
| PlungA | 300.0 | 0.10 | 0.45 | 0.0463 | 0.7845 | 4.535 | 8.27 |
| PlungB | 850.0 | 0.10 | 0.57 | 0.0598 | 9.2771 | 262.529 | 34.47 |
| MKLP10 | 80.0 (0.067) | 0.11 | 0.11 | 0.0114 | 0.3051 | 3.447 | 33.79 |
| MKHP10 | 40.0 (0.233) | 0.09 | 0.06 | 0.0055 | -0.0474 | 2.894 | 39.10 |
| MKLP30 | 36.3 (0.290) | 0.10 | 0.17 | 0.0174 | 0.2264 | 3.093 | 38.29 |
| EROM 4D | 200.0 | 0.13 | 0.61 | 0.0768 | 3.1713 | 18.958 | -5.54 |
| EROM 4S | 350.0 | 0.10 | 0.56 | 0.0558 | 2.7156 | 15.003 | -0.50 |
| Bell 4 | (3.000) | 0.12 | 0.06 | 0.0071 | 0.0747 | 2.743 | 10.67 |
| Bell 6 | (4.200) | 0.12 | 0.12 | 0.0138 | 0.1157 | 2.896 | 7.91 |
| ClearWa | | 0.29 | 0.25 | 0.0748 | 0.9169 | 4.647 | 0.00 |
| PlungA | 700.0 | 0.32 | 0.78 | 0.2496 | 3.2969 | 23.826 | -122.39 |
| PlungB | - | - | - | - | - | - | - |
| MKLP10 | - | - | - | - | - | - | - |
| MKHP10 | 180.0 (0.767) | 0.32 | 0.22 | 0.0706 | 0.4538 | 3.254 | 111.01 |
| MKLP30 | 80.0 (0.720) | 0.30 | 0.23 | 0.0697 | 1.0147 | 5.043 | 240.00 |
| EROM 4D | 300.0 | 0.32 | 1.01 | 0.3271 | 3.2266 | 18.335 | -10.08 |
| EROM 4S | 650.0 | 0.30 | 0.74 | 0.2230 | 3.3706 | 24.366 | -2.40 |
| Bell 4 | (6.000) | 0.22 | 0.10 | 0.0219 | 0.2740 | 2.965 | 29.69 |
| Bell 6 | (18.000) | 0.26 | 0.25 | 0.0664 | 0.8978 | 3.831 | 55.12 |

as fluid. The pressure gradients and statistical moments at bottom stresses of ~0.1 N/m$^2$ and ~0.3 N/m$^2$ are compiled in Tab. 2, with operational details per device given in the text below. In addition, power density spectra of $\tau$ and $\Delta p/\Delta r$, collected in the devices at a mean value of $\tau = 0.1$ N/m$^2$, are compared in Figs. 13 and 15.

The figures on device performance present data in groups comprised of radial distributions of a) time-averaged, local bottom stress, b) time-averaged pressure gradient, together with c) a 10-s window of unsteady bottom stress $\tau$ and d) the time-series histogram of bottom stress $\tau$.

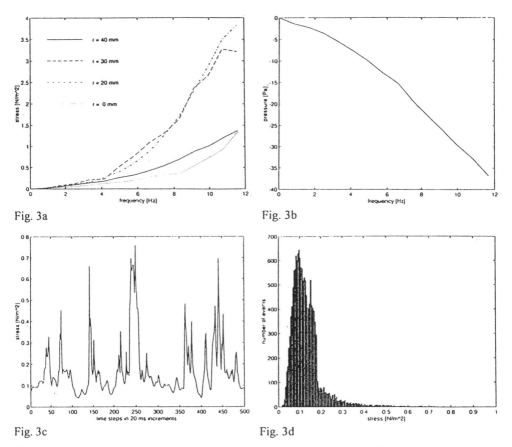

Fig. 3a

Fig. 3b

Fig. 3c

Fig. 3d

**Figure 3.**    Interfacial hydrodynamics of Eromes (device no.3 of Tab.1), forward turning propeller: a) radial distributions of mean bottom stress measured by skin friction probes (60*frequency=rpm), b) mean pressure gradient $\Delta p_{0-40}$; details of unsteady bottom stress $\tau$ with c) sample of time series at $\tau_{40} = 0.13$ N/m$^2$, d) histogram (15000 values, 200 bins). Index values indicate radial positions in mm with r = 0 at center.

*Rotating annuli*

Entrainment data collected in this type of device are presented in the literature by many authors (see, for example, sources in Lavelle et al., 1984; Kuijper et al., 1989). The underlying hydrodynamic forcing has been identified by several researchers, using either LDA for flow velocity measurements in the water column and/or flush-mounted hot-film probes for bottom stress measurements (Graham et al., 1992). Experiments were conducted with clear water over impermeable beds. Values of horizontal pressure gradients have not been reported so far. Modeling activities of the resulting flow fields, the associated secondary circulations, and ensuing bottom stress patterns were reported by Sheng (1989) and others. Efforts are underway to utilize rotating annuli in-situ (Maa, 1991; Amos et al., 1992). The maximum wall shearing stresses generated by these devices range from 0.0 to 1.0 N/m² (see tab. 5.4. in Kuijper, 1993).

*Propeller stirring (Eromes)*

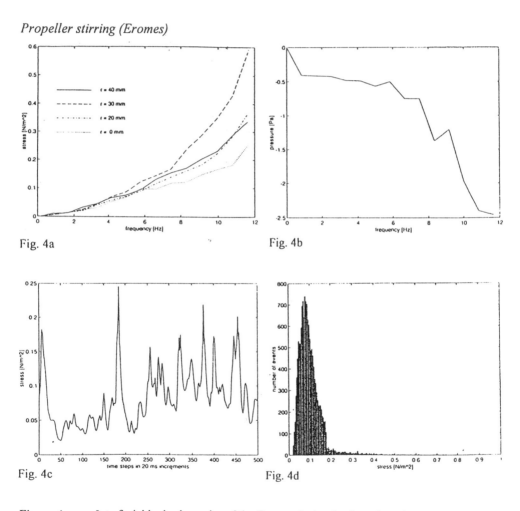

Fig. 4a  Fig. 4b

Fig. 4c  Fig. 4d

**Figure 4.**     Interfacial hydrodynamics of the Eromes device, backward turning propeller: a) radial distributions of mean bottom stress measured by skin friction probes, b) mean pressure gradient $\Delta p_{0-40}$; details of unsteady bottom stress $\tau$ with c) sample of time series at $\tau_{40} = 0.1$ N/m², d) histogram (15000 values, 200 bins). Index values indicate radial positions in mm with $r = 0$ at center.

In collaboration with GKSS, an original Eromes stirring unit (Schünemann and Kühl, 1991) with 3-bladed propeller of 5 cm in diameter, placed 4 cm above the bottom, was investigated in both forward and backward turning modes. In the forward turning mode, both the radial distribution of effective bottom stress based on the threshold of moving quartz grains of known diameter (Unsöld, 1982), and on mean skin friction calibration curves using hot-film anemometry (Fig. 3a) were obtained at different locations as function of propeller turning rate, together with data on the horizontal pressure gradients (Fig. 3b). Distributions of mean skin friction (Fig. 4a) and horizontal pressure gradient (Fig. 4b) were obtained in the reverse turning mode as well. Examples of unsteady time series of τ and histograms of τ₄₀ are shown in Figs. 3c, 3d for the forward turning mode, and in Figs. 4c, 4d for the reverse turning mode. For other statistical parameters, see Tab. 2.

*Vertically oscillating grid (Plunger).*

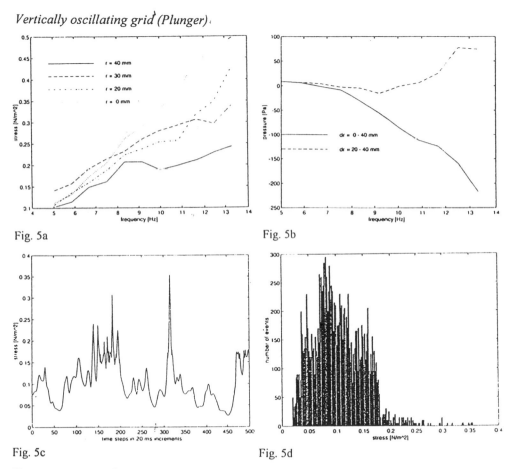

Fig. 5a                    Fig. 5b

Fig. 5c                    Fig. 5d

**Figure 5.** Interfacial hydrodynamics of the Plunger (device no.4 of Tab.1) with hydraulic similarity. a) radial distributions of mean bottom stress measured by skin friction probes, b) mean pressure gradients $\Delta p_{0-40}$ and $\Delta p_{20-40}$; details of unsteady bottom stress τ with c) sample of time series at $\tau_{40} = 0.1$ N/m², d) histogram (15000 values, 200 bins). Index values indicate radial positions in mm with r = 0 at center.

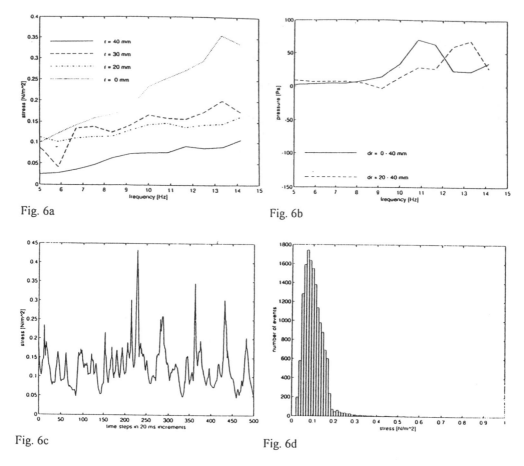

Fig. 6a

Fig. 6b

Fig. 6c

Fig. 6d

**Figure 6.**     Interfacial hydrodynamics of the Plunger device with slightly changed geometry relative to original device. a) radial distributions of mean bottom stress measured by skin friction probes, b) mean pressure gradients $\Delta p_{0-40}$ and $\Delta p_{20-40}$; details of unsteady bottom stress $\tau$ with c) sample of time series at $\tau_{40} = 0.1$ N/m², d) histogram (15000 values, 200 bins). Index values indicate radial positions in mm with r = 0 at center.

Tsai and Lick (1986) developed a device (shown in Fig. 1 of the original article) which erodes sediment by means of an oscillating grid. An 'equivalent shear stress' was introduced by the authors based on the assumption that if concentrations of resuspended sediments are the same in the shaker and in a calibrated laboratory flume under same environmental conditions, then the stresses needed to produce these resuspended sediments are equivalent (ibid., p.315). Data on the hydrodynamics at the bottom were not found in the literature. The inner diameter of the original holding vessel was 11.7 cm. Two series of replication experiments were run in an acrylic tube with 9.9 cm inner diameter at stroke frequencies and amplitudes reported by Tsai and Lick: one with the outer diameter of the shaker disk at 9.3 cm and hole diameter of 1.24 cm (disk porosity of 42.8%), the second with a gap of 0.5 cm between shaker disk and tube wall, disk diameter of 8.9 cm, and hole diameter of 1.2 cm (same disk porosity).

For the first geometry, hydrodynamic similarity was established with the original work by Tsai and Lick (1986). The vertical shaker motions then generated an outwardly pointing turbulent flow at the bottom, with the radial distribution of mean wall shearing stresses shown in Fig. 5a for different oscillation frequencies. The associated mean horizontal pressure gradients $\Delta p$ ($\Delta r$=0-40mm and $\Delta r$=20-40mm) are displayed in Fig. 5b for increasing shaker frequency. Unsteady values of $\tau$ are shown in Figs. 5c (time series) and 5d (histogram). At frequencies > 10 Hz, bubbles appeared in the fluid, the origin of which may be either entrapped air or cavitation. The free surface of the fluid showed standing radial waves of low amplitude < 0.5 cm. The spatial distribution of mean wall shearing stresses produced by the second geometry, with only slight changes from the first case, is distinctly different (Fig. 6a). Associated data on pressure gradients, unsteady bottom stress at $r$ = 40 mm and histogram are given in Figs. 6b, 6c, 6d, respectively.

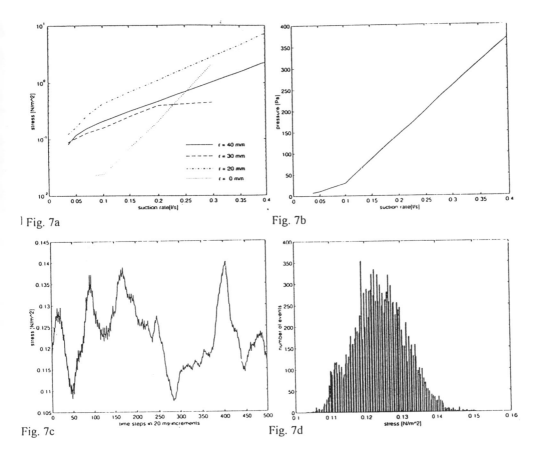

| Fig. 7a

Fig. 7b

Fig. 7c

Fig. 7d

**Figure 7.**    Interfacial hydrodynamics of the Erosion bell (device no.5 of Tab.1) with 4 mm duct height at different suction rates. a) Radial distributions of mean bottom stress measured by skin friction probes, b) mean pressure gradient $\Delta p_{0-40}$; details of unsteady bottom stress $\tau$ with c) sample of time series at $\tau_{40}$ = 0.12 N/m$^2$, d) histogram (15000 values, 200 bins). Index values indicate radial positions in mm with $r$ = 0 at center.

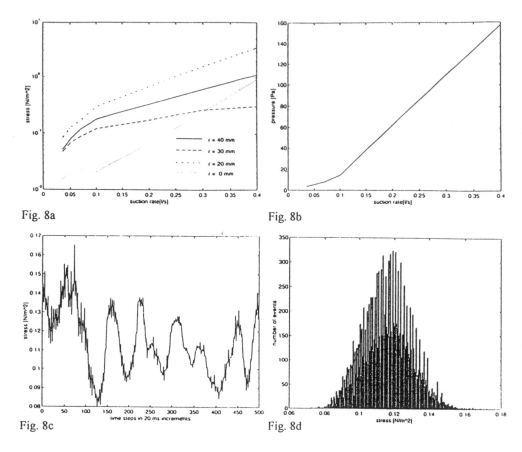

Fig. 8a

Fig. 8b

Fig. 8c

Fig. 8d

**Figure 8.** Interfacial hydrodynamics of the Erosion bell with 6 mm duct height at different suction rates. a) Radial distributions of mean bottom stress measured by skin friction probes, b) mean pressure gradient $\Delta p_{0-40}$; details of unsteady bottom stress $\tau$ with c) sample of time series at $\tau_{40} = 0.12$ N/m², d) histogram (15000 values, 200 bins). Index value indicates the radial position in mm with r = 0 at center.

## Suction head (Erosion bell)

In the suction head device, a particularly shaped, radial-symmetric duct section is formed by placing bell heads of specific geometry (Figs. 8 and 12 of Williamson and Ockenden, 1993) at selected heights over the bottom. Fluid enters the duct area circumfencially through a narrow gap between the bell housing and the bell head, and moves towards the center axis where it is lifted by suction. The fluid can be recirculated or is wasted and replaced from other reservoirs including the environment. Following the original report, where center suction rates varied between 0.035 and 1.0 l/s (with duct gaps adjusted accordingly), the optimized head shape was investigated at gap heights 4 and 6 mm and discharge rates up to 0.4 l/s. With a different design for fluid entry, exit and recirculation sections of the head unit but same shape of the bell head, radial distributions for mean bottom stresses and mean pressure gradients are shown in Figs. 7a,b (4 mm gap) and Figs. 8a,b (6 mm gap). In the development of the final head shape, numerical calculations aimed at optimizing the bottom stress distribution under laminar flow were applied by the original authors. Our data with the replicate heads indicate, though, that the flows were turbulent (Figs. 7c,d for h = 4mm, Figs. 8c,d for h = 6mm). Further details of the hydrodynamics at $\tau = 0.1$ N/m$^2$ and $\tau = 0.3$ N/m$^2$ are presented in Tab. 2; the power spectral densities of bottom stress $\tau$ and of $\Delta p$ at $\tau = 0.1$ N/m$^2$ are incorporated in Figs. 13, 15. The erosion bell of Williamson and Ockenden (1993) exists both as laboratory and field version.

*Suction and stirring (Microcosm)*

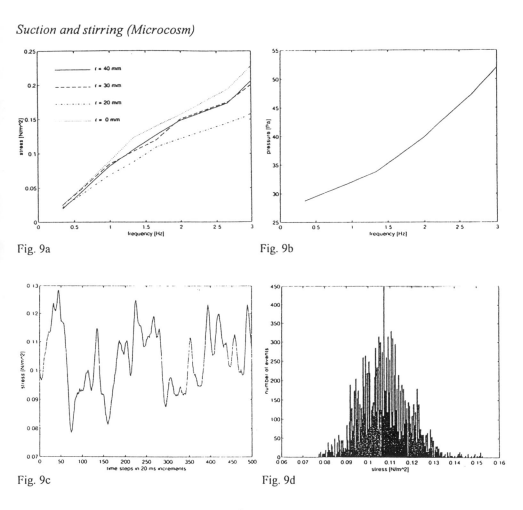

Fig. 9a          Fig. 9b

Fig. 9c          Fig. 9d

**Figure 9.** Interfacial hydrodynamics for the Microcosm (device no.6 of Tab.1, Low-Pumping modus (suction rate = constant = 0.0667 l/min)) with 10 cm in diameter and 5 cm water height. a) Radial distributions of mean bottom stress measured by skin friction probes, b) mean pressure gradient $\Delta p_{0-40}$; details of unsteady bottom stress $\tau$ with c) sample of time series at $\tau_{40} = 0.11$ N/m$^2$, d) histogram (15000 values, 200 bins). Index values indicate radial positions in mm with r=0 at center.

163

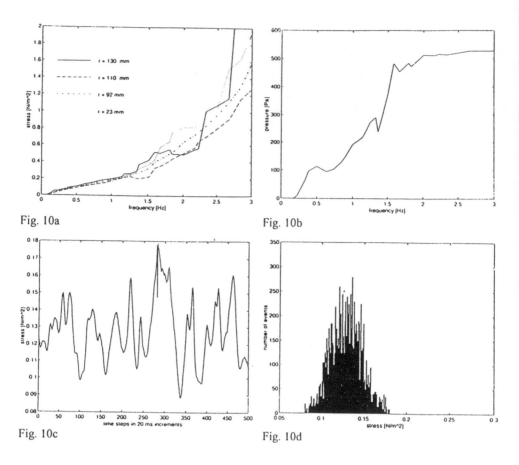

Fig. 10a

Fig. 10b

Fig. 10c

Fig. 10d

**Figure 10.** Interfacial hydrodynamics for the Microcosm device with 30 cm in diameter and 10 cm water height (from Gust and Müller, in prep.). a) Radial distributions of mean bottom stress measured by skin friction probes, b) mean pressure gradient $\Delta p_{0-130}$; details of unsteady bottom stress $\tau$ with c) sample of time series at $\tau_{130} = 0.12$ N/m², d) histogram (15000 values, 200 bins). Index values indicate radial positions in mm with r = 0 at center.

The microcosm generates a spatially homogeneous wall shearing stress by controlling two stirring parameters simultaneously: one is the angular frequency of a stirring disk at selectable heights (currently 3 to 15 cm) over the bottom, the other is the volume of metered, recirculated fluid removed through the rotating axis (Gust, 1987; 1991). Schematics of the device are shown in Figs. 1, 2 of Gust (1990). Design variables include diameter of confinement housing, of stirrer disk, distance d between stirrer and bottom, stirrer angular frequency, and suction rate of recirculated/replaced fluid (Gust and Müller, in prep.). Data on wall shearing stresses and horizontal pressure gradients are presented in Tab. 2 for a microcosm of 10 cm in diameter with 5 cm distance d between bottom and disk for $\tau \leq 0.25$ N/m². The figures show radial distributions of mean $\tau$ and mean $\Delta p$ (Figs. 9a,b), and unsteady $\tau$ at r = 40 mm (time series in Fig. 9c and histogram in Fig. 9d). In Figs. 10a,b, radial mean $\tau$ and $\Delta p$ are shown for a microcosm of 30 cm in diameter with 10 cm water height (d = 8 cm) for mean bottom stress $\leq 0.1$ N/m², with unsteady $\tau$-values at $\tau_{mean} = 0.1$ N/m² given in Figs. 10c,d. Of this device, field versions are in use for intertidal deployments (Humann et al., in prep.). Automated systems, deployable down to the deep sea floor, are under development.

*Entrainment curves of abiotic Kaolinite*

The 30 cm version of the microcosm was used to collect entrainment data for abiotic Kaolinite in the laboratory. Five replicas of freshly stirred sediment cores with 24-h and 48-h consolidation times were exposed to stepwise increased bottom stresses in the range 0.004 to 0.5 N/m². One series was run at 240 h consolidation time. From filtered and weighted suspension samples (250 ml) taken at onset and termination of each time step of 10-min duration, the mass E entrained per unit area and unit time was determined. The resulting mean entrainment curves are compiled in Fig. 11.

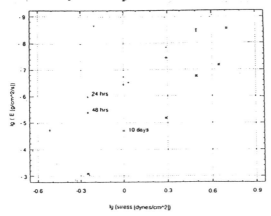

**Figure 11.**    Entrainment curves of abiotic Kaolinite consolidated for 24, 48 and 240 hours in a 30-cm microcosm. Data points are means of 5 replicas for 24 and 48 h consolidation times and from a single experiment for 240 h consolidation time. Erosion time intervals were set at 10 min.

*Entrainment rates with other sediments, devices and stress time histories*

The data base of entrainment rates reviewed here includes Mississippi river sediment eroded in the microcosm (Gust and Müller, in prep.) as well as undisturbed cohesive sediments eroded in-situ in Puget Sound/Seattle by means of a seaflume at water depth < 200m (Fig. 9 of Gust and Morris, 1989). Further entrainment data of cohesive sediments, eroded in rotating annuli, were compiled by Lavelle et al. (1984) and by Kuijper (1993).

Advances in erosion device control permitted experiments in the microcosm under precisely prescribed accelerating and decelerating wall shearing stresses while monitoring on-line the ensuing entrainment rates (Eden et al., in prep.). As a first exercise, the question was investigated if natural cohesive sediments can reach equilibrium concentrations under steady, eroding flows within the 6-h time scales typical for semi-diurnal tidal cycles with direction-reversing ebb- and flood-flows. The concept of equilibrium concentrations had emerged

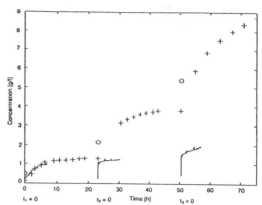

**Figure 12.** Time series of 6 h duration of concentration increase in a 30-cm microcosm (line print) for sieved natural mud from Mühlenberger Loch (Elbe Estuary) at 0.1, 0.196, and 0.25 N/m². In each of the three time series, the sediment had consolidated for 24 h. The constant stress value was reached in each case from zero at $t_1$, $t_2$, $t_3$ in a constant acceleration ramp of 10 min duration. These data are compared with entrainment rates (crosses) of cohesive sediments (China Clay) in the rotating annulus of Kuijper et al. (1989) and with 24-h consolidated abiotic Kaolinite from Fig. 11 (open circles).

from previous work utilizing rotating annuli, yet recent designs of annuli with counterrotating flume section provided data indicating that this erosion behavior may not be universal (Kuijper et al., 1989). Results in a 30-cm microcosm by Eden et al. (in prep.) indicated for 6-h intervals continued constant entrainment rates for $\tau > \tau_{cr}$ for natural sediments from Mühlenberger Loch (Elbe estuary), indicative of absence of equilibrium concentrations (Fig. 12).

## DISCUSSION AND CONCLUSIONS

*Device performance*
Device characteristics have to be judged in context with the interfacial forces prevailing in the natural (or engineering) boundary layer flows. The proper use of devices thus hinges on their ability to reproduce these natural hydrodynamic parameters. For example, both weak and strong erosive forces are required by entrainment devices to simulate those of intertidal flows (see Fig. 1). Three sets of device-specific characteristics determine their application range:

*Fine structure of interfacial bottom stresses*
The entrainment-relevant bottom stress of an erosion apparatus has to compare with that in the field not only in mean (spatially homogeneous) values but also in its fine structure (statistical moments). Representative power spectra of fluctuations of $\tau$ inside the various devices are compared with those of straight-flume flow in Figs. 13a,b at $\tau = 0.1$ N/m². That the latter parameter can affect or even control the entrainment of sediment is demonstrated by a device with forward - turning propeller (Müller et al.,1995), where the skin friction,

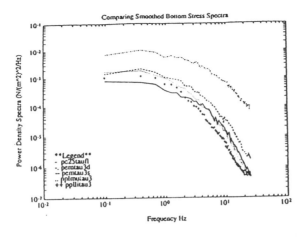

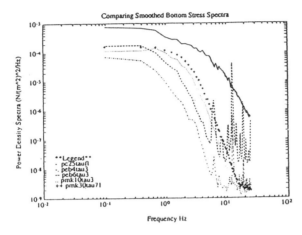

**Figure 13.** Intercomparison of bottom stress spectra collected in the devices 2 to 6 of Tab. 1 at a mean bottom stress of 0.1 N/m$^2$ (5 min time series, 50 Hz). Top figure: Straight flume, clear water: solid line; Eromes, forward turning: dash-dot, backward turning: dash-dash; Plunger, hydraulic similarity: dot-dot, with geometry slightly changed: cross-cross. Bottom figure: Straight flume, clear water: solid line; Erosion bell, optimized head, 4 mm duct height: dash-dot, 6 mm duct height: dash-dash; Microcosm, 10 cm in diameter: dot-dot, 30 cm in diameter: cross-cross.

comprised of a distribution of high-peak unsteady values (different from that of natural flows at same mean $\tau$), moved quartz grains by the top ten percent of the probability function of unsteady $\tau$-values ($\tau_{1/10}$) rather than by the mean value $\tau_{mean}$ or by $\tau_{1/3}$ (Fig. 14).

*Presence of horizontal pressure gradients*
All devices show to some extent horizontal pressure gradients (see last column of Tab.2), with power density functions of fluctuations of $\Delta p$ intercompared in Figs. 15a,b. Even if mean values and fine structure of the bottom stress are close to the natural conditions, $\Delta p$ can affect or control the entrainment of sediment. Such situation is suggested for the plunger by comparing the measured interfacial hydrodynamics with the curve of 'equivalent shear

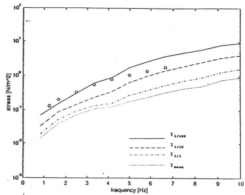

**Figure 14.**    Intercomparison of statistical parameters of bottom stress distribution (probatility of exceedance: $P(\tau < \tau_\alpha) = 1 - \alpha$), measured by skin friction probes in Eromes at forward turning mode of propeller, with an effective threshold bottom stress initiating movement of quartz grains of different diameter (open circles).

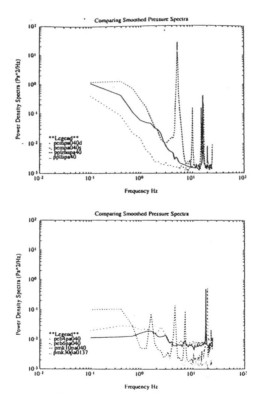

**Figure 15.**    Intercomparison of power spectra of the horizontal pressure gradient $\Delta p_{0-40}$ concurrently sampled with $\tau$-values used to calculate the bottom stress spectra of Fig. 13. Top figure: Eromes, forward turning mode: solid line, backward turning mode: dash-dot; Plunger, hydraulic similarity: dash-dash, with geometry slightly changed: dot-dot. Bottom figure: Erosion bell, optimized head, h = 4 mm: solid line, h = 6 mm: dash-dot; Microcosm, 10 cm diameter: dash-dash, 30 cm diameter with $\Delta p_{0-137}$ : dot-dot.

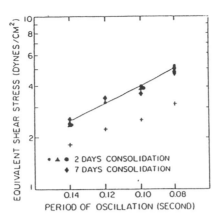

**Figure 16.** Intercomparison of the 'effective entrainment stress' introduced by Tsai and Lick (1986) with mean skin friction data (crosses) of the hydraulically similar Plunger for same shaker frequencies. Note the high value of τ at commencing shaker frequency of 5 Hz. The differences between the calibration curves of effective and actual τ are interpreted as effects of both mean and fine structure of the horizontal pressure gradient generated in the device. Statistical moments of the bottom stress resembled very well those of straight-flume flow at $\tau = 0.1$ N/m$^2$.

stress' introduced in the original paper by Tsai and Lick (Fig. 16). The concentration values in the hydraulically similar plunger device are obtained at smaller values of bottom stress than applied in the calibration flume. The large pressure gradients are proposed as the source of additional erosive forces, with a pronounced spectral peak at the plunger oscillation frequency. This result suggests that in the plunger each sediment investigated may require independent calibration efforts.

*Adherence to geometry and/or operational mode*
In most devices the repeatability of experiments and stability of the results critically depends on a strictly observed experimental protocol. Minute changes in geometry and/or operational mode changed the spatial and temporal stress patterns to both better and worse, with underlying reasons not always evident or obvious. Examples are presented for 1. the propeller-driven device, where switching to a backward-turning mode largely improved the bottom stress characteristics but reduced the magnitude of τ (compare Figs. 3,4), 2. the plunger, where by merely changing the gap space between disk and sidewall by a few millimeters, the wall shearing stress, direction of the internal flow cell and pressure gradient characteristics switched to completely different patterns (compare Figs. 5,6), 3. the erosion bell, where our data (at same duct shape but different design of the recirculation path) improved the radial homogeneity of the mean bottom stress over those given in the original report (Fig. 17), and 4. for the microcosm, where a high central fluid removal rate, when enacted instead of the prescribed low-pumping mode, generates a high bottom stress in the center of the sediment core with localized erosion.

Of the devices tested, it appears that for laboratory simulations of entrainment processes as well as for field applications at bottom stresses at $\tau < 1$ N/m$^2$, the microcosm offers the most suitable calibration characteristics with a reasonable spatially homogeneous bottom stress distribution (Fig. 18 a), and a bottom stress fine structure (Fig. 18 b) which variance

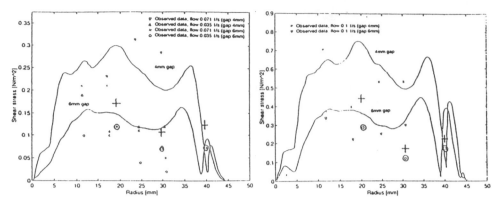

**Figure 17.** Intercomparison of a) measured (symbols) and modeled (solid line) bottom stress by Williamson and Ockenden ('1993) with data from the replicate head at 4 mm (crosses) and 6 mm (double open circle). Left panel: Fluid recirculation rate 0.05 l/s, right panel: Fluid recirculation rate 0.1 l/s.

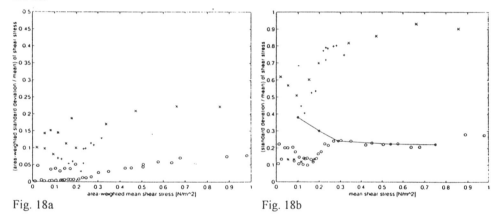

Fig. 18a                                        Fig. 18b

**Figure 18.** Evaluation of hydrodynamic characteristics of the various erosion devices tested (EROMES (x), plunger (+), erosion bell (*), microcosm(o)): a) Spatial homogeneity of the mean bottom stress of the devices tested. For a perfectly homogeneous bottom stress the ordinate value would be zero. b) Intensity of unsteady bottom stress compared with the reference values of straight, smooth flume flow ( o ). Data from radial position r = 40 mm except for 30cm microcosm where r = 130 mm.

is very similar to that of natural flows at $\tau > 0.3$ N/m². These features come with a simple and stable operational mode, and reproducible performance features for steady and time-dependent erosion forcing functions. Attention has to be paid to possible localized erosion in the center area, which can be controlled by covering the center section of 5 cm² with a solid disk in lab simulations or optimizing the stirring/suction ratio at the expense of a slightly reduced spatial homogeneity. Entrainment and deposition simulations of cohesive bottoms are possible with this device as long as the sediment interface remains smooth or transitional. The fluid recirculation path permits inclusion of additional measurement and processing units (designed to either investigate suspension characteristics and biogeo-chemical parameters or to control the amount of suspended material retained in the eroding fluid, for example). As a consequence, investigations into the effects eroded cohesive sediment at various concentrations may have on the calibration features of erosion devices, as well as aspects of water quality, become feasible. A fully computerized microcosm with an on-line concentration recording unit for repetetive tidal and other unsteady time histories of the bottom stress has been designed and is currently tested (Eden et al., in prep.). A deep-sea entrainment module is under construction as well. Furthermore, independent measurements of $\tau$ and $\gamma$ are feasible at the interface between an impermeable bottom and a fluid (suspension) in the device, thus effectively providing a turbulent-flow suspension rheometer.

*Entrainment functions*

Relative errors of mean entrainment rates of abiotic Kaolinite, consolidated for 24 h, ranged for the 5 replicate measuring sequences from 25 to 60%. Similar error margins were found for the 48-h consolidation time experiments and for all Mississippi river mud data. The uncertainty in experimental procedure was determined by the material remaining in the fluid volume of the system (~7 l) during preparation of the core (mg-range). Detection thresholds of entrainment E lay at approximately $10^{-8}$ g/cm$^2$/s. Nearly identical masses were collected in experiments with 10 and 100 min duration at same low bottom stresses < 0.03 N/m$^2$ . This experimental evidence established erosion thresholds with $\tau_{crit.} \geq 0.03$ N/m$^2$ for abiotic Kaolinite and Mississippi river mud. Equilibrium concentrations of suspended material, eroded at bottom-stresses exceeding this value, were not reached for natural mud from Mühlenberger Loch (Elbe Estuary) within the 6-h run time (Eden et al., in prep.). At bottom stresses lower than this threshold, concentrations typically had dropped at the end of 10-min intervals beneath values of blanks taken at the beginning of experiments, suggesting active deposition of material.

The entrainment rates of abiotic Kaolinite were not statistically different from the Mississippi river mud data with same 24-h consolidation time, except for the the highest value at 0.31 N/m$^2$ as determined by utilizing the Statgraphics software package.

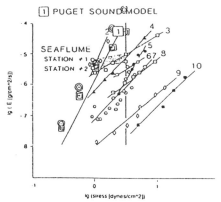

**Figure 19.**    Intercomparison of entrainment data from various devices, sediments and erosion time increments. The newly added data are 24-h consolidated Kaolinite (double open circle) and Mississippi river mud (double open square) over and beyond those compiled and identified already in Fig. 9 of Gust and Morris (1989).

171

The 24-h consolidated Kaolinite and Mississippi river entrainment data are compared with the field entrainment data of a seaflume (Gust and Morris, 1989), and results from rotating annuli (Lavelle et al., 1984) in Fig. 19. Most entrainment rates increase by about on order of magnitude for the range $\tau_{threshold} < \tau_{erod} < 0.5$ N/m$^2$. It is not clear if the differences observed for E in Fig. 19 can be attributed to differences in experimental techniques employed (equilibrium values vs. entrainment data from 10-min intervals), since the data of Fig. 12 show rather good agreement between entrainment rates of annuli and microcosms for abiotic Kaolinite. The different slope of the seaflume entrainment data is attributable either to the fact that unsteady values of E were recorded, or to biological stabilization effects of the undisturbed in-situ sediment beds. The data set is currently too small to permit any firm conclusions.

The differences in entrainment rates between the different techniques applied, and partly even among individual devices within the same generic type of device, require investigation into the possibility of suspension effects on flow and device performance. Flows could switch from Newtonian to non-Newtonian behavior for some or all of the suspensions, affecting device performance by providing then a different drag law (for example, between rotating lid and suspension in an annulus). To answer these questions, the relationship between $\gamma$ and $\tau$ has to be obtained for suspensions, both in natural flows and for erosion devices. By such approach validity of clear-water calibration relationships of $\tau$ (or corrections thereof) will be determined. Once these issues are clarified, more refined experiments will become possible to resolve the minimum set of parameters from particular physico-chemical, rheological and biogeochemical processes (Verreet and Berlamont, 1986) which control the entrainment of natural cohesive sediments.

## ACKNOWLEDGEMENTS

Funding for parts of this research was provided by Waterways Experiment Station, Estuaries Program at Vicksburg, MS, and by Technical University of Hamburg-Harburg. Staff members, students and scientists who provided data for this review are currently preparing their articles as identified in the literature list. The collection of hitherto unpublished data from the former flume laboratory of G.G. at the University of South Florida, Marine Science Dept. (St. Petersburg, USA) was assisted by E. Wysk-Koch and Z. Li. Our thanks go to all of them for their contributions.

## REFERENCES

*Amos, C., Grant, J., Daborn, G. R. and K. Black*: Sea Carousel - A benthic annular flume. Estuarine, Coastal and Shelf Science, 34, 557-577, 1992.

*Bellhouse, B. J. and D. L. Schultz*: Determination of mean and dynamic skin friction, separation and transition in low-speed flow with a thin-film heated element. J. Fluid Mech. 24, 379-400, 1966.

*Best, J. L. and M. R. Leeder*: Drag reduction in turbulent muddy seawater flows and some sedimentary consequences. Sedimentology 40, 1129-1137, 1993.

*Cardoso A. H., Graf, W. H. and G. Gust*: Uniform flow in a smooth open channel. J. Hydr. Res. <u>27</u>, 603-616, 1989.

*Eckelmann, H.*: The structure of the viscous sublayer and the adjacent wall region in a turbulent channel flow. J. Fluid Mech., 439-459, 1974.

*Eden, H., Gust, G., Müller, V. and W. Dasch*: Entrainment functions and underlying hydrodynamic forcing by wind and tides in shallow regions of the Elbe estuary (in preparation).

*Einstein, H. A. and N. Chien*: Effects of heavy sediment concentration near the bed on velocity and sediment distribution. Univ. of Calif., Berkeley, and U.S. Army Corps of Engr., Missouri River Div., Report No. <u>8</u>, 96 pp., 1955.

*Emmerling, R*: Mitteilungen aus dem Max-Planck-Institut fhr Strömungsforschung und der Aerodynamischen Versuchsanstalt, Göttingen, Nr. <u>56</u>, 1973.

*Graham, D. I., James, P. W., Jones, T. E. R., Davies, J. M. and E. A. Delo*: Measurement and prediction of surface shear stress in annular flume. J. Hydr. Engrg., ASCE, 118(9), 1270-1286, 1992.

*Grant, J. and G. Gust*: Prediction of coastal sediment stability from photopigment content of purple sulfur bacterial mats. Nature <u>330</u>, 244-246, 1987.

*Gust, G.*: Observations on turbulent drag reduction in a dilute suspension of clay in sea water. J. Fluid Mech. <u>75</u>, 29-47, 1976.

*Gust G.*: Discussion on 'Velocity profiles with suspended sediment' by N. L. Coleman. J. Hydr. Res. <u>22</u>, 263-275, 1984.

*Gust, G.*: Skin friction probes for field applications. J. Geophys. Res. <u>93</u>, 14121-14132, 1988.

*Gust, G.*: Method of generating precisely-defined wall shearing stresses, U. S. Patent No. 4,973,165, 1990.

*Gust, G. and J. B. Southard*: Effects of weak bed load on the universal law of the wall. J. Geophys. Res. <u>88</u>, 5939-5952, 1983.

*Gust, G. and G. L. Weatherly*: Velocities, turbulence and skin friction in a deep-sea logarithmic layer. J. Geophys. Res. <u>90</u>, 4779-4792, 1985.

*Gust, G. and V. Müller*: A device for interfacial solute and material fluxes with calibrated, user-selectable bottom stress fields (in prep.).

*Gust, G. and M. Morris*: Erosion thresholds and entrainment rates of undisturbed in-situ sediments. J. Coastal Res., Spec. Issue <u>5</u>, 87-99, 1989.

*Hinze, J. O.*: Turbulence, 2nd ed., McGraw-Hill, New York, 1975.

*Hüttel M. and G. Gust*: Impact of bioroughness on interfacial solute exchange in permeable sediments. Marine Ecology Progress Series <u>89</u>, 253-267, 1992.

*Humann, K., Kies, L. and G. Gust*: The effects of benthic algae on the stability of cohesive sediments in an intertidal region (in prep.).

*Krishnappan B. G.*: Rotating circular flume. J. Hydr. Engrg., ASCE 119(6), 758-767, 1993.

*Kuijper, C., Editor*: On the methodology and accuracy of measuring physico-chemical properties to characterize cohesive sediments. G6M Coastal Morphodynamics, Report of MAST-I Programme, 1993.

*Kuiper,C., Cornelisse, J.M., and J.C. Winterwerp*: Research on erosive properties of cohesive sediments. J. Geophys. Res. <u>94</u>, 14341-14350, 1989.

*Lavelle, J. W., Mofjeld, H. O. and E. T. Baker*: An in situ erosion rate for a fine-grained marine sediment. J. Geophys. Res. <u>89</u>, 6543-6552, 1984.

*Ludwieg, H. and W. Tillmann*: Untersuchungen über die Wandschubspannung in turbulenten Reibungsschichten. Ing. Arch. <u>17</u>, 288-299, 1949.

*Maa, J. P., Shannon, T. S., Li, C., and C. H. Lee*: In situ measurements of the critical bed shear stress for erosion. Environmental Hydraulics, Lee and Cheung, Ed., Balkema, Rotterdam, 1991.

*Müller, V. and G. Gust*: Hochauflösende Strömungsmessungen in meeresgeologisch relevanten Zonen. Proc. 2nd Workshop Meeresgeowissenschaften, Rostock, pp 64-68, 1994

*Müller, V., Vorrath, D., Werner, A. and G. Witte*: Skin-friction characteristics of the EROMES-system - Hydrodynamic measurements and erosion experiments with Kaolinite. GKSS Report 95/E/43, 1995

*Partheniades, E. and J. F. Kennedy*: Depositional behaviour of fine sediment in a turbulent fluid motion. Proc., 10th Conf. on Coastal Engrg., Tokyo, 2, 707-724, 1966.

*Sanford, L. P., Panageotou, W. and J.P. Halka*: Tidal resuspension of sediments in northern Chesapeake Bay. Marine Geology 97, 87-103, 1991.

*Sattel, H.*: Wandschubspannung an umströmten Körpern. Univ. der Bundeswehr München, Dissertation, 1994.

*Schlichting, H*: Grenzschicht-Theorie, Karlsruhe, Braun, 1979.

*Schünemann, M. and H. Kühl*: A device for erosion-measurements on naturally formed, muddy sediments: the EROMES-System. GKSS Report 91/E/18, 1991.

*Sheng, Y. P.*: Consideration of flow in rotating annuli for sediment erosion and deposition studies. J. Coastal Res., Spec. Issue 5, 207-216, 1989.

*Stewartson, K.*: On the flow between two rotating coaxial disks. Proc. Cambridge Phil. Soc. 49, 333-341, 1952.

*Tsai, C. H. and W. Lick*: A portable device for measuring sediment resuspension. J. Great Lakes Res. 12(4), 314-321, 1986.

*Unsöld, G.*: Der Transportbeginn rolligen Sohlmaterials in gleichförmigen turbulenten Strömungen: Eine kritische Überprüfung der Shields-Funktion und ihre experimentelle Erweiterung auf feinstkörnige, nicht-bindige Sedimente. Ph. D. Dissertation der Math. Nat. Fakultät, Universität Kiel, 1982.

*Vanoni, V. A.*: Transportation of suspended sediment by water. Trans. ASCE, III, 67-133, 1946.

*Verreet, G. and J. Berlamont*: Rheology and non-Newtonian behavior of sea and estuarine mud. Encyclopedia of Fluid Mechanics 7, 135-149, 1986.

*Williamson, H. J. and M. C. Ockenden*: In situ erosion of cohesive sediment. Report ETSU TID 4112, HR Wallingford, 1993.

# 11  On the development of instruments for *in situ* erosion measurements

**J. M. CORNELISSE[1], H. P. J. MULDER[2], E. J. HOUWING[3], H. J. WILLIAMSON[4] and G. WITTE[5]**

[1]*Delft Hydraulics, P.O. Box 177, 2600 MH Delft, The Netherlands*
[2]*Rijkswaterstaat/RIKZ, P.O. Box 207, 9750 AE Haren, The Netherlands*
[3]*Utrecht Univesity, P.O. Box 80115, 3058 TC Utrecht, The Netherlands*
[4]*HR Wallingford, Howbery Park, Wallingford, Oxon OX10 8BA, England*
[5]*GKSS, Research Centre, D-21502, Geesthacht, Germany*

## ABSTRACT

Instruments to measure the in situ erosion rate are under development all over the world, though applicable for different purpose and under different circumstances. Some instruments have been designed to measure the effects of bio-turbation, while others are meant to establish the threshold of erosion and/or the erosion rate as a function of depth. In this paper four instruments, designed by European institutes are described and compared. The comparison is done by a series of reproduction tests with each instrument under identical conditions. These tests were carried out with China clay (Kaolinite) and fresh water; all institutes used Kaolinite sub-samples from the same main Kaolinite sample. A further analysis was carried out by comparing the results with similar experiments carried out in the annular flume of Delft Hydraulics. The deviations in concentration between the repetitive experiments vary from 25 % to 100 %, depending on the apparatus. The reproduction error depends on the reproducibility of both the hydrodynamics and the spatial distribution of the bed strength. The erosion rate quantified with any device will be as inaccurate as the reproduction error. Space and time dependent variations in the bed shear stress and the bed strength are important for the erosion rate. The paper is completed with a short discussion on the need to carry out reproduction tests. It is concluded that small errors in reproducibility of the bed properties and the applied erosive stress can result in large errors in the measured erosion rate. This effect is larger for smaller instruments.

## 1  GENERAL INTRODUCTION

Many years of laboratory experiments on the erosive properties of (natural) cohesive sediment beds have not led yet to a full understanding and an adequate description of the erosion process. Especially the combination of physical, chemical and biological processes makes a successful systematic laboratory research program very difficult To know the erosive properties of natural beds one should carry out field measurements. Various instruments have been developed and reported in literature.

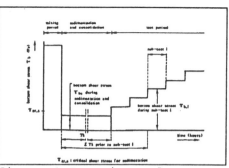

figure 1 Experimental setup

*Cohesive Sediments.*
Edited by Neville Burt, Reg Parker and Jacqueline Watts

These instruments are called sea-flumes (straight or annular) or particle entrainment simulators (e.g. Young, 1977, Lavelle and Davis, 1987, Tsai and Lick, 1986, Wilkinson and Jones, 1989, Houwing, 1992). Unfortunately no information is available on the reproducibility of these instruments. Reproduction tests however are essential for a correct interpretation of field data in order to know whether the experimental result is reliable. Therefore several in situ erosion instruments from different institutes were tested in the laboratory, viz. EROMES (propeller in vertical cylinder; GKSS Geesthacht), PES-1 and PES-2 (vertically oscillating grids in cylinder and square chamber respectively; Delft Hydraulics), ISIS (funnel with radial current; HR Wallingford) and ISEF (vertical standing annular flume, Univ. of Utrecht). The results were compared to the carousel of Delft Hydraulics. The experiments were carried out using the same sediment (kaolinite) and following the same procedure (figure 1) for the forming of the bed, the consolidation period and the steps in increasing erosive forces.

## 2 CAROUSEL Annular flume (DELFT HYDRAULICS/ DELFT/ The Netherlands)

**PRINCIPLE:** The Delft Hydraulics Carousel consists of a circular channel (figure 2) with a mean diameter of 2.1 m and a channel width of 0.2 m. The water depth during the experiments was 0.3 m. A circular lid at the surface drives the water by friction. In this way a Couette- like longitudinal velocity profile is generated, with a high velocity close to the lid (approximately 3 times the mean velocity in the channel). The shape of this longitudinal velocity profile results in an secondary current, comparable with the secondary current in the bend of a river. In order to minimize the effect of this secondary current, the channel itself is rotated in the opposite direction. In this way the secondary current can be minimised and even optimised in such a way that at the bottom of the channel no lateral shear stress is present.

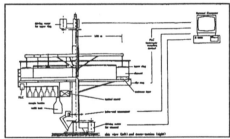

figure 2 Carousel DELFT HY-DRAULICS

**CALIBRATION:** The operational speeds of the lid and the channel are experimentally determined using small spheres with a density slightly larger than that of the water. Furthermore, velocities above the bottom were measured with an electro magnetic current meter. The proportion of rotational speeds is found to vary between 2.8 and 2.9, which is almost the same as the value of 3.1 derived by Mehta and Partheniades in a somewhat smaller flume. For the experiments described a value of 2.8 is used. The optimal ratio of the angular velocities of the upper lid and the channel is also a function of the water

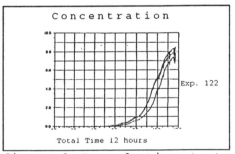

figure 3 Reproduction tests in the Carousel

176

depth. In the case of an erosion experiment, when a sediment layer is present, the water depth in the annular flume is less than 0.30 m and changing during erosion. The operational speeds of the upper lid and the channel are corrected slightly to account for this effect. The bed shear stresses in the carousel are calculated with an expression as derived by Mehta and Partheniades, where $\tau_b$ is the bed shear stress [Pa], $V_1$ is $[|\omega_1|+|\omega_2|]r$ [m/s], $\omega_1$ is the angular velocity of the annular lid [rad/s], $\omega_2$ is the velocity of the channel [rad/s], ris the mean radius of the annular flume [m], a = 0.275 and b = 1.37:

$$\tau_b = aV_1^{\nu}$$

To measure the concentration a small amount of the suspension is pumped out of the flume at mid depth (0.14 m above the channel bottom) and through a small optical cell, in which the light adsorption is measured. The suspension is then returned into the flume. At regular time intervals samples are taken from the flume to calibrate the optical probe for measuring the suspension concentration. The sampling as well as the rotational speeds of the lid and the channel are controlled by a computer.

DEPLOYMENT: For many years erosion, deposition and tidal experiments have been performed in this carousel. Initially, the experiments were done with the artificial mud, Kaolinite. Over the last 5 years only natural mud from different locations in the Netherlands have been investigated. Some of these experiments also contained repeated tests, which produced the same reproducibility as with Kaolinite bed experiments.

REPRODUCTION: The annular flume is located inside a climate controlled room. A temperature of 20 oC and a humidity of 80 % guarantee almost constant test conditions. Each reproduction experiment is carried out on a new sediment layer formed from a Kaolinite suspension of 60 g/l at a water depth of 0.3 m and a consolidation period of 24 hours. The erosion behaviour of each layer is determined by increasing the bed shear stress in successive small steps: between 0.05 and 0.47 Pa. The results of those experiments are presented in figure 3. During experiment 122 an error was made in the start-up procedure, which caused the differences over the first three hours. Depending on the way of defining reproducibility (see Discussion) a reproducibility between 5 - 20 % is found for establishing the erosion parameters for instance in the erosion formulation of Parchure or Partheniades. The reproducibility in establishing the critical stress for erosion is better than 0.02 Pa.

3  PES (Particle Entrainment Simulator) (DELFT HYDRAULICS/ Rijkswaterstaat/ The Netherlands)

PRINCIPLE: In order to quantitatively predict sediment transport in nature a Particle Entrainment Simulator (PES) was built. The same type of device has been developed or used by other investigators as Tsai and lick (1986) with good results. The PES (PES-1) consists of a round prismatic vertical chamber inside of which a horizontal grid oscillates vertically. A relatively thin sheet of the sediment is deposited at the bottom of the chamber while the remaining part of the chamber is filled with water. The chamber of the PES is built in such a way that the erosion parameter of undisturbed natural sample can be measured.
The grid oscillates in the water and creates turbulence and pressure fluctuations in the

water above the bed which penetrate down to the sediment water interface and causes sediment resuspension. The turbulence and hence the amount of sediment resuspended is proportional to the frequency ( at constant amplitude) of the grid oscillations.

CALIBRATION: It can be easily recognized that the turbulence together with the pressure fluctuations is quite different from that generated by currents or waves, which are the prime natural elements for sediment entrainment. However, the idea is that the PES can be calibrated against another device, i.e. an annular flume. The calibration parameter is the erosion rate, giving a relation between amplitude and frequency of the grid and the shear stress in the annular flume. Of course, it is important that the PES itself reproduces its results under identical tests conditions. Because of the bad reproducibility of erosion experiments performed in PES-1, a second PES (PES-2) was built with a square prismatic chamber. PES 2 was built in such a way that the dimensional differences between the experiments is less then 0.1 mm during a whole experiment and over all the components. Figure 4 shows a front view of the PES-1. Because the reproducibility ⹁was not acceptable reproducibility (see figure 5), the experiments with the PES were stopped, so there is no calibration present.

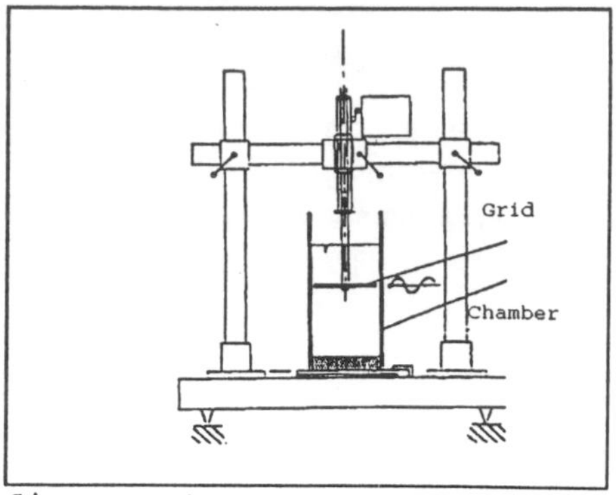

figure 4 PES-1 DELFT HYDRAULICS_

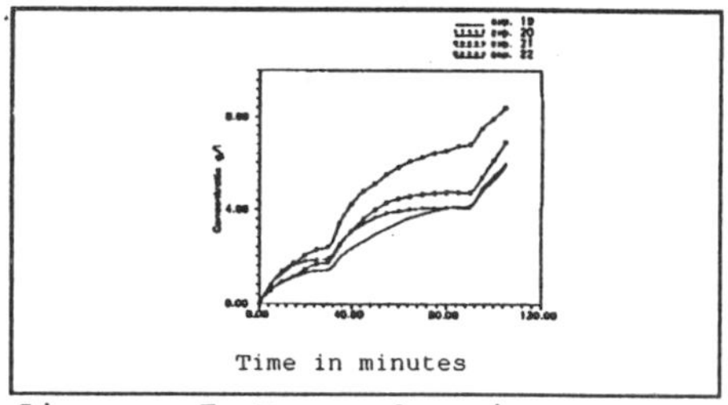

figure 5 Reproduction tests in the PES-2

DEPLOYMENT: In total more than 22 experiments done mostly in groups of four experiments were carried out. Most of the time two or three experiments gave almost the same results, but always there was at least one experiment with an inexplicable result.

REPRODUCTION: The reproducibility of the experiments was between 50 and 100 %. To find the reason for these poor results video recordings were made of the turbulence, erosion and water-bed interactions. Those recordings showed that in the PES vortices were present with the size of the water depth, which where not spatially stable, (sometimes) the vortices relocated, and which resulted in an unreproducible experiment.

### 4 ISIS (Instrument for Shear stress in-Situ) (HR Wallingford/ Wallingford/ England)

PRINCIPLE: The ISIS consists of a circular inverted, curved bell-shaped funnel of 84 mm diameter which fits inside a circular perspex tube with a small annular space of 3 mm around the edges (Figure 6). This is positioned just above the sediment bed, at a typical distance of 2 to 8 mm from the lowest part of the bell head to the mud bed (hereafter called the 'gap'). Water is drawn up through the centre of the bell by smooth pumping, and is replaced by water drawn down the sides of the bell. The bell is shaped so that the water flow across the surface of the bed is laminar and flows radially towards the bell centre, exerting an even shear stress across the whole of the bed. The water removed through the bell is replaced into the column by recirculation and flows through a diffuser, to minimise other circulations in the column.

During operation, the applied shear stress is increased by adjusting the flow and gap in user defined steps and time intervals (according to the sediment being eroded). An optical backscatter probe is used to measure turbidity, and two ultrasonic probes are used to measure the gapsize. Turbidity, flow and gapsize are logged during a test.

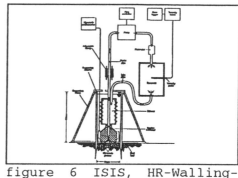

figure 6 ISIS, HR-Wallingford

CALIBRATION: ISIS was calibrated at the University of Plymouth against hot film shear stress probes. The instrument was set up within a calibration rig, whereby two flush-mounted hot wire shear stress probes could be mounted at any one time in specially made calibration bottom plates in the support column. Two plates were used which allowed the applied shear stresses from the radial flow through the bell head to be measured at different radii from the centre of the bell head. The bell head was calibrated for applied shear stress on the bottom plate at four radii for a range of gaps and flows. A total of 235 data points of mean shear stress against gap, flow and radius were measured for the calibration. Plots of shear stress against flow and gap

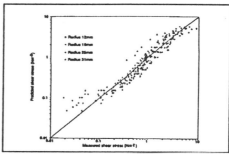

figure 7 Calibration ISIS

were produced to investigate the form of the relationship of shear stress with gapsize, flow and radius. It was found that the relationship was of the form:

$$\log_{10}\tau = a \log_{10}gap + b (flow)^2 + c (flow) + d$$

where $\tau$ is the applied shear stress (Nm$^{-2}$), gap is the bell head to mud surface in (mm), flow is the flow through bell head (l/s) and a,b,c,d are constants. A multiple regression was obtained using a statistical processing computer package. Figure 7 shows the predicted shear stress from the derived equation against the measured shear stresses from the hot wire shear stress probes. There is a minor change in the relationship with radius, but overall the shear stress is mostly dependent on the gap and flow.

DEPLOYMENT: ISIS can be used in laboratory and field conditions. The basic technique is the same for each application, with some minor modifications. The ISIS apparatus is first pumped through with clear, ambient water, or saline solution in the laboratory and bled to remove air. In the field the bell head section is positioned on the test site within a supporting bed frame. The column is filled with water via the diffuser. The bell head is then lowered close to the mud surface (typically a 6 mm gap), and the baseline gap measurements are taken. Flow is slowly initiated through the bell head. Measurements of flow, gap, turbidity and visual erosion characteristics are taken every three minutes from the start. Surface shear stress measurement is taken at the point where the surface mud is just beginning to be eroded by the action of the applied shear stress and the concentration meter starts to register a turbidity increase. The exact point of erosion is established by analysis of the signals after an erosion test. The flow and gap measurements at the point of erosion are used to calculate the surface shear stress, according to the calibration equation.

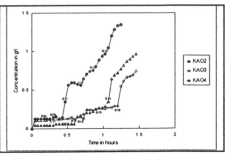

figure 8 Reproduction tests in the ISIS

REPRODUCIBILITY: With the ISIS three experiments were done with a consolidation period of 24 hours, the results are presented in fig 8. The shear stresses given in the plot (by labels) are the shear stresses applied at equilibrium conditions for each step. The labels are located at the start of a sub-test. At the equilibrium point the erosion has stopped.

## 5  EROMES (GKSS/ Geesthacht/ Germany)

PRINCIPLE: The EROMES-System is a patented development of the GKSS Research Centre for erosion measurements on naturally formed, muddy sediments Kühl and Puls (1990), Schünemann and Kühl (1991), Schünemann and Kühl (1993). After sampling, the sediment core remains confined in a 10 cm diameter perspex sampling tube with a 30 cm water column above the sample. The turbulence and hence erosion is induced by a propeller 4 cm above the sediment surface. Special mounting baffles prevent rotation of the water column and vertical balancing currents.

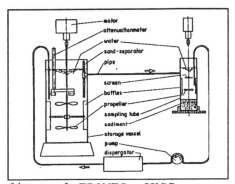

figure 9 EROMES, GKSS

CALIBRATION: The applied bottom shear stress is a function of the propeller speed and is calibrated by quartz sands of specified grain size and known critical shear stress. The increasing distance between propeller and sediment surface during the erosion experiment is negligible if that increase does not exceed about 10 mm.

**DEPLOYMENT:** During an experiment the suspension in the sampling tube is continuously pumped into a second much larger storage container with its own baffles, propellers and turbulence, which ensure a homogeneous distribution of the eroded suspended matter. On its way to the storage container the suspension passes a dispersing device. This is necessary for cohesive and biologically consolidated soils which sometimes erode in pellets. A return-pipe

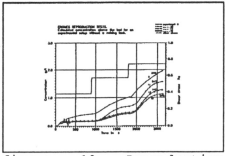

figure 10 Reproduction tests in the EROMES

carries the suspension back to the sampling tube. Thus, there is a continuous exchange of suspension with different concentrations between storage container and sampling tube. The volume of the storage container ensures that the suspended matter concentration will not exceed natural conditions. The suspension-concentration in the container is monitored by measuring the light attenuation of a beam of light of constant intensity. The principal components of the EROMES-System are shown in fig. 9. The erosion rate $\varepsilon(t)$ as a function of time is calculated from the measured concentration $C_c(t)$ in the storage container. The evaluation formula is obtained by estimating the mass of the suspended matter in the loop as follows:

$$\varepsilon(t) = \frac{\dfrac{d^2C_c(t)}{dt^2}\dfrac{V_c V_T}{\dot{V}_r} + \dfrac{dC_c(t)}{dt}[V_c + V_T]}{A_T}$$

where $\varepsilon$ is the erosion rate [g m$^{-2}$ s$^{-1}$], t is the time [s], $C_c$ is the suspended matter concentration in the storage container [g m$^{-3}$], $V_c$ is the storage container volume [m$^3$], $V_t$ is the sample tube volume [m$^3$], $\dot{V}_r$ is the flow between sample tube and container [m$^3$/s] and $A_T$ is the bed surface in [m$^2$].

**REPRODUCIBILITY:** In a coorporation between GKSS and Delft Hydraulics two series of experiments were performed with the EROMES. With the first series the reproducibility of the critical erosion stress for erosion was measured. The instrument gave a good reproducibility for this type of measurement. The aim of the second type of experiments was establishing erosion parameters in the same way as has been done in the Carousel and PES. The results of those measurements are presented in figure 10 (concentration versus time). From the four experiments two are almost equal. The greatest difference between the four experiments is in the order of 100%

## 6 ISEF (In Situ Erosion flume)(Utrecht-University/Urecht/ The Netherlands))

**PRINCIPLE:** In 1992, a new in-situ erosion meter was been developed to measure the bed shear strength of a cohesive bed. The ISEF is a circulation flow system in the vertical plane (figure 11). It consists of a lower horizontal test section, two bend sections and an upper section where the flow is generated by a propeller.

The horizontal test section and the two bend sections have a rectangular cross-section with a height of 0.1 meter and a width of 0.2 meter. The bottom part of the horizontal test section is open over a length of 0.9 meter. When the Erosion Flume is resting on the sediment bed, the surface of the bed will be in line with the steel bottom plates of the flume on both ends of the horizontal section. The total weight of the erosion flume is about 50 kg. The volume of water in the flume is 85 litres. The propeller can

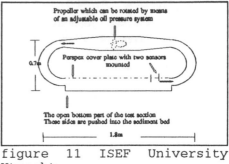

figure 11 ISEF University Utrecht

be rotated at various speeds by means of an adjustable oil pressure system. The flow velocity in the horizontal section is measured by means of a small disc-type electro-magnetic flow meter placed at 0.55 meter from the entrance of the horizontal test section. The suspended mud concentration is measured by an optical turbidity meter

**CALIBRATION:** The erosion process of sediment particles of the bed in the test section is related to the prevailing bed-shear stress. The bed-shear stress can be determined from the measured velocity profile assuming a logarithmic distribution in vertical direction. However, due to the short length of the test section an equilibrium flow will not establish and the logarithmic velocity profile will deviate. To account for this effect a coefficient $\alpha$ is introduced. This coefficient is determined, for non-cohesive material, by measuring velocity profiles above a flat bed of moving sand and gravel particles at conditions just beyond the initiation of motion (Houwing, 1992). The value of $\alpha$ varies from 0.9 to 0.74 as function of increasing Reynolds number. The formula which determines the minimal shear stress needed for erosion reads:

$$\tau_b = \rho \kappa^2 (\alpha u_m)^2 (\ln \frac{\delta}{0.033 k_s + 0.11 \upsilon [\frac{\tau_b}{\rho}]^{-0.5}})^{-2}$$

where $u_z$ is the flow velocity at a height z above the bed [m/s], $u_*$ is $(\tau_b/\rho)^{0.5}$ [m/s], $\tau_b$ is the bed shear stress [Pa], $\rho$ is the fluid density [kg/m$^3$], $\kappa$ is the constant of von Karman (0.4) [-], z is the height above the sediment bed [m], $k_s$ is the effective bed roughness height of Nikuradse [m], $\upsilon$ is the kinematic viscosity [m$^2$/s] and $\alpha$ is acalibration coefficient [-]

**DEPLOYMENT:** The ISEF instrument was first used on a mud flat in the salt marsh area near the Groninger Wadden Sea Coast to get experience with the equipment. The results of those measurements were presented at ICC Kiel 1992.

**REPRODUCIBILITY:** Three reproductivity tests have been carried out according to the tests performed with a laboratory carrousel (De Jong, 1991). An artificial bed was constructed in the test-section for each experiment. The bed consisted of Kaolinite and was formed by deposition from still water. The mean density of the initial bed used in the ISEF was somewhat smaller than the one used in the test performed by De Jong. During the experiments, the current velocity in the ISEF was increased in steps of one hour. Within these hours the sediment concentrations were determined.

Differences found between the three reproduction tests were probably due to differences in local erosion which resulted in
small differences in turbulence and the velocity profile above the bed. However, the
Kaolinite bed was eroded smoothly over the
entire test section in all three reproduction
tests; figure 12. Results not yet used in the
discussion.

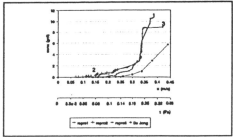

figure 12 Reproduction tests
in the ISEF

## 7 DISCUSSION

**GENERAL DISCUSSION:** Natural beds erode under the action of
waves and currents. In both cases forces act which the bed cannot resist and erosion
takes place. By current erosion these forces are generated by the turbulent flow and
consist of relative small random fluctuating shear stresses and pressure. Sometimes
strong variations occur by bursts. Under (not breaking) waves the turbulence generated
by orbital velocities is generally small. Under waves not only shear stress but also
pressure fluctuations act on the bed. Due to these pressure fluctuations, the effective
stresses within the bed can decrease. As a result, the bed may liquify and/or the
resistance against erosion by currents can decrease. Each natural bed is different, in size
distribution , homogenity, density profiles, mineralogical chemical- biological and
physical characteristics resulting in different erosion characteristics. Relative small
changes in those characteristics can have a big effect on erosion. Over the depth not
only the composition of the bed is changed but also the strength, which means that
when a bed is eroded in layers different strength levels are encountered. In general all
the instruments are able to establish an accurate and reproducible critical shear stress for
erosion, in particular the EROMES was very accurate. This is partly caused by the
possibility of visual observation. So all the instruments generated a reproducible
situation for erosion threshold. For establishing the erosion parameters, the bed must be
eroded slowly over approximately 5-20  mm. None of the instruments were able to do
so. In most of the cases a pit in the bed is eroded, which means that layers of different
strength are encountered. Another problem with these pits is that the resulting uneven
bed topography strongly effects the flow pattern in most instruments. As a result, the
driving erosion force alters and no reproducibility can be obtained. This problem is more
serious for smaller instruments.

**REPRODUCTION ERROR:** The results of the erosion meters show
deviations in concentration when an experiment is repeated. To quantify these variations
a reproduction error is defined as $R = 2*s/m*100$ %, where s is the standard deviation
and m the mean value of the concentration. The parameter R is determined for each
experimental step, see table 1. The reproducibility of the ISIS is estimated roughly to be
100 % (figure 8). For the ISEF R is estimated as 20 % (figure 12)

183

| | $\tau_b$ [Pa] | T [min] | c [kg m⁻³] for exp. no. | | | | s | m | R | A [cm⁻²] |
|---|---|---|---|---|---|---|---|---|---|---|
| | | | 1 | 2 | 3 | 4 | | | | |
| PES 1 disk 1 | – | 30 | 2.32 | 3.56 | 2.37 | | 0.70 | 2.75 | 51.0 | 98.5 |
| disk 2 | – | 30 | 4.52 | 4.03 | 2.97 | 3.37 | 0.69 | 3.72 | 36.8 | |
| PES 2 disk 2 | – | 30 | 1.42 | 1.89 | 1.74 | 2.42 | 0.42 | 1.87 | 44.7 | 645 |
| | – | 60 | 4.21 | 4.11 | 4.79 | 6.84 | 1.27 | 4.99 | 51.0 | |
| | – | 15 | 5.84 | 5.95 | 6.90 | 8.39 | 1.18 | 6.77 | 34.9 | |
| EROMES | 0.38 | 15 | 0.43 | 0.17 | 0.17 | 0.20 | 0.13 | 0.24 | 104.4 | 78.5 |
| | 0.57 | 15 | 1.01 | 0.43 | 0.70 | 0.48 | 0.27 | 0.65 | 81.3 | |
| | 0.74 | 14 | 1.96 | 1.12 | 1.57 | 1.29 | 0.37 | 1.49 | 49.5 | |
| CAROUSEL | 0.12 | 60 | 0.03 | 0.03 | | | 0.00 | 0.03 | 0.0 | 13200 |
| | 0.15 | 60 | 0.16 | 0.16 | | | 0.00 | 0.16 | 0.0 | |
| | 0.19 | 60 | 0.40 | 0.46 | | | 0.04 | 0.43 | 19.7 | |
| | 0.24 | 60 | 1.01 | 1.00 | | | 0.01 | 1.01 | 1.4 | |
| | 0.30 | 60 | 2.89 | 2.92 | | | 0.02 | 2.91 | 1.5 | |
| | 0.37 | 60 | 5.49 | 5.74 | | | 0.18 | 5.62 | 6.3 | |
| | 0.47 | 90 | 7.68 | 8.10 | | | 0.30 | 7.89 | 7.5 | |

Table 1 Results of reproduction experiments. c = concentration; $\tau_b$ = bed shear stress; T = duration of (sub)test; sd = standard deviation; m = mean concentration; R = reproduction error; A = bed surface area.

The third experiment (122) of the carousel is excluded (see figure 3), however including the final result of this test would lead to R = 12.3 %. Although the number of experiments is small (2 - 4), R seems to decrease with increasing surface area. A plausible explanation for this is the fact that spatial variations in the bed properties and the turbulent stresses are averaged out when the erodible area is larger. Statistically the error in the average value of a series of n stochastic independent values with the same error, say $\sigma$, can be approximated by $\sigma/\sqrt{n}$. This means

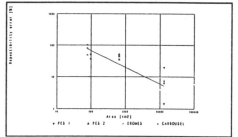

Fig.13    Log-log plot of reproduction error versus bed surface area.

R is related to $A^{-0.5}$, where A is the surface area of the bed. From a fit with the data in table 1 (without tests where R = 0) it follows R = 769*$A^{-0.528}$ (n = 13, $r^2$ = 0.72), see fig.13 . The exponent of A is very close to the theoretical value. According to this relation the R-values for the ISIS (A = 66.5) and ISEF (A = 1800) are 84 % and 15 % respectively.

MODEL SIMULATIONS: The effect of small changes in the bed shear stress, the bed density or the bed strength can be demonstrated with an erosion model. This model was applied by Winterwerp et al (1993) for the analysis of the carousel results on the behaviour of natural mud from several locations in the Netherlands. The erosion rate is of the form:

$$E = E_f e^{\alpha (\tau_b - \tau_{ce}(z))} \quad \text{for } \tau_b > \tau_{ce}$$
$$E = E_f \quad \text{for } 0 < \tau_b \leq \tau_{ce}$$
$$E = 0 \quad \text{for } \tau_b = 0$$

184

where E the erosion rate [kg m$^{-2}$ s$^{-1}$], E$_f$ the floc erosion rate [kg m$^{-2}$ s$^{-1}$], $\alpha$ a constant, $\tau_b$ the bed shear stress [Pa] and $\tau_{ce}$ the bed strength [Pa] as a function of the depth z in the bed layer. The suspended concentration is derived from:

$$\frac{d\overline{c}}{dt} = \frac{E}{h}$$

where $\overline{c}$ is the depth-averaged concentration [kg m$^{-3}$], t is time [s] and h is the water depth [m]. While Winterwerp et al (1993) found no evident relation between bed strength and the bed density, they suggested:

$$\tau_{ce}(z) = \beta \frac{z}{H} + \gamma$$

where H is the thickness of the erodible bed (surface at z = H) and ß and $\gamma$ are constants. The erosion depth is calculated from the amount of eroded mass and the integrated dry density of the bed. The dry density profile of the bed is given by a top layer (above z/H = $\zeta$) with a linear profile and a lower layer with a power law profile:

$$c_b(z) = \overline{c_b}\left(a - b\left(\frac{z}{H}\right)\right) \quad \text{for } \zeta \leq \frac{z}{H} \leq 1$$
$$c_b(z) = \overline{c_b}k\left(\frac{z}{H}\right)^{-n} \quad \text{for } 0 \leq \frac{z}{H} < \zeta$$

where $c_b$ is the dry density [kg m$^{-3}$], and a, b, k and n are empirical constants. For given values for the depth-averaged $c_b$, $\zeta$, a, b and n the value of k is calculated from the equality of $c_b$ at $\zeta$ for both profiles.

With the model a simulation is made for a reference situation and three runs in which the main parameters are varied with 10 %, which is regarded as realistic. Figure 14 shows the effect of an increase of 10 % in the bed shear stress, in the shear strength (ß and $\gamma$) and in the density of the top layer only (a and b). Each run consists of 3 subsequent steps of 30 minutes with $\tau_b$ = 0.2, 0.4 and 0.6 Pa respectively. In the reference

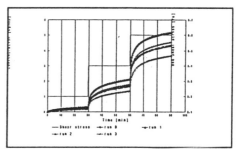

Fig.14 Model result for ref. run 0, 10 % higher shear stress (run 1), 10 % higher shear strength (run 2) and 10 % higher density in top layer (run 3).

case the model parameters have the typical values: E$_f$ = 10$^{-7}$, $\alpha$ = 30, h = 0.25, H = 0.05, ß = -3, $\gamma$ = 3.08, a = 6, b = 6, $\overline{c_b}$ = 350, $\zeta$ = 0.9, n = 0.4 and k = 0.575. In run 0 the final erosion depth is 7.6 mm, so mainly the top layer of 5 mm is eroded. A 10 % larger bed shear stress causes a 20 - 30 % higher concentration at the end of a step. However this effect can be higher at much lower shear stresses. The effect of a 10 % stronger bed results in a 15 - 20 % lower concentration, while a 10 % higher density in the top layer results in 5 - 8 % higher concentrations. The model simulation and the experimental results show that variations in the concentration can be expected depending on the capacity of the apparatus to reproduce the eroding force and the bed properties. The reliability will be higher when the bed area is larger. The maximum error can be as large as 100 %.

**Acknowledgement**

Our thanks are extended to all involved institutes and contractors, for providing the financial support to enables us to carry out this investigation.

REFERENCES
CARROUSEL
Cornelisse, J.M., Kuijper, C., Winterwerp, 1993. Evaluation of in situ erosion devices. (in Dutch), Report 45, Rijkswaterstaat, Delft Hydraulics, Cohesive Sediments Research, Z161-36, march 1993.
Jong, P. de, 1991 Reproduction-research carrousel. (in Dutch). Internal report Delft Hydraulics, note z161-94.
Kuijper, C., Cornelisse, J.M., Winterwerp, J.C., 1989. Research on erosive properties of cohesive sediments, Journal of Geophysical Research, Vol. 94, N0 C10, Pages 14341-14350, oktober 15 1989.
PES
Tsai, Cheng-Han, Lick, Wilbert, 1986 A Portable device for measuring sediment resuspension, J.Great Lakes Res. 12(4): 314 -321, Internatonal Association Great Lakes Res. 1986
ISIS
Graham D.I, James P.W., Jones T.E.R., Davies J.M., Delo E.A, 1992. "Measurement and prediction of surface shear stress in annular flume". ASCE Journal of Hydraulic Engineering. Vol 118, No 9, September, 1992.
ETSU, (1992) "In-Situ Erosion of Cohesive Sediment". Energy Technology Support Unit, Harwell, UK, Report No. ETSU-TID-4112.
EROMES
Kühl, H.; Puls, W.,1990 Offenlegungsschrift DE 3826044 A1. Deutsches Patentamt,1990
Schünemann, M.; Kühl, H. 1991 A device for erosion-measurements on naturally formed,muddy sediments: The EROMES-System. GKSS Report GKSS 91/E/18, Geesthacht, 1991
Schünemann, M.; Kühl, H.,1993 Experimental Investigations of the Erosional Behaviour of Naturally Formed Mud from the Elbe-Estuary and the Adjecent Wadden Sea, Germany.Coastal and Estuarine Studies 42 (ed. by A. J. Mehta), American Geophysical Union, Washington DC, 1993, p. 314-330
Raudkivi, A.J., 1976 Loose Boundary Hydraulics. Pergamon Press, Oxford, 1976
ISEF
Houwing, E.J. and L.C. van Rijn, 1992 In situ erosion flume (EROSF): determination of bedshear stress and erosion of a kaolinite bed. First concept, 7/7/1992.
DISCUSSION
Winterwerp, J.C, Cornelisse, J.M. and Kuijper, C. (1993). Erosion of natural sediments from the Netherlands. Model simulations and sensitivity analysis. Rijkswaterstaat, Delft Hydraulics, Report 43 of Cohesive Sediments Research, February 1993.

# 12 Developments in the combined use of acoustic doppler current profilers and profiling siltmeters for suspended solids monitoring

**J. M. LAND**[1], **R. KIRBY**[2] **and J. B. MASSEY**[3]
[1]*Dredging Research Ltd, UK*
[2]*Dredging Research Ltd and Ravensrodd Consultants Ltd, UK*
[3]*Geotechnical Engineering Office, Civil Engineering Office, Hong Kong*

## INTRODUCTION

The Territory of Hong Kong, at the mouth of China's third largest river, the Pearl, has a land area of only a little over 1,000 square kilometres and a population of almost six million people. Its steep, rugged terrain makes much of the land area unsuitable for development and for over a hundred years land has been formed by reclamation from the sea. Hong Kong's proximity to southern China and its sheltered deep water harbour make it the busiest container port in the world in terms of annual container throughput. Cargo forecasts indicate a need to double container handling capacity by the end of the decade and double it again by 2011. Hong Kong also needs a new international airport and associated infrastructure. Implementation of the Port and Airport Development Strategy has generated a demand for fill material this decade of 650 Mm$^3$.

Traditionally, the fill material required for such development has come from cutting into hillsides, but with the recent greatly increased pace of development, reliance has shifted to dredging of sand from the seabed. Not only is this much cheaper and quicker when modern high production dredgers are used, but it is also preferable environmentally, while not free of environmental concerns.

All major construction projects in Hong Kong are subjected to a rigorous process of environmental impact assessment. The scale of the dredging has given rise to much public concern, principally from environmental groups, capture fishermen and mariculturalists. The chief concerns are the removal of areas of the seabed and the potential effects of sediment put into suspension by the sand dredging and mud disposal operations. Direct measurements of suspended sediment throughout the water column and areally are required in the vicinity of the marine borrow areas and mud disposal grounds to measure the concentration of sediment plumes and to track their movements.

In order to provide the data required to assess the effects of the dredging and disposal operations, two sediment measurement techniques have been extensively used, both of which have required a significant degree of development and refinement:

a)  acoustic doppler current profilers (ADCPs) have been used to estimate concentrations in the main part of the water column;

b)  profiling, rapid-response pulsed infra-red siltmeters have been used, particularly to study near-bed sediment suspensions at concentrations of up to 50,000 mg/l.

*Cohesive Sediments.*
Edited by Neville Burt, Reg Parker and Jacqueline Watts
©1997 John Wiley & Sons Ltd.

This paper is an updated version of that which was presented at the conference and includes modifications based on experience gained during broadly similar work undertaken in Germany and the UK in 1995-6.

## ACOUSTIC DOPPLER CURRENT PROFILERS

ADCPs were developed to measure water currents throughout the water column in a non-intrusive manner. They measure the doppler shift of acoustic pulses emitted from four highly directional transducers, arranged in a Janus configuration, and backscattered from suspended particulate matter in the water column. The data are used to calculate the particle velocity and, by inference, the water velocity, in the direction of each acoustic beam. By combining measurements from three beams with the known orientation of the ADCP and its speed over the seabed, the current speed and direction relative to earth co-ordinates can be derived. The measurements are made at depth intervals which may be as small as 0.10 metre, depending on the specification of the ADCP used, by 'range-gating' the emitted pulse.

The intensity of the backscattered energy has been used for several years as a means of measuring the concentration of suspended particulates, notably plankton (Flagg and Smith, 1989). ADCPs are becoming increasingly common as a means of measuring suspended sediment and they have been used for the measurement of material put into suspension during dredging and dredged material disposal operations, particularly in the USA (eg. Thevenot and Kraus, 1993; Ogushwitz, 1994). They have also been used to study plumes from wastewater outfalls (Dammann *et al.*, 1991).

The theory on which the acoustic measurement of suspended solids rests has been described in a number of recent works (eg. Sheng and Hay, 1988; Thorne *et al.*, 1991). The work described here has not focused on acoustic theory, although some minor refinements have been made. The main thrust of the investigations has been directed towards three practical aspects:

a)   establishing reliable methods of compensating for the environmental parameters which contribute to acoustic attenuation during transmission;

b)   controlling or compensating for the instrument-dependent parameters that affect the measurement of acoustic backscatter intensity;

c)   development of a powerful software package that can process the very considerable amount of data which is gathered during surveys.

All of the ADCPs used for this work were manufactured by RD Instruments and deployed from moving vessels. They include several 1200 kHz narrowband instruments and 300, 600 and 1200 kHz broadband instruments. Data acquisition was controlled by means of Transect© software supplied by the instrument manufacturer and installed on portable computers.

### Acoustic Attenuation

During transmission through the water column, the intensity of the acoustic pulse diminishes due to beam spreading, absorption by the water and absorption and scattering by the suspended solids in the water column. The spherical expansion of the beam is a simple geometric function and can easily be accounted for during data processing.

It is necessary to pay great attention to variations of temperature and salinity in the water column if accurate compensations are to be made for acoustic absorption by the water. Salinity gradients in Hong Kong's wet season can be as much as 5 ppt/m and temperature gradients can be as high as 1°/m. These have a marked effect on the acoustic absorption and relatively small deviations of temperature and salinity from the assumed values give rise to large errors in concentration estimates. Field procedures include the detailed measurement of the temperature and salinity profiles; these are also used to compute the variation of speed of sound through the water column in order to derive very accurate ranges to the measurement intervals.

Acoustic energy is lost due to absorption and scattering by the sediment. The principle controlling factor is the size of the sediment but its density and compressibility are also important. The potential consequences of this are that errors may occur:

a) in areas where mixed sediment types occur in varying proportions

b) if the degree of flocculation of the sediment changes due to variations of concentration, water chemistry and turbulence;

The work described here has been concerned mainly with fine sediments with particle diameters in the range 5 - 75 μm and with a mean diameter of typically 10 μm. During each deployment, samples of suspended sediment are taken for particle size analysis. The calculation procedures can accommodate variation of particle size with depth. Both time-variable and current speed-variable calibrations have been used to overcome problems arising from the flocculation of fine sediments during periods of slack water.

## Instrument Parameters

Each ADCP has unique performance characteristics which must be taken into account if accurate estimates are to be obtained. The most important are:

a) the power output of the transducers;

b) the noise of the signal amplification circuits;

c) the performance and temperature sensitivity of the signal amplification circuits.

The power output of each transducer is unique and varies with the supply voltage. Observed relative backscatter intensity can be corrected to account for variations of supply voltage but this has not been done during the work described here. The procedure which has been adopted is to carefully control the supply voltage and to ensure that it is exactly the same each time the instruments are used.

The noise of the signal amplifiers is an unwanted component of the measurement of relative backscatter intensity and varies from one instrument to another. Field measurements of the amplifier noise are made and data processing incorporates this value in the computation of quasi-absolute backscatter intensity and the identification of data which are likely to have been corrupted by noise.

The amplification circuits which boost the received signal have unique performance characteristics. The relationship between instrument counts and dB can be measured in the laboratory and, for each data ensemble, is varied according to the temperature recorded at the ADCP transducers. The errors which are introduced if this factor is ignored can be considerable.

There are other characteristics which affect the measurements but, for the most part, these are reasonably constant for a given instrument. Errors arising from these factors have been

189

largely eliminated by carefully calibrating each instrument. In addition, verification data are obtained during each deployment in order that instrument stability can be checked.

## Data Processing

The successful use of the ADCP is reliant on good data-processing software. The amount of data which can be collected by an ADCP is very considerable and data files exceeding 250 MB have been obtained. The main functions of the data processor are as follows:
1) data import and editing of unwanted data;
2) combination with navigation data obtained by electronic positioning systems;
3) calculation of the range to each measurement interval and of the length of each interval using measured speed of sound;
4) computation of dB values using the measured characteristics of each transducer receiver board;
5) correction for acoustic absorption in water using temperature and salinity data;
6) calculation of sediment attenuation coefficients;
7) estimation of the particle concentration using instrument-specific calibration curves.

The calculation procedure is iterative and is carried down through the water column from the uppermost measurement interval. This is necessary because the attenuation due to the sediment is a function of the sediment concentration which is the unknown. The concentration in the first bin is estimated using the calibration data and a value for sediment attenuation is derived. This is used to establish a new concentration and the calculation is repeated until the estimated concentration changes by less than 0.001 mg/l per iteration. This procedure is carried out in a step-wise manner down through the water column. It is recognised that the procedure can lead to significant accumulated errors but none have yet been detected in water depths of up to 60 metres and at concentrations in excess of 2,000 mg/l.

## Limitations of Measurement

Measurements using ADCPs are subject to several limitations which, in some cases, may be severe. One of the most important is that air bubbles and any particulate matter, in addition to sediment, will reflect acoustic energy and it is rare that the water column is entirely free of air bubbles or particulates other than sediment. Air bubbles have been a frequent problem during our studies and are caused either by other vessels or by wave action. The affected records are usually easily recognisable but there is nothing which can be done to correct the data. Even when the air bubbles are only present in the upper part of the water column, the data from the lower part cannot accurately be processed because the acoustic pulse must pass twice through the affected part of the water column and because the calculations are top-down.

The instrument itself is also a source of error. The statistical uncertainty of the measured backscatter varies according to the type of instrument being used and from one instrument to another. In principle, the uncertainty is least when using broadband instruments but all instruments which are used must be investigated to determine the actual uncertainty.

*Comparison between Concentrations Estimated using Backscatter Data and Other Methods*

Many thousands of data sets have been obtained which can be used to compare the concentrations estimated using acoustic backscatter data and those measured using conventional techniques.

Great care is required when obtaining calibration data in order to ensure close temporal and spatial matching. Both pumped and bottle water samplers have been used in combination with turbidity meters (white light transmissometers and infra-red backscatter devices). Despite painstaking precautions, it is almost impossible to obtain data from all three measurement techniques which do not exhibit considerable scatter.

The areas where most of the data presented here were obtained were close to areas of intensive dredging activity and were also subject to rapid natural variations of concentration. Independent observations made with both the ADCPs and the siltmeters showed that concentrations could vary by 50% or more over a period of a few seconds and over distances of a few metres. Figures 1-4 show typical sets of comparative data.

The degree of correlation between the ADCP estimates and the water sample/turbidity meter data has been observed to be very similar to, or better than, that between the siltmeter observations and the water samples over four orders of concentration magnitude (< 1 mg/l to > 2,000 mg/l).

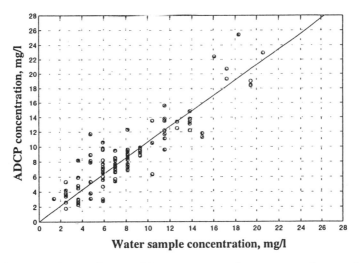

**Figure 1.** Concentration estimates Vs. water sample concentrations - 1200 kHz narrowband ADCP - Hong Kong

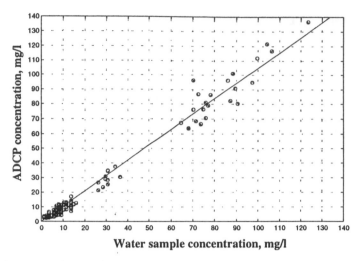

**Figure 2.** Concentration estimates Vs. water sample concentrations - 1200 kHz narrowband ADCP - Hong Kong

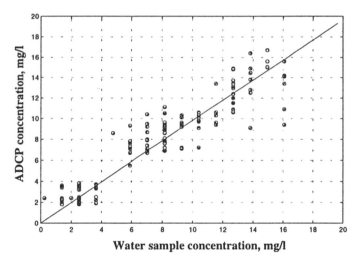

**Figure 3.** Concentration estimates Vs. water sample concentrations - 1200 kHz broadband ADCP -Hong Kong

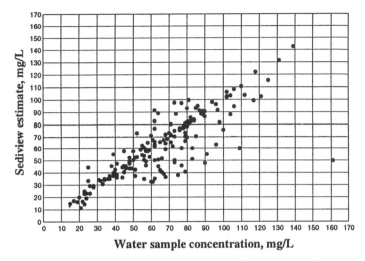

**Figure 4**. Concentration estimates Vs. water sample concentrations - 1200 kHz broadband ADCP - Germany

## PROFILING SILTMETERS

*Calibration, Deployment & Data Recording*

The instrument array is calibrated using silt indigenous to the survey locality. A 70 point, including 10 double point, calibration has been carried out in order to validate the internal electronic linearisation. Pre-survey set up and a post-survey check is performed with a small number of well homogenised suspensions of the local sediment. The array is mounted on a frame and deployed such that it can be towed at 2 m/s and vertically profiled at 1 m/s. During vertical profiling the data capture rate is such that only raw data can be collected and stored on the down-going leg with conversion to real parameters made during the up-going recovery phase. Graphical display of these is available on a monitor as soon as the instrument reaches the surface. In about 20m of water a maximum of two vertical down-going profiles can be accomplished in one minute.

## SURVEY OF OPEN-WATER DREDGING AND DISPOSAL SITES

When used in combination, the ADCP permits whole cross-sections and complex, highly dynamic events to be detected and surveyed rapidly on a relatively gross scale. Whereas the profiling siltmeters permit small-scale diagnostic internal anatomical characteristics of suspensions, such as lutoclines, and the vital near-bed dense layers to be resolved. In June 1993 surveys were undertaken around mechanical and trailer suction dredgers to establish the disturbance to the local environment at dredging and disposal sites. The subsequent migration of fines out of the areas was also investigated in order to determine possible engineering and environmental impacts and transport processes.

At the dredging site horizontal traverses in the wake of a trailer suction dredger overflowing fines in calm weather from weirs on the port and starboard sides revealed twin descending plumes. In the example shown one plume is more turbid due to the angular relationship between the tracks of the dredger and survey vessel. Figure 5.

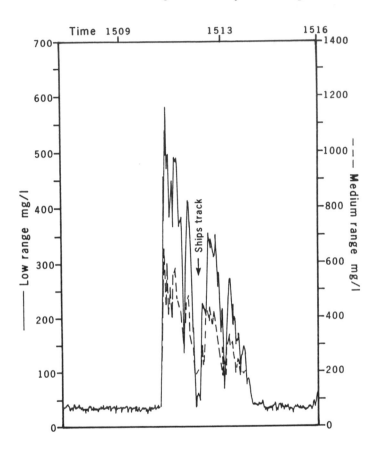

**Figure 5.** Horizontal siltmeter traverse at 3.0m towing depth across the wake of a trailer suction dredger showing the twin overflow weir/ALMOB plumes with low turbidity ships track between. Plume turbidity increased with depth and proximity to the dredger.

On reaching the bed a dense ephemeral layer formed, which extensive profiling revealed to be localised to the dredging site. At the disposal site discharge of the barge loads from grab dredgers led to uneven deposits veneered in places by thin (0.3 - 1.0 m) dense fluid mud layers. On the timescale of the surveys there was no evidence that these dense layers were reworked. The loads carried by trailers were very likely differentiated into a dense and less dense phase in the hopper in the dredging process. On discharge the less dense phase expanded rapidly and could be tracked by the ADCP and siltmeter for long periods as it was advected out of the area by tidal currents. Such low concentration phases appeared to be diluted rapidly by mixing and reduced by settling onto the bed. A second high concentration phase often 7 - 8 m deep on discharge was detected by the profiling siltmeter. These dense near-bed layers were tracked for 10 - 12 km. away from the discharge site. They thinned rapidly due to spreading and migrated along contrasted pathways to the shallower low concentration phase. Their transport was determined by a combination of tidal current and seabed slope. At times, especially after their initial differentiation, the dense near-bed suspensions develop a lutocline sharp enough for the ADCP to regard them as the seabed.

These dense layers contain a large fraction of the discharged load indicating further the necessity for ADCP and profiling siltmeter to be used as a complementary pair. Figure 6. Neither the shallower dilute phase nor the dense near-bed phase of the discharged material were tracked into environmentally sensitive or important engineering sites identified prior to monitoring operations, where they might have given rise to serious problems

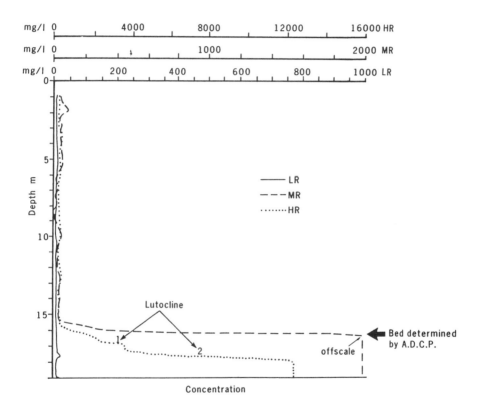

**Figure 6.**  Vertical siltmeter profile showing low and high turpidity phase suspensions arising from trailer disposal. The dense near-bed layer is 2.65m deep, shows multiple layering and reaches a concentration of 12,500 mg/l in the near-bed zone. The layer was not detected by the ADCP.

Several surveys of a similar nature have been carried out including two in which over 250 million ADCP sediment concentration observations were collected in an area of 1,000 km$^2$ over periods of four days in order to provide an overview of the natural sediment regime in Hong Kong and to identify perturbations of the regime caused by dredging and dredged material disposal.

## CONCLUSIONS

The combination of the ADCP and profiling infra-red siltmeters permits the whole of man-induced events to be visualised and monitored, with the consequence that entire sediment losses from the several mechanisms involved can be established. As important, losses can be verifiably and unequivocally established.

The ADCP has successfully been used to measure sediment concentrations along almost 10,000 km of survey line in Hong Kong. In addition, it has been used to study transport of sediment from dredged material disposal sites in Germany and sediment transport in the River Mersey, UK.

The work described in this paper is enabling a far greater understanding of natural sediment processes and processes related to dredging and marine disposal in Hong Kong than have hitherto been possible. Surveys completed so far indicate that suspended sediment plumes from marine borrow areas can have significant effects in the immediate vicinity but diminish rapidly with time and distance, and probably have less effect on marine biology than natural disturbances such as tropical cyclones (typhoons), energy and sediment inputs from the nearby Pearl River, strong tidal currents, temperature variations and unnaturally high predation of some species due to heavy commercial fishing. At marine disposal sites, during and after dumping operations, very little sediment is released into suspension, although dense near-bed suspensions do appear to form. However, the ongoing programme of surveys has so far indicated that the environmental impacts of dredging and marine disposal in Hong Kong are within acceptable limits.

### ACKNOWLEDGEMENT

This paper is published with the permission of the Director of Civil Engineering, Hong Kong Government.

### REFERENCES

Dammann, W.P., Proni, J.R., Craynock, J.F. and Fergen, R. (1991). Oceanic wastewater outfall plume characteristics measured acoustically. Chemistry and Ecology, 5, 75-84.

Flagg, C.N. and Smith, S. (1989). Zooplankton abundance measurements from acoustic doppler current profiler. Proceedings of Oceans '89, IEEE and MTS, Seattle, WA, pp1318-1323.

Ogushwitz, P.R. (1994). Measurements of acoustical scattering form plumes of sediment suspended in open waters. Journal of Marine Environmental Engineering., 1, pp119-130.

Shen, J. and Hay, A.E. (1988). An examination of the spherical backscatter approximation in aqueous suspensions of sand. Journal of the Acoustical Society of America, 83, pp598-610.

Thevenot, M.M. and Kraus, N.C., (1993). Comparison of acoustical and optical measurements of suspended material in the Chesapeake Estuary. Journal of Marine Environmental Engineering., 1, pp65-79.

Thorne, P.D., Vincent, C.E., Hardcastle, P.J., and Pearson, N. (1991). Measuring suspended sediment concentration using acoustic backscatter devices. Marine Geology, 98, pp7-16.

# 13 A laboratory study of cohesive sediment transport

M. CRAPPER[1] and K. H. M. ALI[2]

[1]*Department of Civil and Environmental Engineering, University of Edinburgh, UK*

[2]*Department of Civil Engineering, University of Liverpool, UK*

## INTRODUCTION

This paper describes the current status of research into cohesive sediment transport at Liverpool, including insights into cohesive sediment behaviour obtained from laboratory experiments and the first results of a comparison between the Wallingford FLUIDMUDFLOW-2D fluid mud transport model and the authors' experimental data.

It is well known that suspended sediment affects the internal dissipation of energy of a flow. Also, the presence of large concentrations of suspended sediment has a damping influence on vertical mixing (O'Connor 1991).

Vanoni (1946) and Hino (1963) considered the dynamic interaction between the turbulent fluid and the suspended sediment. Their analyses were based on simple turbulence concepts involving mixing length calculations. Hino's theoretical analysis showed that for neutrally buoyant particles, the von Kármán constant $\kappa$ decreased and turbulent intensity increased with the increase in volume concentration. For cohesionless sediment particles, the theory again predicted a decrease in $\kappa$, a small decrease in turbulence intensity and a rather rapid decrease in the life of the eddies.

The present authors have obtained a series of detailed near bed velocity distributions and measurements of turbulent shear velocity in dense suspensions of cohesive sediment in a laboratory flume. These have been used to investigate the conclusions of Hino (1963) and their applicability to cohesive sediment regimes.

Numerical Computer Models are the most convenient and economical tools with which engineers can predict the behaviour of the marine environment and the impact of new engineering works. Many such models, for example that by Nicholson and O'Connor (1986), have already been developed for cohesive sediment. These models have not generally taken into account the effects of fluid mud.

Fluid mud has been described as a suspension of flocculated mud dense enough to change significantly the physical properties of the mud/water mixture compared with those of clear water of the same temperature and salinity. Fluid mud can form under a variety of conditions and can then flow under the influence of gravity and hydrostatic forces and settle out and dewater in navigation channels and berths. This causes rates of siltation significantly greater than those resulting from settling directly from suspension (Odd and Rodger 1986).

*Cohesive Sediments.*
Edited by Neville Burt, Reg Parker and Jacqueline Watts
© 1997 John Wiley & Sons Ltd.

Attempts to model the behaviour of fluid mud have been developed at Hydraulics Research Wallingford Ltd. by Odd and Rodger (1986) and by Odd and Cooper (1989), the latter authors having described the application of their FLUIDMUDFLOW-2D model to a prototype situation in the Severn Estuary. However, these model investigations have been limited by the dearth of reliable data against which to evaluate the performance of the model.

The present authors are attempting to rectify this situation by obtaining in-depth laboratory data on the behaviour of fluid mud. The results of the Wallingford FLUIDMUDFLOW-2D model will then be compared with these detailed data.

## THEORETICAL CONSIDERATIONS - THE WALLINGFORD FLUIDMUDFLOW-2D MODEL

The Wallingford FLUIDMUDFLOW-2D Model uses two-dimensional, vertically averaged transport models that solve the diffusion-advection equation with empirical erosion and deposition functions as source and sink terms and a 'fluid mud layer' interposed between the settled bed and the mud/water suspension proper. This fluid mud layer can change its thickness due to exchange of mud with the bed and the suspension, and can flow due to gravitational and hydrostatic effects.

The motion of the fluid mud is determined by solving the momentum and continuity equations for the layer. The momentum equation for the x-direction is given in equation 1 below and the continuity equation in equation 2.

$$\frac{\partial u_m}{\partial t} + \frac{1}{d_m \rho_m}(\tau_0 - \tau_i)_x - \Omega v_w + \frac{\rho_w}{\rho_m} g \frac{\partial \eta_w}{\partial x}$$

$$+ \frac{g}{\rho_m}(\rho_m - \rho_w)\frac{\partial \eta_m}{\partial x} + \frac{g d_m}{2}\frac{\partial}{\partial x}(\rho_m - \rho_w) = 0 \tag{1}$$

$$\frac{\partial}{\partial t}(d_m c_m) + \frac{\partial}{\partial x}(u_m d_m c_m) + \frac{\partial}{\partial y}(v_m d_m c_m) = \frac{dm}{dt} \tag{2}$$

Here u and v refer to velocities in the x and y-directions, $\rho$ to densities and $\eta$ to surface elevations. The subscript $_m$ refers to the fluid mud layer and the subscript $_w$ to the overlying water. $d_m$ represents the thickness of the fluid mud layer, $c_m$ its concentration, and $\Omega$ is the Coriolis parameter. $\tau_0$ and $\tau_i$ are the shear stresses on the bed and the fluid mud/water interface respectively. In the model, $c_m$ is assumed to be a constant and is related directly to $\rho_m$. Values of $\tau_0$ and $\tau_i$ are derived from a friction factor related empirically to the depth-mean velocities of fluid mud layer and the overlying water.

The term dm/dt in equation 2 represents the sources and sinks from the fluid mud layer, given by settling into the fluid mud, dewatering of the fluid mud, erosion of the underlying bed by the fluid mud, erosion of the fluid mud by the overlying water, and entrainment of overlying water into the fluid mud layer itself.

In the FLUIDMUDFLOW-2D programs, the settling velocity of the suspended mud is derived from the equation

$$\omega = \omega_s C$$

$$\omega \geq 0.0001 ms^{-1}$$

(3)

(Odd and Cooper 1989). Here $\omega$ is the settling velocity, $C$ is the concentration of mud in suspension close to the bed/fluid mud and $\omega_s$ is a constant. Equation 3 takes no account of hindered settling effects which cause the fall velocity to reduce with increasing concentrations for values of concentration above about $15kgm^{-3}$. This important phenomenon appears to have been excluded from the FLUIDMUDFLOW-2D programs because suspended sediment concentrations in excess of $15kgm^{-3}$ were considered sufficiently rare in the field for their consideration to be unnecessary.

For the authors' application of the FLUIDMUDFLOW-2D model, Coriolis effects were assumed to be negligible and the parameter $\Omega$ was set to zero.

## EXPERIMENTAL ARRANGEMENTS

The experiments so far used for comparison with the FLUIDMUDFLOW-2D model were carried out using mud from Eastham Ferry in the Mersey Estuary in a perspex tank as described in Ali and Georgiadis (1991). This apparatus allowed the measurement of fluid mud thickness, elevation, velocity and concentration for various bed slopes and initial average concentrations. All these results were obtained in still water.

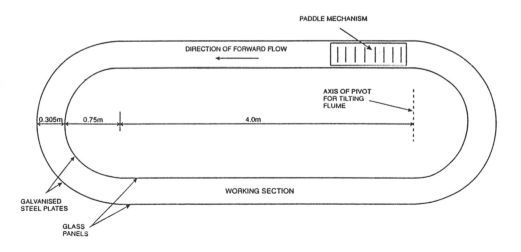

Figure 1  Plan of the Liverpool Race Track Flume

**Figure 2** The Liverpool Race Track Flume

**Figure 3** The Toshiba SDL-50A Ultrasonic Imaging Machine with SDL-01A Pulsed-Doppler Unit

The authors have also conducted a series of fluid mud experiments in the Race Track Flume (figures 1 and 2). These tests were carried out using mud dredged from Seaforth Dock in the Port of Liverpool.

The Race Track Flume consists of two semi-circular bends of internal radius 0.75m joined by two straight sections of length 4m. The channel width is 0.305m and the maximum working depth of water is about 0.6m. Flow in the flume is driven by a paddle mechanism consisting of a toothed belt positioned level with the water surface. The paddle mechanism is situated at the upstream end of the straight section opposite the flume's working section. The Race Track Flume can be tilted to provide a uniform bed slope over the working section, and the authors' experiments were conducted using a bed slope of 1:15.612.

A series of experiments was carried out in the Race Track Flume in order to determine the effect of varying concentrations of suspended mud on the near bed velocity distribution and to ascertain the thickness, elevation, velocity and concentration of fluid mud under various conditions of water flow and initial suspended sediment concentration. The flume was prepared by filling it with saline water of density approximately $1021 kgm^{-3}$. The correct amount of mud was then introduced so as to give the desired average concentration when all the mud was in suspension. The paddle was then set to full speed and the flume was run for half an hour to ensure that all the mud was in suspension and to allow it time to flocculate. Manual stirring with brooms was also used to ensure that no mud remained on the flume bed. During the stirring and flocculation process, measurements of the near bed velocity and concentration profiles were taken. Finally the paddle speed was reduced to the desired value for the fluid mud test, and measurement of the fluid mud flow commenced as soon as the mud flocs began to settle.

## USE OF MEDICAL DIAGNOSTIC ULTRASOUND TECHNOLOGY

The study of cohesive sediment in the race track flume gives rise to problems in the location of the mud bed and the fluid mud layer and in the measuring of flow velocities. Mud in suspension is completely opaque, making Laser Doppler Anemometry impossible and rendering visual examination impractical, except adjacent to the glass sides of the flume's working section.

These problems have been partly resolved by the use of medical diagnostic ultrasonic imaging together with pulsed-Doppler velocity measurement, an idea originally suggested by Dr. W.R. Parker of Blackdown Consultants Ltd. A Toshiba SAL-50A ultrasonic imaging machine with an SDL-01A pulsed-Doppler unit was obtained on loan from the Clinical Engineering Department of the University of Liverpool (figure 3).

The application of this instrument in the Race Track Flume is fully described in Ali and Crapper (1993). With it, it was possible to obtain precise measurements of the location of sharp density interfaces such as a settled mud bed or hindered settling lutocline, and to measure flow velocities with a spatial resolution of 2mm.

The pulsed-Doppler unit allows velocities to be measured in only one dimension, in the direction of the Doppler beam. In the case of the Race Track Flume tilted at a slope of 1:15.612, this direction was at an angle of 58.665° to the horizontal.

The authors had previously conducted a detailed investigation of flows in the Race Track Flume. This included measurements taken using a two-dimensional Laser-Doppler Anemometer (LDA) system in clear water and a Computational Fluid Dynamics (CFD) simulation. This investigation showed that mean vertical and transverse flow velocities in the flume's working section were insignificant, being never greater than 10% and typically less than 5% of the longitudinal flow velocity (i.e. the flow along the axis of the flume's working section in a horizontal direction). It was therefore concluded that the ultrasonic-Doppler readings taken in the direction of the Doppler beam could be said to correspond to $u\cos\theta$, where u is the longitudinal flow velocity and $\theta$ is the Doppler angle of 58.665°.

The pulsed-Doppler unit is limited in that it has a wall-motion filter which filters out all Doppler frequencies below 100Hz, meaning that the lowest longitudinal velocity that can be measured is approximately $0.055 \text{ms}^{-1}$. This meant that the instrument was useless for measuring fluid mud flow, for which the maximum velocity encountered was only about $3 \text{mms}^{-1}$. However, during the stirring and flocculation period of each experimental run, when the flume paddle was set to full speed, flow velocities at heights greater than 5mm above the flume bed were greater than $0.055 \text{ms}^{-1}$. Thus, by using the pulsed-Doppler unit within this period, it was possible to obtain a series of near-bed velocity distributions for different concentrations of mud in suspension.

## MEASUREMENT OF TURBULENCE USING THE ULTRASONIC DOPPLER

An analysis of clear water LDA results taken at a sampling frequency of 1000Hz was carried out using a Fast Hartley Transform with a Hanning window. This showed the only significant frequencies occurring in the flow to be below 0.1Hz. The ultrasonic pulsed-Doppler unit, having a minimum pulse repetition frequency of 4kHz, is capable of measuring flow velocity with a temporal resolution well in excess of this and could therefore be used, at least in theory, to obtain direct measurements of turbulent fluctuations as well as of mean flow velocity in suspensions of mud.

The turbulence information of interest in the study of the velocity distributions under various suspensions of mud is the shear velocity $\sqrt{[u'w']}$, where u' is the fluctuating component of the longitudinal velocity u and w' the fluctuating component of the vertical velocity w. The brackets [] indicate a temporal mean quantity. Results obtained from the authors' investigation of flows in the Race Track Flume indicate that the turbulence in the flume is reasonably isotropic, so it may be assumed that there is a consistent relationship between the one-dimensional fluctuating component, q', measured with the pulsed-Doppler unit and the principal fluctuating components u' and w'. Therefore, it may be supposed that there will be a consistent relationship between the shear velocities $\sqrt{[q'q']}$ and $\sqrt{[u'w']}$.

A plot of shear velocity $\sqrt{[u'w']}$ measured with the two-dimensional LDA system against shear velocity $\sqrt{[q'q']}$ measured with the ultrasonic pulsed-Doppler unit is shown in figure 4. It can be seen that, although there is scatter, most of the points are grouped around a line of slope equal to 1.331.

The ultrasonic instrument is designed to measure relatively steady laminar flows. No specification is available as to its design limits in the measurement of turbulent fluctuations. The instrument is susceptible to external noise and scattering of results and requires careful set-up in order to maximize its signal to noise ratio. Nevertheless, it has not previously been possible to measure turbulent quantities in dense suspensions of mud, and the authors believe that the application of this technology to the problem offers a significant advance.


## OTHER MEASURING TECHNIQUES IN THE RACE TRACK FLUME

The inability of the pulsed-Doppler unit to measure slow speeds meant that it was necessary to resort to visual inspection of the fluid mud phenomenon close to the glass sides of the flume's working section.

Rhodamine-B dye was injected in lines to enable fluid mud velocity profiles to be measured and a CCD video camera with a zoom lens and SVHS recorder were used to record the process for later analysis. Further measurements of fluid mud velocity, thickness and elevation were made manually during the course of each experiment.

The video equipment was also used to study particle size and fall velocity, the camera being fitted with a 135mm lens and a bellows extension, as described in Dearnaley (1991). This gave a magnification of up to 50 times, allowing mud flocs of diameter 20μm to be resolved. Samples of mud flocs were extracted from the race track flume as follows: a perspex tube was aligned horizontally with the water flow and then sealed at the ends, enclosing a volume of mud suspension. The tube was then withdrawn from the flume and turned upright to enable close up filming of the flocs settling under gravity. The recorded information was analyzed at Hydraulics Research Wallingford Ltd. using a Micro Eye TC frame grabber and Micro Scale TC image analysis software developed by Digithurst Ltd.

Concentration profiles were measured using a FOSLIM probe manufactured by the Delft Hydraulics Laboratory. This instrument relates the transmission of infra-red light to the concentration of suspended sediment, and can be adjusted to measure linearly up to concentrations of about 60kgm$^{-3}$, though it does have to be re-calibrated frequently using the same sediment as that to be measured. The probe intrudes into the flow, though it is small and does not appear to introduce much disturbance. In order to measure vertical profiles of concentration, it was connected to a vernier allowing it to be moved up and down.

Concentrations of fluid mud, which were generally too high to be measured using the FOSLIM probe, were determined by sucking samples from the fluid mud layer using a syringe connected to a thin glass tube. The glass tube was attached to a vernier and set at the height of the centre of the moving layer of mud. Samples were either stored in bottles and later weighed using a specific gravity bottle or passed through a Paar oscillating-U-tube density meter. Measurement of concentration/bulk density by this method is fraught with difficulty, since it is impossible to determine whether a sucked sample is truly representative of the flow from which it was extracted. For concentrations in excess of about 10kgm$^{-3}$, which is much lower than the upper limit of the FOSLIM concentration probe, it seems likely that the suction process entrains water, causing the measured concentration to underestimate the real value.


## EXPERIMENTAL RESULTS - TURBULENCE

Detailed results obtained using the pulsed-Doppler unit and corresponding concentration measurements obtained with the FOSLIM are included in the appendix. It will be noted that the velocities measured are in general well above the 0.055ms$^{-1}$ lower measuring limit of the pulsed-Doppler system.

Figure 5 shows the dimensionless turbulent shear velocity in the Race Track Flume against mud concentration for measurements at sections, B and E. Section B is located 1.0m into the working section of the flume and section E 3.0m into the working section. At each section, values are averaged over the lower 50mm of the flow.

It is possible to identify a trend of shear velocity reducing with increasing mud concentration, this being given for the respective sections, by:

$$\left(\frac{u^*}{U_{mean}}\right) = 0.201 - 0.001\,C \qquad (4)$$

where $u^*$ is a shorthand notation for the shear velocity $\sqrt{[u'w']}$, $U_{mean}$ represents the mean longitudinal flow velocity and C is the concentration.

The difference in relationship between C and $u^*/U_{mean}$ at the two sections may in part be influenced by the differing degrees of experimental scatter. It is also probably influenced by the flume's semicircular bend, as section B is only 1.0m downstream of this and the flow there is still somewhat affected by that feature.

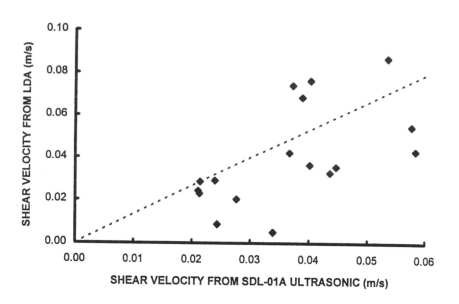

**Figure 4** Plot of turbulent shear velocity u* measured directly with the LDA against that measured with the Toshiba SDL-01A ultrasonic Pulsed-Doppler unit

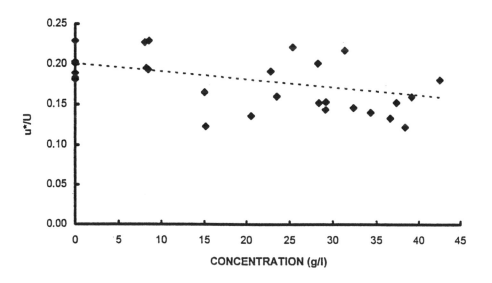

**Figure 5** Plot of u*/U against mean concentration of suspended mud including data from measuring sections B and E

## EXPERIMENTAL RESULTS - VELOCITY DISTRIBUTIONS

Using mixing length concepts (Rouse, 1962, Daily and Harleman 1973), it can be seen that the velocity distribution near smooth and rough boundaries is given by:

$$\frac{u - U}{u^*} = m \log\left(\frac{z}{Z}\right) + C_1 \tag{5}$$

where U is the maximum velocity at $z = Z$ and m is the slope given by $2.303/\kappa$. The von Kárman parameter $\kappa$ is about 0.4 for clear water and the constant $C_1$ is about 5.5. Rearranging equation 5 gives:

$$\frac{u}{U} = m_1 \log\left(\frac{z}{Z}\right) + 1 \tag{6}$$

where $m_1 = mu^*$.

Rouse (1961) gives experimental velocity distributions for a 'wide' channel with a width/depth ratio of 1.724 and for a 'narrow' channel with a width/depth ratio of 0.600. Rouse's results were given in the form u/U against z/H where H is the water depth. The present authors re-plotted these curves using z/Z instead of z/H and the results are given in figure 6.

Some of the authors' dimensionless experimental velocity distributions are given in figure 6 for sections B and E (width/depth ratios 0.616 and 0.589 respectively). Clearly there is considerable scatter in the results. Nevertheless, for a given value of z, the experimental velocities are consistently lower than Rouse's 'wide' and 'narrow' channel values, especially near the bed.

Semi-logarithmic plots of the experimental velocity distributions resulted in reasonable straight lines in most cases. The slopes of these lines together with the measured shear velocities near the bed were used to determine values of $\kappa$ in table 1. Values of the von Kárman parameter in table 1 were averaged over sections B and E and show a general reduction with the increase in mud concentration. This trend confirms the findings of Hino (1963).

**Table 1** Variation of the von Kárman Parameter $\kappa$ with Sediment Concentration

| Mean Concen-tration (g/l) | $\kappa$ Newtonian | $\kappa$ Non-Newtonian | Mean Concen-tration (g/l) | $\kappa$ Newtonian | $\kappa$ Non-Newtonian |
|---|---|---|---|---|---|
| 0.0 | 0.47 | - | 29.7 | 0.32 | 0.28 |
| 8.4 | 0.37 | 0.35 | 40.8 | 0.25 | 0.22 |
| 24.0 | 0.30 | 0.26 | 80.9 | 0.10 | 0.05 |

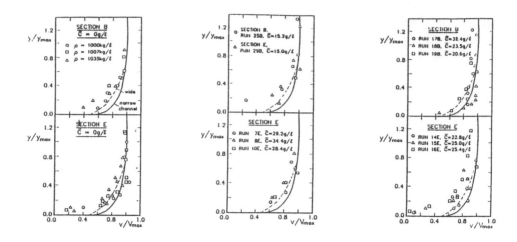

**Figure 6** Near-bed velocity distributions in various concentrations of mud in the Race Track Flume together with 'narrow' and 'wide' channel profiles after Rouse (1962)

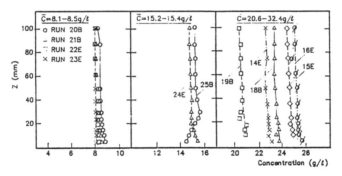

**Figure 7** Near-bed concentration profiles in the Race Track Flume

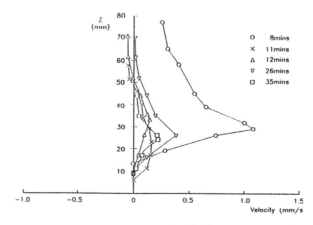

**Figure 8** Fluid mud velocity profiles in the Race Track Flume

High concentration mud suspensions can be treated as non-Newtonian fluids with an apparent Bingham shear strength $\tau_B$ of 0.1 to 0.4Nm$^{-2}$ (Odd and Rodger 1986). Assuming a value of $\tau_B = 0.1$Nm$^{-2}$, the modified values of $\kappa$ for a non-Newtonian fluid are included in table 1. The table shows that the change in $\kappa$ is about 5% for a concentration of 8.4kgm$^{-3}$ and about 50% for a concentration of 80.89kgm$^{-3}$.

Variation of concentration with z is given in figure 7. Clearly, in most cases, there is very little variation with z. This trend confirms the conclusions of Sleath (1984).

## EXPERIMENTAL RESULTS - FLUID MUD

Figure 8 shows a series of fluid mud profiles observed and recorded on video through the glass side of the Race Track Flume. These results were obtained at section E with an initial average concentration in the flume of 80.89kgm$^{-3}$ and a flow speed of zero in the overlying water. The Reynolds Numbers of the fluid mud flow given by $R_e = u_m d_m / v_m$ with $v_m$ equal to $2.6 \times 10^{-6}$m$^2$s$^{-1}$ are all of the order of 100, indicating the fluid mud flow to be laminar. Equation 20 in Ali and Georgiadis (1991) gives a value for the mean velocity of a fluid mud layer flowing under laminar conditions as:

$$u_m = \sqrt{\left( \frac{8g(\rho_m - \rho_w)d_m S}{\rho_m f(1 + \alpha)} \right)} \tag{7}$$

where g is the acceleration due to gravity, S is the channel slope, f is the friction factor, $\alpha$ is a factor assumed by Harleman (1961) to be 0.64 and the other symbols are as previously defined. Comparison of the observed profiles with this equation suggests that the friction factors involved have very high values of the order of 1.

The video recordings show complex behaviour with the whole mass of mud beneath the hindered settling lutocline behaving as a continuum, some of which flows downslope. There is, however, is no clearly visible boundary between the overlying suspension and the moving fluid mud.

Table 2 shows values of particle size and fall velocity obtained from the video imaging system for a mud concentration of 40kgm$^{-3}$. Samples were taken at various times during a long run in order to determine whether there was any change in floc structure over time. However, as can be seen from the table, there was no such trend. The average fall velocity of 0.81mms$^{-1}$ corresponds to a value obtained from Thorn (1981) of 1.32mms$^{-1}$, and is therefore of the correct order of magnitude.

Analysis of the video record is difficult due to residual turbulence in the settling column. An allowance must be made for this when determining particle motion between successive frames. There is also great difficulty in identifying individual flocs within the dense mass of mud particles shown on the video tape. The implementations of the video imaging analysis method described in Dearnaley (1991) were typically only for concentrations up to 0.3kgm$^{-3}$, for which this latter problem does not occur.

**Table 2** Particle Size and Fall Velocity Information Obtained Using Video System

| Sample Number | Time from Start of Test (hh:mm) | Floc Area (sq µm) | Maximum Diameter (µm) | Minimum Diameter (µm) | Equivalent Circle Diameter (µm) | Fall Velocity (mms$^{-1}$) |
|---|---|---|---|---|---|---|
| 2.1 | 00:02 | 9850 | 124 | 59 | 67 | 0.64 |
| 1.1 | 00:10 | 8760 | 112 | 52 | 58 | 0.72 |
| 2.2 | 00:35 | 7230 | 108 | 59 | 55 | 0.47 |
| 1.2 | 01:03 | 5870 | 100 | 49 | 51 | 0.90 |
| 2.3 | 01:31 | 11160 | 122 | 64 | 68 | 0.95 |
| 1.3 | 02:07 | 11020 | 118 | 56 | 64 | 1.00 |

## MODELLING RESULTS

Application of the Wallingford FLUIDMUDFLOW-2D Model to the authors' results from the tilting flume experiments showed that the model results were highly sensitive to the sediment fall velocity used in the calculations. A comparison of fluid mud velocities predicted using different values of $\omega_s$ in equation 3 is shown in figure 9. The results shown are for an initial average concentration of 30kgm$^{-3}$ and a bed slope of 1:5. A value of $\omega_s$ of $5 \times 10^{-5}$m$^4$kg$^{-1}$s$^{-1}$ in equation 3 gives a fall velocity for C = 30kgm$^{-3}$ of 1.5mms$^{-1}$, which corresponds exactly to the value obtained from Thorn (1981).

The high degree of sensitivity to fall velocity is to be expected since this parameter controls the amount of sediment settling into the fluid mud layer.

Figure 10 shows modelled fluid mud velocities for various bed slopes with a 15.6kgm$^{-3}$ initial average concentration. These are compared with experimental data from Ali and Georgiadis (1991). It is clear that the FLUIDMUDFLOW-2D model is producing fluid mud velocities of the correct order of magnitude and which vary over time in a broadly realistic way. However, the fit between modelled and measured velocities is not ideal. Modelled peak velocities do not coincide with measured ones, and the modelled velocities for slopes of 1:10 and 1:20 are shown to decay to zero at approximately 16 minutes, whereas the experimental data appear to approach a steady-state value. The former discrepancy is attributable to the way in which vertical settling into the fluid mud layer is modelled using equation 3, whilst the latter problem appears to be due to the fact that the model assumes a constant concentration for fluid mud, whereas Ali and Georgiadis (1991) found the concentration to increase throughout their experiments.

The figures show only preliminary results, and modelling work is continuing. However, it appears at present that accuracy of the FLUIDMUDFLOW-2D software's predictions is significantly affected by the programs' sensitivity to mud settling behaviour.

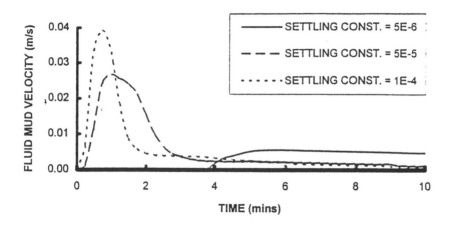

**Figure 9** Effect of variations in settling constant $\omega_s$ on modelled fluid mud velocity

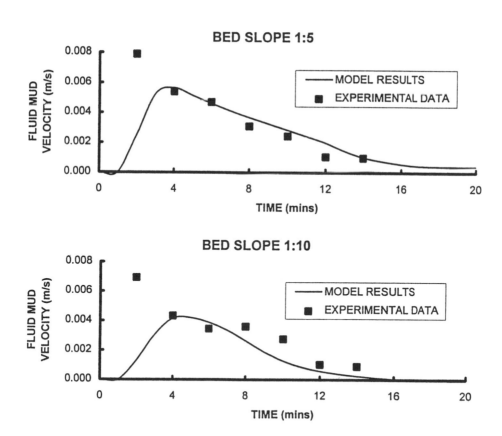

**Figure 10** Model results for various bed slopes compared to fluid mud velocity data from Ali and Georgiadis (1991)

## CONCLUSIONS

The study described in this paper is continuing, but the main conclusions reached so far may be summarised as follows:

1. It is possible to measure turbulent features in the flow of dense suspensions of cohesive sediment using ultrasonic Doppler technology. Use of this technology represents a significant advance in the laboratory study of cohesive sediment.

2. concentrations of suspended cohesive sediment in the Race Track Flume cause a reduction in turbulence intensity and a decrease in the von Kármán constant $\kappa$ in line with the theoretical concepts of Hino (1963).

3. Fluid mud flow in the Race Track Flume is generally laminar in nature with a high friction factor of the order 1. The behaviour of the mud is complex and it is difficult to identify a clearly defined 'fluid mud layer'

4. Results of the Wallingford FLUIDMUDFLOW-2D model are highly sensitive to the way in which the fall velocity of suspended sediment is modelled, but under certain circumstances the FLUIDMUDFLOW-2D programs are capable of modelling experimental fluid mud velocities with a reasonable degree of accuracy.

## ACKNOWLEDGEMENTS

The work described in this paper was funded under UK Science and Engineering Research Council Grant No GR/H14991 for which the authors express their gratitude.

Thanks are due to Prof. B.A. O'Connor, Professor of Maritime Civil Engineering at the University of Liverpool for valuable advice; to Dr. W.R. Parker of Blackdown Consultants Ltd., Taunton U.K. for suggesting the use of ultrasound technology and for valuable advice; to Dr. Thien How of the Clinical Engineering Department, University of Liverpool, U.K. for the loan of the Toshiba instrument; to Dr. W. Roberts, Dr. A.J. Cooper and N.V.M. Odd of Hydraulics Research Wallingford Ltd. for providing the FLUIDMUDFLOW-2D software and advising on its use; to Dr. M.P. Dearnaley of Hydraulics Research Wallingford Ltd. for arranging and helping with the use of the Digithurst video analysis system; and to the Mersey Docks and Harbour Company for allowing access to recover mud from the dredger 'Mersey Mariner'.

## REFERENCES

Ali K.H.M. and Georgiadis K., (1991), *Laminar Motion of Fluid Mud*, Proc. I.C.E., Part 2, vol. 91, December, pp.795-821

Ali K.H.M. and Crapper M., (1993), *Measuring techniques including the application of medical ultrasound technology to the laboratory study of fluid mud*, B-5-4, Proc. XXVth I.A.H.R. Congress, Tokyo, Japan, Aug.-Sept., pp.166-173

Daily J.W. and Harleman D.R.F., (1973), *Fluid Dynamics*, Addison Wesley Publishing Co., USA, pp.227-244

Dearnaley M.P., (1991), *Flocculation and settling of cohesive sediments*, Report no. SR 272, Hydraulics Research Wallingford Ltd.

Harleman D.R.F. (Streeter V.L. (ed.)), (1961), *Stratified flow, Handbook of fluid dynamics*, McGraw-Hill, New York, ch. 26

Hino M., (1963), *Turbulent Flow with Suspended Particles*, Proc. A.S.C.E., Journal of the Hydraulic Division, vol. 89, no. HY4, July, pp.161-185

Nicholson J. and O'Connor B.A., (1986), *A Cohesive Sediment Transport Model*, A.S.C.E. Journal of Hydraulic Engineering vol. 112 no. 7, July, pp.621-640

O'Connor B.A., (1991), *Suspended sediment transport in the coastal zone*, Proc. Int. Symp. on the Transport of Suspended Sediments and its Mathematical Modelling, Firenze, Italy, pp.17-63

Odd N.V.M. and Cooper A.J., (1989), *A two-dimensional model of the movement of fluid mud in a high energy turbid estuary*, Journal of Coastal Research, Special Issue no. 5, Summer

Odd N.V.M. and Rodger J.G., (1986), *An analysis of the behaviour of fluid mud in estuaries*, Report no. SR 84, Hydraulics Research Wallingford Ltd., March

Rouse H., (1961), *Fluid Mechanics for Hydraulic Engineers*, Dover Publications, Inc., New York, pp.274-281

Rouse H., (1962), *Elementary Mechanics of Fluids*, John Wiley and Sons Inc., New York, pp.176-199

Sleath J.F.A., (1984) *Sea Bed Mechanics*, John Wiley and Sons, Inc., New York, pp.279-281

Thorn M.F.C., (1981), *Physical Processes of siltation in tidal channels*, Hydraulic modelling applied to maritime engineering problems, Paper no. 6, Institution of Civil Engineers

Vanoni V.A., (1946), *Transportation of suspended sediment by water*, Transactions, A.S.C.E., vol. 111, pp.67-133.

# DEPOSITION AND EROSION

# 14 Biological mediation of sediment erodibility: ecology and physical dynamics

**D. M. PATERSON**
*Gatty Marine Laboratory, University of St Andrews, UK*

## INTRODUCTION

Ecology is broadly defined as the study of the interactions between organisms and their environment (Colinveaux 1993). The "environment" of an organism incorporates all physical and biological aspects of the surroundings, including other organisms, the physical setting and even the influence of local air and water motion. Advective movements of the medium are often exploited by organisms but also have the potential to be disruptive, harmful or even lethal. Exploitation often includes feeding strategies and the dispersal of sperm and propagules (Denny 1993). In ecological studies, strong interactions between the physical habitat and the biological community have long been recognised but recently more detailed attention has been paid to how organisms utilise or adapt to the physical dynamics of their specific environment (Vogel 1981, Denny 1988 & 1993). Traditional sedimentology has viewed the environment as one where the factors of first order importance are the physical forces and the abiotic properties of the substratum. The possibility that the biology of a sediment system might mediate the response of the bed to physical forcing was historically a factor that received little attention. Most experimental studies (Miller *et al.* 1977) were designed to examine the fundamental behaviour of particles and actively preclude biotic effects. This approach is entirely reasonable if the goal is to establish the physical basis behind the erosion of particles but, as with many autecological studies in biology, the problem arises as to the relevance of these studies, conducted under highly defined conditions, to the natural world. The interactions of biological processes with sediments introduces an entire suite of variables that may or may not be important in the study of sediment erodibility (Montague 1986). The object of this short review is not to provide a comprehensive coverage of the literature (see Heinzelmann & Wallisch 1991, Black 1992) but to highlight some of the milestones in the examination of the biological stabilisation of sediments, with a bias towards studies concerning the ubiquitous microbiota of surficial sediments.

*Cohesive Sediments.*
Edited by Neville Burt, Reg Parker and Jacqueline Watts
©1997 John Wiley & Sons Ltd.

## THE EROSION OF COHESIVE SEDIMENTS

Discussion of the biological mediation of cohesive sediment erosion is complicated by difficulty in describing the erosion process. Erosion of a bed can be considered as taking place when the shear stress at the surface exceeds the inherent strength of the bed (Delo & Ockenden 1992). Therefore, erosion can be defined in terms of a "threshold stress" followed by a subsequent erosion rate. Within these boundaries, several types of erosion can occur (e.g. Type Ia, Type Ib & Type II *sensu* Amos *et al.* 1992a). However, erosion can also be considered as one side of a deposition/entrainment equilibrium which is always occurring at the bed surface. Under these circumstances, the stress at which the removal of particles first exceeds deposition is the point of interest and no "threshold" for erosion exists. For the purpose of this review, the mechanism of erosion is not critical, the fact that the bed becomes more resistant to erosion is the critical factor.

In theory, organisms inhabiting sediments may have three potential influences on sediment erodibility: *neutral* (no effect); *negative* (decreasing stability); or *positive* (increasing stability). Neutral effects are where no *measurable* changes in sediment erodibility, as compared with controls, are found (the problem of control systems is discussed later). The most common negative effect considered in the literature is the result of the reworking or packaging of the sediment by organisms, known as bioturbation. It should be noted that bioturbation *per se* does not always lead to sediment destabilisation (Meadows & Tait 1989) and that other activities by organisms may also change the resistance of a sediment to erosion by purely physico-chemical mechanisms (Montague 1986). A positive effect on sediment stability is often described as the *biogenic stabilisation* or *biostabilisation* of the bed and has been defined as *"a decrease in sediment erodibility caused directly or indirectly by biological action"* (Paterson & Daborn 1991). This definition does not infer the nature of the erosion process and includes the influence of detrital biomass (old root systems & senescing filaments) and the production of extracellular polymeric substances (EPS).

## EARLY EVIDENCE OF BIOGENIC STABILISATION

As early as 1868, Huxley observed an all-pervasive slime or mucilage on the ocean bed. With his colleague Haekel, he considered this "slime" to be a living entity and an important agent in sediment stabilisation and conferred the generic name *Bathybius haeckelii* to this ubiquitous "organism" (cited by Krumbein *et al*, 1994). Although incorrect in interpretation, Huxley had recognised the wide spread occurrence of organic material on the ocean floor. Since this time, the potential binding effects of cyanobacteria, bacteria and algae has continued to receive sporadic attention. These observations were generally unsupported by quantitative evidence partly due to the technological difficulty in measuring "sediment stability" (see Gust this volume,

Cornelisse *et al.* this volume). One of the first, and still most innovative, attempts to examine the phenomenon of biogenic stabilisation came from the study of subtidal mats in the Bimini Lagoon, Bahamas (Scoffin 1970, Neumann *et al.* 1970). Their observations ranged from mangroves to microbial mats and included the use of an *in situ* benthic flume. Mats composed of different assemblages of cyanobacteria and algae were found to have a different resistance to erosion and this was related to the microstructure and composition of the mats (*Enteromorpha>Lyngbya > Schizothrix > Laurentia*, in order of decreasing resistance). *Enteromorpha* mats (a filamentous green alga) were five times more resistant to erosion than bare areas of sediment. The algal mats were found to cause less sediment accumulation and stabilisation than macrophytes (e.g. sea grasses) and were more ephemeral in nature (Scoffin 1970).

## MECHANISMS OF BIOGENIC STABILISATION

Observations that biota might hinder the erosion of sediments abound in the literature (Montague 1986, Heinzelmann & Wallisch 1991). Few studies make quantitative measurements and fewer still address the mechanism by which the stabilisation occurs, an aspect which is fundamental towards the understanding of biogenic stabilisation. A number of the mechanisms of biogenic stabilisation (Table 1) are discussed below and shown in Figure 1.

### *Alteration of flow above the bed*

A number of biogenic effects influence the flow over the substratum. On the microbial scale, organisms may influence the flow across the bed surface in several ways. Firstly, the stress experienced by a particle on the surface of a bed is composed of elements of lift and drag forces (Allen 1985). Particles protruding above the bed experience greater stress than those more closely associated with the surface. This bed *roughness* is expressed as the hydraulic roughness (Zo) which, for a flat bed, is related to the size of the bed particles (Nowell & Jumars 1984, Denny 1988). As Zo increases, the effective stress experienced by the particles increases for the same free stream velocity and particle entrainment becomes more likely. Hydraulically rough beds can become smoothed where particles become covered by a microbial mat or if the inter-particle voids are filled by microbes and their secretions (Manzenrieder 1983). This has the effect of decreasing the stress experienced by the particles and raising the critical velocity for particle entrainment. In addition to smoothing effects, microbial development may lead to increasing elasticity in the bed, even for sediments normally considered to be non-cohesive. Oscillation of the surface of a diatom-colonised cohesive bed in response to perpendicular jets has been previously noted (Paterson 1989) while similar but more extreme behaviour for a microbial mat on a mixed tidal flat has also been found (Yallop *et al.* 1994).

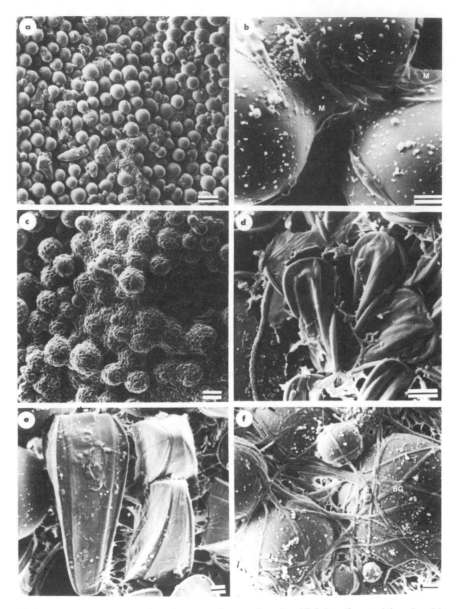

**Figure 1**. Low-temperature scanning electron micrographs of artificial sediment (glass beads) and the mechanisms of sediment stabilisation by bacteria, diatoms and cyanobacteria. a: The glass bead cultures at an early stage in colonisation (bar marker 100 um). b: The development of cohesive bonds between the beads by bacteria and diatoms. M = mucilage. N & C = different diatom species (*Navicula* sp. and *Cylindrotheca* sp.) (bar marker = 30 um). c: Further development of the assemblage. Beads are now covered by diatoms and bacteria Cohesion has increased and roughness (Zo) is beginning to be affected (bar marker = 40 um). d: Cohesive binding between diatom cells (*Amphiprora* sp.) (bar marker = 20 um). e: Detail of extracellular polymeric substances (EPS) forming links between diatom cells (*Surirella* sp. ) (bar marker = 10 um). f: Network formation by cyanobacterial filaments (BG) across the surface of the glass bead culture (bar marker = 10 um).

**Table 1.** Mechanisms responsible for biogenic stabilisation

| Mechanism | Organism | Selected references |
|---|---|---|
| *Influence on flow above the bed* | | |
| a. Topography (Zo) | Microbiota | Heinzelmann & Wallisch 1991 |
| b. Skimming flow/flow reduction | Faunal tubes, Sea grasses Algae | Luckenbach 1986, Gambi *et al.* 1990 |
| *Burrow structures* | Infauna | Meadows & Meadows 1991 |
| *Network effects* | Filamentous forms, roots, cyanobacteria, fungi | Meadows *et al.* 1994 Yallop *et al.* 1994 |
| *Extracellular polymeric substances (EPS)* | | |
| a. Particle cohesion | Bacteria/diatoms/infauna | Manzenrieder 1983, Underwood |
| b. Organic matrix | Bacteria/diatoms/infauna | & Paterson 1993a & 1993b |

A more recent and intriguing suggestion for an effect of microbes on sediment erosion is through rheological changes brought about by microbial secretions. Polymers are frequently used in industry to enhance flow and reduce drag. An analogous situation can occur at the sediment/water interface. The exchange of solutes between the bed and the water column has been demonstrated many times (Huettel & Gust 1992) and almost all sediment-inhabiting microbes produce polymeric materials, many of which become suspended (Denny 1988, Decho 1990). Entry of these polymers into the boundary layer flow field is likely to influence the nature of the stress experienced at the bed through the *"Toms effect"* in a similar manner to the mucus secreted by the epithelial cells of rainbow trout (Daniel 1981). This secretion can reduce the drag experienced by a swimming fish by up to 50%. The potential of microbial polymers to influence the boundary flow conditions through alterations in viscosity seems less improbable in the context of other studies such as Deacon (1979) who cited mucilage secretions from coral polyps as the active agents in inhibiting the formation of wind-induced capillary waves in coral atolls. The use of controlled stress rheometry has great potential towards the investigation of the effects of biopolymers on stress at the surface of cohesive beds (see Jones, this volume).

The corollary of boundary layer smoothing is the increase in roughness which can be caused by bioturbation (Nowell *et al.* 1981). This effect is not straightforward since although the increase in Zo might be expected to reduce the critical erosion threshold, faecal pellets and mounds also incorporate mucilage which can have a stabilising influence (Black 1992). A more extreme example of biologically increased boundary roughness is the formation of tubes which protrude from the sediment surface. In this case the density of the protruding elements (e.g. animal tubes) is of critical importance to the stress imposed on the bed (Eckman *et al.* 1981). Rigid structures will tend to entrain flow and increase local turbulence leading to scour around the element (Heinzelmann & Wallisch 1991). However, once the density of elements surpasses a critical level, the flow field is altered to produce an area of high turbulent intensity at the boundary of the element field leading to a reduced stress at the actual bed (skimming flow). An analogous mechanism is suggested for macrophytes such as sea (Gambi *et al.* 1990). The problem is made more complex by the fact that the presence of tubes or blades is not independent of other biological factors such as microbial growth and the production of faecal pellets. Tubes at high density are effective in the reduction of surface stress but other biogenic factors seem to be involved in the stability of sediments around the tubes (Eckman *et al.* 1981, Luckenbach 1986).

*Network effects*

Network effects may influence sediments on a variety of spatial scale passing through several orders of magnitude. At the largest scale are the effects of trees such as the mangrove (Scoffin 1970) while at an intermediate level are the roots of aquatic macrophytes (van Eerdt 1985) and the filaments of macrophytic algae which ramify through the surface of sediments. At the microbial level, all filamentous organisms including fungi, bacteria, cyanobacteria and algae may be implicated in increasing the integrity of the bed (Paterson & Daborn 1991). For larger forms, a reasonable attempt can be made to measure the biomass and tensile strength of the effective elements (Waldron 1977, van Eerdt 1985). On a microbial level, the influence of the network is less easily interpreted since the binding elements are microscopic and cannot be easily separated out from the rest of the biomass. A common example of the efficiency of such binding is the filamentous red algae, *Audouinella floridula* (Dixon & Irvine 1977) which can often be found under very exposed conditions and yet maintains a layer of non-cohesive sediments bound to the surface of exposed intertidal rocks. Evidence that binding by network formation is important tends to be qualitative but is sometimes supported by visualisation of the substratum through the use of various forms of microscopy (Paterson 1995). Interpretation of the quantitative efficacy of microbial network formation suffers from the microscale heterogeneity of microbial assemblages (Yallop *et al.* 1994, Paterson *et. al.* 1994). Evidence that network formation is often highly effective on a microbial scale is often circumstantial but convincing. Quantitative evidence concerning increased slope stability induced by bacteria and fungi has been given by Meadows *et al.* (1994). These worker found that network formation by fungal hyphae (*Penicillium chryogenum*) significantly increased slope stability. In other studies, filamentous cyanobacteria were found to increase the stability of cultures much more than coccoid forms (Krumbein *et al.* 1994) while the invasion of experimental diatom cultures by filamentous cyanobacteria greatly increased the stability of the experimental systems (Paterson 1990). The shear strength of the surficial region of inter-tidal mud is also greatly increased by the growth of the filamentous algae, *Vaucheria* sp. (Paterson, unpublished data). Factors that effect the binding potential of network formations include the tensile strength and length of the elements, which may be single filaments or bundles of filaments (e.g. the cyanobacterium *Microcoleus*, Yallop *et al.* 1994), the amount of branching and the tenacity of the attachment to the surrounding matrix.

### Animal burrows

The presence of animal burrows within sediments can also lead to the strengthening of the sediment bed. A considerable amount of work on the influence of infauna on sediments has been carried out by Meadows and co-workers (Meadows & Tait 1985 and 1989, Meadows *et al.* 1991, Meadows & Meadows 1991). This work has amply demonstrated the potential of burrowing organisms to increase the local shear strength of sediments but less is known about how shear strength may be interpreted in terms of sediment entrainment. Studies by Rhoads *et al.* (1978) indicated that tube-forming polychaetes (*Heteromastus filiformis*) could increase the critical rolling velocity for silty-clay sediments by 80%. However, these workers point out that the mechanism of stabilisation may be a combination of effects including: tubes acting as a barrier to bioturbation; polychaete activity stimulating microbial activity (Luckenbach 1986); and projection of tube fields reducing the shear stress at the bed surface.

### Extracellular polymeric substances

The production and degradation of extracellular polymeric substances (EPS) plays a pivotal role in sediment ecology (Decho 1990). EPS is implicated in a multitude of functions (Hoagland *et al.* 1993) including cell attachment (Rosenberg *et al.* 1989), protection from desiccation and buffering against toxins (Decho 1990), locomotion in diatoms (Edgar & Pickett-Heaps 1984), as a food resource (Decho & Moriaty 1990. The ubiquitous nature of EPS explains the observations of Huxley (1868) and, with the realisation of the ecological importance of EPS, the relative quantities found in nature (> 17 mg g$^{-1}$ in intertidal cohesive sediments by Underwood & Paterson 1993a) become less surprising. Field observations in which EPS was cited as the "active ingredient" of stabilisation by microbial mats (Frankel & Mead 1973) were followed by the laboratory study of Holland *et al.* (1974). In this study, 6 species of diatom were examined and the conclusion reached that sediment stabilisation was related to mucilage production. Cells producing little mucilage were less effective at retarding erosion than those which produced copious amounts. This innovative study lacked the hydrodynamic rigour of later studies but made the valuable point that biogenic stabilisation was likely to be highly variable and species (or by inference, assemblage) specific. A modern analogue of this work (de Jonge & van den Bergs 1987) showed that diatoms enter suspension in response to hydrodynamic stress but not necessarily in the proportions found on the sediment but this result was variable and dependant on the sediment type. Several authors confirm the variability of binding through the production of EPS (e.g. Grant *et al.* 1986a, Paterson *et al.* 1986) and Grant *et al.* (1986a) suggested that the importance of diatom binding by mucilage was on a spatial scale of millimetres and a temporal scale of days. Vos *et al.* (1988) and de Boer (1981) considered diatom binding to be of more general importance and the former examined the organisms and the mechanisms responsible for sediment stabilisation. They described a *"cohesive effect"* due to an increase in the surface cohesion of particles caused by an organic coating and a *"network effect"* generated by the formation of mucilage "threads". The interpretation of the mucilage secreted by diatoms and bacteria as threads requires some care because of artefacts that may be produced in the preparation of materials for scanning electron microscopy (SEM, see discussion in Westall & Rince 1994, Paterson 1995). The stabilisation of sandy sediment by bacterial EPS was confirmed in an elegant series of experiments by Dade *et al.* (1990). This laboratory study examined the influence of bacterial isolates grown in culture on the entrainment of the sediments. In addition to measuring the effect of bacteria *in situ*, EPS was extracted from the cultures and added to clean sediments. This work confirmed that EPS *per se* is capable of binding sediment although the efficiency of the binding was reduced in comparison with undisturbed cultures. In addition to stabilisation, EPS may enhance accretion (Grant *et al.* 1986b). Micro-algal populations were implicated as important in the development of salt-marshes (Coles 1979) and this has been supported by recent studies which show that diatom populations and their associated mucilage enhance accretion of intertidal mudflats (Underwood & Paterson 1993a).

## COMPARISON OF DATA: METHODS AND CONTROLS

Not all work on biogenic stabilisation has been included in this review but tabulations which attempt to compare the results of such studies are available (Heinzelmann & Wallisch 1991, Paterson & Daborn 1991). Comparisons (Tables 2) are often of dubious because of the variety of techniques employed and the differences between laboratory and field measurements. The current trend is towards the *in situ* measurement of as many sediment parameters and processes as possible (Amos *et al.* 1992b, Black & Paterson, in press) but many varied techniques are under development (Williamson & Ockenden 1996, Gust this volume, Cornelisse *et al.* this volume). It seems unlikely that a general consensus will be easily reached since the objectives of each study and the habitat involved will greatly constrain the methodology.

The expression of results as an increase in stabilisation of sediment over an abiotic control (Table 2) is sometimes used to produce a coefficient of biological stabilisation (Manzenrieder 1983) which has been adopted by a number of researchers (Yallop *et al.* 1994). This approach is more straightforward when non-cohesive sediments are involved since the sediments can be cleaned of organic material and then used as the control. However, the packing structure of a bed has an influence on erodibility and therefore disturbance of the bed may bias results. Such a simple control system is not possible with cohesive sediments because of the complex nature of the interactive forces between fine particles (Allen 1985). Controls for biological and particularly microbiological effects are therefore difficult unless using abiotic mineral sediment in the laboratory (Parchure 1984). Poisoning of the bed has been used (de Boer 1981, Black 1992, Daborn *et al.* 1993, Underwood & Paterson 1993a) but the resulting changes in the sediment are not well understood. Black (1992 & this volume) compared the effect of different poisons on a cohesive bed and significantly different results were produced. Biological processes may be halted but organics are not always removed and may even be "fixed" by the addition of some chemicals (e.g. formalin).

Table 2. Selected measurements of biogenic stabilisation from the literature.

| Substratum | Date | Biota | Relative stabilisation % increase over control |
|---|---|---|---|
| *Non-cohesive* | | | |
| Neumann et al. | 1970 | Algae/cyanobacteria | 500 |
| Holland et al. | 1974 | Diatoms | (100) |
| Manzenrieder | 1983 | Bacteria/algae | 300-700 |
| Grant & Gust | 1987 | Purple sulphur bacteria | 390 |
| | | Cyanobacteria | 350 |
| Vos | 1988 | Diatoms | 100 |
| Dade et al. | 1990 | Bacteria | 200 |
| Paterson | 1990 | Diatoms | 400 |
| Madsen et al. | 1993 | Diatoms/bacteria | 300 |
| Yallop et al. | 1994 | Diatoms/cyanobacteria | >960 |
| | | | |
| *Cohesive* | | | (>350) |
| Holland et al. | 1974 | Diatoms | 45 |
| Rhoads et al. | 1978 | Unidentified microbes | 300 |
| Parchure | 1984 | Unidentified microbes | 200 |
| Black | 1992 | Diatoms/bacteria | 500 |
| Yallop et al. | 1994 | Diatoms/bacteria | 300 |

() = best estimate from available data.

## THE SEARCH FOR A PREDICTIVE RELATIONSHIP

One of the critical objectives in the field of sediment transport research is to increase predictive capabilities for the natural transport of sediments and for the implications of anthropogenic change to the environment. With this goal in mind, the influence of biological and chemical factors on sediment erosion and transport requires further attention (Dyer 1989). A relationship between the biological status of the bed and sediment erodibility would be a major step forward. However, since it is the *mechanism* of stabilisation rather than the biomass *per se* that is important, it is only if biomass is directly related to binding influence that such a relationship is produced. This is more likely to occur under laboratory conditions when microbial growth is controlled and assemblage diversity is low such as in the study of Rhoads *et al.* (1978) who found a good correlation between a measure of microbial biomass (ATP) and the mean rolling velocity of glass bead cultures. Grant and Gust (1987) showed an impressive relationship, for a natural assemblage, between photo-pigment content (BChl *a* and Chl *a*) and critical erosion (friction) velocity ($U_{*crit}$) of mats of purple sulphur bacteria (Chromatiaceae). The maximum stabilisation produced was a $U_{*crit}$ of 4.06 cm s$^{-1}$ as opposed to 1.04 cm s$^{-1}$ for a control (cleaned) core with a correlation of 0.80 between BChl *a* and $U_{*crit}$. However, high correlations only indicate covariation and do not suggest a causal relationship. Photo-pigment content is not a direct measure of sediment binding. Grant and Gust (1987) found that a more direct measure of polymer content (colloidal carbohydrate) was slightly less well correlated with sediment stability than photo-pigment content (r = 0.77). Colloidal carbohydrate only represents a fraction of the total polymer load within the mat system (Grant *et al.* 1986a) and the sediments contained filamentous cyanobacteria which may also have a binding effect. This study also included the important information that even apparently "clean" sediments found in the field were more difficult to erode ($U_{*crit}$ = 1.28 cm s$^{-1}$) than abiotic controls ($U_{*crit}$ = 1.04 cm s$^{-1}$).

Several other attempts have been made to find a specific biological parameter of use in the prediction of sediment erosion. Madsen *et al.* (1993) found that sub-tidal mats of algae did stabilise sediments but correlations of stability versus colloidal polysaccharide (*sensu* Grant *et al.* 1986a), Chl *a* or cell biovolume were not significant. The best correlation (0.59) was with a measure of algal cell volume at the sediment surface.

## ECOLOGICAL PERSPECTIVE

The data available on the biogenic mediation of sediments appears patchy and confused. For studies that promise much in the way of predicting sediment behaviour (Manzenrieder 1983, Grant & Gust 1987, Dade *et al.* 1990) there are an equal number studies that suggest stabilisation cannot be predicted from a biological parameter (Grant *et al.* 1986a, Madsen *et al* . 1993) or are better predicted from physical parameters such as wind speed (de Jonge 1992). In biological terms, this variation is unsurprising and from an ecological perspective many of the studies outlined are hardly comparable even if the technologies employed were similar (see Gust this volume, Paterson 1994). Ecosystems under investigation vary from soil systems to subtidal ecosystems. Few ecologists would be brave enough to make many generalisations about the nature and dynamics of such varied communities. It is only where a single mechanism of biostabilisation is dominant that a single parameter for a biological effect is likely to be the best predictor of sediment behaviour. Studies of this type of assemblage are rare but give rise to the most "promising" results. An example of this type of community would be a well-developed mat dominated by a particular group of organism (c.f. Grant & Gust 1987) or experimental models where populations can be controlled (Paterson 1990, Dade *et al.* 1990). Individual populations may well have more than one binding effect (e.g. EPS and filaments) and even the production of EPS is not consistent but may depend on local conditions and upon the physiology of the organisms.

It is clear that more information linking the nature of the substratum, the type of assemblage, the mechanisms of stabilisation and the response of both to hydrodynamic forcing is required to model the influence of biogenic effects on sediment transport. A community specific/site specific approach may be required initially which takes into account aspects of the ecology of the environment including spatial and temporal variability. Episodic events are of great importance to the transport of sediments and few biological systems will resist the full force of a storm or flood event. However, the disturbance of the environment paves the way for a new biological succession and several studies suggest that the recovery of the system is enhanced by colonisation by biota (Underwood & Paterson 1993a). Not enough is known about the seasonal and spatial variability of benthic communities to predict the nature and effects of community succession. Seasonal aspects of the influence of biology on sediment transport must be expected and have been shown in the literature in terms of patterns of sedimentation (Frostick & McCave 1979) but are not always as clear as postulated (Rhoads *et al.* 1978). The seasonal and spatial variation of natural biological communities must be superimposed on the physical background. Evidence for biogenic stabilisation as an important factor at some sites on some occasions is overwhelming. Interdisciplinary work has led to some interesting associations being discovered including that between birds and sediment erosion (Daborn *et al.* 1993). Such interdisciplinary approaches were called for by Montague in 1986 and there is still a requirement for more studies to examine the ecology of benthic ecosystems in relation to the physical dynamics of the environment, an interdisciplinary approach which should result in significant advance in both fields.

## ACKNOWLEDGEMENTS

The contribution of funding from the EC Environment award "PRO-MAT" (contract EV5V-CT94-0411) to work reported in this review is gratefully acknowledged while thanks is also due to Alison Miles for comments on the manuscript.

## REFERENCES

Allen, J.W.R., 1985. *Principles of physical sedimentology*. Allen & Unwin.

Amos, C.L., Daborn, G.R., Christian, H.A., Atkinson, A. & Robertson, A., 1992a. *In situ* erosion measurements on fine-grained sediments from the Bay of Fundy. *Marine Geology*, **108**, 175-196.

Amos, C.L., Grant, J., Daborn, G.R. & Black, K., 1992b . Sea Carousel- A Benthic Annular Flume. *Estuarine, Coastal and Shelf Science*, **34**, 557-577.

Black, K.S. 1992. *The erosion characteristiçs of cohesive estuarine sediments: some in situ experiments and observations*. Ph.D. Thesis, University of Wales.

Black, K.S. & Paterson, D.M. Measurement of the erosion potential of cohesive marine sediments: A review of current *in situ* technology. *Journal of Marine Environmental Engineering* (in press).

Coles, S.M., 1979. Benthic microalgal populations on intertidal sediments and their role as precursors to salt marsh development. In *Ecological Progress in Coastal Environments*. Jefferies, R.L. & Davey, A.J.(eds). First European Ecological Symposium, Oxford, Blackwell, p 24-42.

Colinveaux, P., 1993, *Ecology 2*. Wiley, New York.

Daborn, G.R., Amos, C.L., Brylinsky, M., Christian, H., Drapeau, G., Faas, R.W., Grant, J. Long, B., Paterson, D.M., Perillo, G.M.E. & Piccolo, M.C., 1993. An ecological cascade effect: Migratory birds affect stability of intertidal sediments. *Limnology and Oceanography*, **38(1)**, 225-231.

Dade, B.W., Davies, J.D., Nichols, P.D., Nowell, A.R.M., Thistle, D., Trexler, M.B. & White, D.C., 1990. Effects of bacterial exopolymer adhesion on the entrainment of marine sand. *Geomicrobiology Journal*, **8**, 1-16.

Daniel, T.L., 1981. Fish mucus: *In situ* measurements of polymer drag reduction. *Biological Bulletin*, **160**, 376-382.

de Boer, P.L., 1981. Mechanical effects of micro-organisms on intertidal bedform migration. *Sedimentology*, **28**, 129-132.

de Jonge, V.N. & van den Bergs, J., 1987.  Experiments on the resuspension of estuarine sediments containing benthic diatoms. *Estuarine, Coastal and Shelf Science*, **24**, 725-740.

de Jonge, V.N., 1992. *Physical processes and dynamics of Microphytobenthos in the Ems estuary (The Netherlands)*. Ministry of Transport, Public works and Water Management. pp. 176.

Deacon, E.L., 1979. The role of coral mucus in reducing wind drag over coral reefs. *Boundary-Layer Meteorology*, **17**, 517-522.

Decho, A. W., 1990. Microbial exopolymer secretions in ocean environments: Their role(s) in food webs and marine processes. *Oceanography and Marine Biology Annual Review*. **28**, 73-153.

Decho, A.W. & Moriarty, D.J.W., 1990. Bacterial exopolymer utilization by a harpacticoid copepod: A methodology and results. *Limnology Oceanography*, **35(5)**, 1039-1049.

Delo, E.A. & Ockenden, M.C., 1992. *Estuarine muds manual.* Report SR 309, HR Wallingford.

Denny, M.W., 1988. *Biology and the Mechanics of the Wave-Swept Environment.* Princeton University Press. USA.

Denny, M. W., 1993. *Air and Water: The Biology and Physics of Life's Media.* Princeton University Press. USA.

Dixon., P.S. & Irvine, L.M. 1977. *Seaweeds of the British Isles. Vol. 1. Rhodophyta. Part 1.* Introduction, Nemaliales, Gigartinales. British Museum, London.

Dyer, K.R., 1989. Sediment processes in estuaries: future research requirements. *Journal of Geophysical Research*, **94**, 14.327-14.339.

Eckman, J.E., Nowell, A.R.M., & Jumars, P.A., 1981. Sediment destabilisation by animal tubes. *Journal of Marine Research*, **39**, 361-374.

Edgar, L.A. & Pickett-Heaps, J.D. 1984. Diatom Locomotion. In *Progress in Phycological Research.* Round, F.E. & Chapman, D.J. (eds). **3**, 47-88.

Frankel, L. & Mead, D.J., 1973. Mucilaginous matrix of some estuarine sands in Connecticut. *Journal of Sedimentary Petrology*, **43**, 1090-1095.

Frostick, L.E. & McCave, I.N., 1979. Seasonal shifts of sediment within an estuary mediated by algal growth. *Estuarine and Coastal Marine Science*, **9**, 569-576.

Gambi, M.C., Nowell, A.R.M. & Jumars, P.A., 1990. Flume observations on flow dynamics in *Zostera marina* (eelgrass) beds. *Marine Ecology Progress Series*, **61**, 159-169.

Ginsberg, R.N. & Lowenstam, H.A., 1958. The influence of marine bottom communities on the depositional environment of sediments. *Journal of Geology*, **66**, 310-318.

Grant, J. & Gust, G., 1987. Prediction of coastal sediment stability from photopigment content of mats of purple sulphur bacteria. *Nature,* **330**, 244-246.

Grant, J., Bathmann, U.V. & Mills, E.L., 1986a. The interaction between benthic diatom films and sediment transport. *Estuarine, Coastal and Shelf Science*, **23**, 225-238.

Grant, J., Mills, E.L. & Hopper, C.M., 1986b. A chlorophyll budget of the sediment-water interface and the effect of stabilising biofilms on particle fluxes. *Ophelia*, **26**, 207-219.

Heinzelmann, C. H. & Wallisch, S., 1991. Benthic settlement and bed erosion. A Review. *Journal of Hydraulic Research*, **29**, 355-371.

Hoagland, K.D., Rosowski, J.R., Gretz, M.R. & Roemer, S.C., 1993. Diatom extracellular polymeric substances: function, fine structure, chemistry, and physiology. *Journal of Phycology*, **29**, 537-566.

Holland, A.F., Zingmark, R.G. & Dean, J.M., 1974. Quantitative evidence concerning the stabilization of sediments by marine benthic diatoms. *Marine Biology,* **27**, 191-196.

Huettel, M. & Gust, G., 1992. Solute release mechanisms from confined sediment cores in stirred benthic chambers and flume flows. *Marine Ecology Progress Series*, **82**, 187-192.

Huxley, T.H., 1868. On some organisms living at great depths in the North Atlantic Ocean. *Quarterly Journal of Microscopy*, **VIII**, 203-212.

Krumbein, W.E., Paterson, D.M., Stal, L.J. 1994. *Biostabilisation of sediments*. Oldenburg University Press, Germany.

Luckenbach, M.W., 1986. Sediment stability around animal tubes: The roles of hydrodynamic processes and biotic activity. *Limnology and Oceanography*, **31(4)**, 779-787.

Madsen, K.N., Nilsson, P. & Sundback, K., 1993. The influence of benthic microalgae on the stability of a subtidal sediment. *Journal of Experimental Biology*, **170**, 159-177.

Manzenrieder, H., 1983. Retardation of initial erosion under biological effects in sandy tidal flats. *Leichtweiss, Inst. Tech. University Braunschweig*. 469-479.

Meadows, A., Meadows, P.S., Muir-Wood, D., & Murray, J. M. H. 1994. Microbiological effects on slope stability: an experimental analysis. *Sedimentology*, **41**, 423-435.

Meadows, P.S. & Meadows, A. 1991. *The environmental impact of burrows and burrowing animals*. Proceedings of a Symposium of the Zoological Society of London, Clarendon Press, Oxford.

Meadows, P.S. & Tait, J., 1985. Bioturbation, geotechnics and microbiology at the sediment-water interface in deep-sea sediments. *Proceedings of the 19th European Marine Biology Symposium*. Gibbs, P.E.(ed). Cambridge University Press, p 191-199.

Meadows, P.S. & Tait, J., 1989. Modification of sediment permeability and shear strength by two burrowing invertebrates. *Marine Biology*, **101**, 75-82.

Meadows, P.S., Tait, J. & Hussain, S.A., 1991. Effects of estuarine infauna on sediment stability and particle sedimentation. *Hydrobiologia*, **190**, 263-266.

Miller, M.C., McCave, I.N. & Komar, P.D., 1977. Threshold of sediment motion under unidirectional currents. *Sedimentology*, **24**, 507-527.

Montague, C.L., 1986. Influence of biota on the erodibility of sediments. In *Lecture notes on coastal and estuarine studies*. Mehta, A.J. (ed). **14**, 251-269.

Neumann, A.C., Gebelein, C.D. & Mehta, A.J., 1970. The composition, structure and erodibility of sub-tidal mats, Abaco, Bahamas. *Journal of Sedimentary Petrology*, **40**, 274-297.

Nowell, A.R.M. & Jumars, P.A., 1984. Flow environments of aquatic benthos. *Annual Review of Ecology and Systematics*, **15**, 303-28.

Nowell, A.R.M., Jumars, P.A. & Eckman, J.E., 1981. Effects of biological activity on the entrainment of marine sediments. *Marine Geology*, **42**, 133-153.

Parchure, T.M., 1984. *Erosional behaviour of deposited cohesive sediments*. Ph.D. Thesis, Coastal & Oceanographic Engineering Department, University of Florida., USA.

Paterson D.M. & Daborn, G. R., 1991. Sediment stabilisation by biological action: Significance for coastal engineering. In *Developments in coastal engineering*. Peregrine, D.H. & Loveless, J.H. (eds). University of Bristol Press. p 111-119.

Paterson, D.M. 1990. The influence of epipelic diatoms on the erodibility of an artificial sediment. In *Proceedings of the 10th International Symposium on Living and Fossil Diatoms, 1988.* Simola, H. (ed). Joensuu, Finland p. 345-355.

Paterson, D.M., 1989. Short-term changes in the erodibility of intertidal cohesive sediments related to the migratory behaviour of epipelic diatoms. *Limnology and Oceanography*, **34(1)**, 223-234.

Paterson, D.M., Yallop, M.L. & George, C. 1994. Spatial variability in sediment erodibility on the island of Texel. In *Biostabilisation of sediments*. Krumbein, W.E., Paterson, D.M. & Stal , L.J. (eds), Oldenburg University Press, Germany.

Paterson, D.M. 1995. The biogenic structure of early sediment fabric visualised by low-temperature scanning electron microscopy. *Journal of the Geological Society, 152, 131-140*.

Paterson, D.M. 1994. Microbiological mediation of sediment structure and behaviour.    In *Microbial Mats* NATO ASI, Vol G 35. Caumette, P. & Stal, L.J. (eds), Springer-Verlag, pp 97-109.

Rhoads, D.C., Yingst, J.Y. & Ullman, W.J., 1978.  Seafloor stability in central Long Island sound. Part 1. Temporal changes in Erodibility of fine-grained sediment. In *Estuarine Interactions* Wiley, M.L. (ed). Academic Press, p 221-244.

Rosenberg, E., Rosenberg, M., Shoham, Y., Kaplan, N. & Sar, N., 1989. Adhesion and desorption during the growth of *Acinetobacter calcoaceticus* on hydrocarbons.  In *Microbial Mats: Physiological ecology of benthic microbial communities* . Cohen Y. & Rosenberg, E. (eds). American Society for Microbiology, Washington, D.C. p 219-227.

Scoffin, T.P., 1970. The trapping and binding of subtidal carbonate sediments by marine vegetation in Bimini Lagoon, Bahamas. *Journal of Sedimentary Petrology, 40,* 249-273.

Underwood, G.J.C. & Paterson, D.M., 1993a.  Recovery of intertidal benthic diatoms after biocide treatment and associated sediment dynamics. *Journal marine biological Association, U.K., 73,* 25-45.

Underwood, G.J.C. & Paterson, D.M., 1993b. Seasonal changes in diatom biomass, sediment stability and biogenic stabilisation in the Severn Estuary. *Journal marine biological Association, U.K., 73,* 871-887.

van Eerdt, M.M., 1985. The influence of vegetation on erosion and accretion in salt marshes of the Oosterschelde, The Netherlands. *Vegetatio, 62,* 367-373.

Vogel, S. 1981. *Life in moving fluids*. Willard Grant Press, Boston.

Vos, P.C., de Boer, P.L. & Misdorp, R., 1988. Sediment stabilization by benthic diatoms in intertidal sandy shoals; Qualitative and quantitative observations. In *Tide-Influenced Sedimentary Environments and Facies* ,De Boer, P.L., Van Gelder, A. & Nio, S.D., Eds. 511-526.

Westall, F. & Rince, Y. 1994. Biofilms, microbial mats and microbe-particle interactions: electron microscope observations from diatomaceous sediments. *Sedimentology, 41,* 147-162.

Williamson, H.J. & Ockenden, M.C. 1996. ISIS: an instrument for measuring erosion shear stress in situ, *Estuarine, Coastal and Shelf Science,* **42:** 1-18.

Yallop, M.L., de Winder, B., Paterson, D.M. & Stal, L.J.,   Comparative structure, primary production and biogenic stabilisation of cohesive and non-cohesive marine sediments inhabited by microphytobenthos. *Estuarine, Coastal and Shelf Science.* 39: 565-582.

# 15 Microbiological factors contributing the erosion resistance in natural cohesive sediments

## K. S. BLACK

*Gatty Marine Laboratory, University of St Andrews, UK*

## INTRODUCTION

Deposited estuarine muds are a complex, multi-component system. They consist of closely spaced silt and clay grains embedded within an amorphous matrix of detrital and bonded organic material, and are usually associated with a variety of benthic biological communities (Montague, 1986). Traditionally (e.g. Raudkivi, 1991) the cohesive strength of these sediments, which governs directly the erosion resistance, has been solely attributed to electrostatic interactions between the charged clay flakes. However, there is an increasing body of research to suggest that cohesive strength may also be a function of organic factors, and in some cases may be dominated by organic factors (Paterson, this volume).

Young and Southard (1978) first reported a positive relationship between the organic content of sub-tidal muds and the critical shear velocity necessary for sediment resuspension ($U\cdot_{crit}$). Since then a variety of manipulative laboratory experiments on both sands and muds (Paulic *et al.*, 1986; Grant *et al.*, 1986; Dade *et al.*, 1990; Montague *et al.*, 1992; Madsen *et al.*, 1993), in addition to a number of field studies (de Boer, 1981; Daborn *et al.*, 1993; Paterson *et al.*, 1990, Underwood and Paterson, 1993 a,b), have demonstrated a similar dependance. Paterson and Daborn (1991) provide a useful review of this and other work and cite a number of studies specifically on microbial mediation of sediment resuspension. Organic matter in many of these cases appears to act as a "glue" between mineral particles, and the cohesive strength is to a greater or lesser extent a function of the non-physical characteristics of the bed.

The present study was undertaken to investigate the potential influence of naturally occuring organic material on the resuspension potential of an estuarine mud. As anchoring of mineral grains to the bed by organics relates to the sediment structural properties (Paterson, 1995), disturbance through sampling is likely to disrupt the organic-mineral bonds. Therefore, experiments were performed directly on undisturbed inter-tidal muds in the field. This was achieved using a small re-circulating flume capable of remote deployment (Black and Cramp, 1995).

*Cohesive Sediments.*
Edited by Neville Burt, Reg Parker and Jacqueline Watts
© 1997 John Wiley & Sons Ltd.

## METHODOLOGY

### In situ techniques

The field flume is shown in Figure 1. It is a small, race-way shaped channel able to be emplaced directly on subaerially exposed muds. Ambient seawater, dribbled slowly into the channel during flume filling, is recirculated within a 0.10 m wide channel at a series of pre-determined speeds over an area of exposed mud (0.02 m$^2$). Erosion is assessed from the time-evolution of suspended sediment in the flume water measured using an optical backscatter turbidity sensor (Downing, 1983). Estimates of the boundary shear stress at the mud surface were obtained from laboratory-based, spatially distributed measurements of the vertical velocity structure in the channel using a laser doppler anemometer. Shear stresses up to 0.6 Nm$^{-2}$ can be imposed on the sediment surface using the flume. Further details of the flume are contained in Black and Cramp (1995).

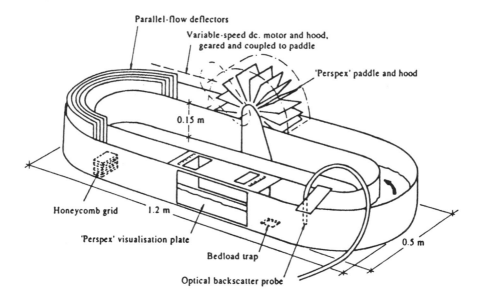

**Figure 1** The portable, *in situ* flume. The area of exposed mud is between the perspex windows. The motor and drive are not shown.

The initial erosion rate ($\varepsilon_i$) was computed from the differential sediment concentration over the first one minute of erosion under a constant bed stress of 0.25 Nm$^{-2}$. Peak erosion rates ($\varepsilon_p$) correspond to the maximum erosion rate measured under this bed stress held constant for ten minutes. The surface critical erosion stress (CES) was computed from the intercept on the x-axis of a plot of bed shear stress (x) against $\varepsilon_i$ (y) i.e. the point corresponding to an erosion rate of zero.

The flume was deployed at nine locations over a 2-day period during June, 1990, along the upper inter-tidal region of a small estuarine mudflat at Carew, S. Wales, U.K. (Fig. 2; median dispersed grain diameter ~12μm). Nine smooth, planar areas were selected for experimentation. Six of these were natural, undisturbed areas of mud as far as possible devoid of surface macrofauna and their tracks and trails (control plots). The remaining three were each chemically treated with either 500ml of 1 Molar copper sulphate ($CuSO_4$), hydrogen peroxide ($H_2O_2$, 30%v/v) or mercuric chloride ($HgCl_2$, 1 M) using a household plant sprayer. These treatments are referred to as *biocides*. A single tide was allowed to inundate the mudflat to restore the chemistry of the porewater prior to an erosion test. A number of additional experiments were also performed using the flume as a conventional laboratory channel on remoulded, organic-free muds from the estuary. Organic-free muds were created by passing natural sediment through a series of acid digestions until the organic content was measured to be zero.

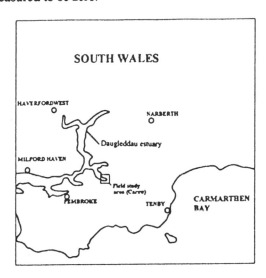

**Figure 2** Location map of study area, Carew estuary, S. Wales, UK.

*Sediment Sampling*
Measurement of a suite of physical and biological sediment properties were made alongside each deployment of the flume. Small brass micro-cores (8mm in diameter) were used to take samples which were then sliced into millimetre thick laminae. Grain size of non-dispersed sediment was measured carefully using a Malvern® Laser Particle Sizer. Water content and bulk density measurements were calculated from oven dried samples. Standard methods were used to determine sediment chlorophyll *a*, colloid carbohydrate and bulk organic content. The units for carbohydrate are glucose equivalents (GE) per gram dry sediment. Sediment ATP content was measured on muds pre-sieved through a 0.5 mm steel gauze using the luciferin-luciferase method of Bancroft *et al.*, (1976) with modifications by Bulleid (1976). Five replicate samples were taken for each property for each flume deployment. Results reported here are from the uppermost millimetre of the sediment only.

## RESULTS AND DISCUSSION

*Erosion Rate Time-Series*

Smoothed profiles of the response of selected sediment beds to an imposed bed stress of 0.25 $Nm^{-2}$ are shown in Figure 3. Qualitatively, both treated (+ biocide) and untreated (control) sediment beds appear to behave in the same fashion - each bed erodes rapidly at first, up to a peak erosion rate ($\varepsilon_p$), and then after ~1 minute the erosion rate slows, eventually to a negligible rate (~$10^{-6}$ $kgm^{-2}s^{-1}$) after ten minutes. This asymptotic erosion pattern corresponds to 'Type I' erosion as described by Mehta and Partheniades (1982), and has been widely reported from laboratory studies (e.g Johansen *et al.*, this volume; Spork *et al.*, this volume) and other field areas (e.g. Amos *et al.*, 1992). In quantitative terms, however, it is clear that there is a difference between biocide treated areas and control plots. The influence of the chemical treatments in each case is to weaken the bed with the result that peak erosion rates are substantially higher and more sediment is ultimately entrained. For example, $\varepsilon_p$ is 7.78x$10^{-4}$ $kgm^{-2}s^{-1}$ for sediments treated with copper sulphate, which is approximately 6-7 times the corresponding rates for control muds (~1.10x$10^{-4}$ $kgm^{-2}s^{-1}$), and around three times as much sediment is scoured into suspension (9.07 g vs. 6.31 g). Copper sulphate appears to have the greatest impact resulting in the most serious erosion. Hydrogen peroxide and mercuric chloride both increase sediment erosion in relation to control plots, but to a lesser degree ($\varepsilon_p$~2x$10^{-4}$ $kgm^{-2}s^{-1}$).

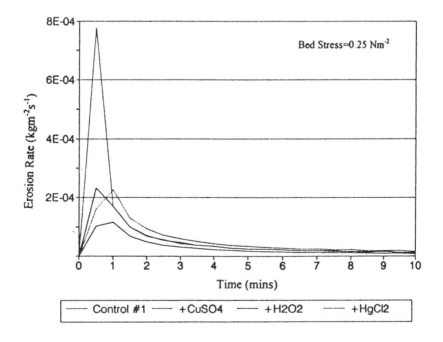

**Figure 3** Erosion through time profiles of biocide treated beds under a bed shear stress of 0.25 $Nm^{-2}$. The profile of control bed #1 (lowermost trace) is included for comparison.

*The Critical Erosion Stress (CES)*

The critical erosion stress corresponds to the point at which bed sediments are entrained into suspension by the fluid flow. A mean value for the CES from the undisturbed control plots of mud is 0.13 $Nm^{-2}$. The effect of application of each of the chemical treatments is to reduce the critical stress by a single order of magnitude (Table I), although the effect of hydrogen peroxide is less marked. Critical erosion stresses of muds treated with mercuric chloride and copper sulphate are in fact close to the predicted value for non-cohesive quartz grains of approximately the same median diameter ($\sim$0.02-0.03 $Nm^{-2}$; Mantz, 1977), and fall either side of the result for organic-free Carew muds of similar density tested in the laboratory ('abiotic muds untreated', CES 0.021 $Nm^{-2}$, Table I).

**Table I** Critical erosion stress and wet bulk density of experimental beds.

| | Critical Erosion Stress ($Nm^{-2}$) | Wet Bulk Density ($kgm^{-3}$) |
|---|---|---|
| *Field Experiments* | | |
| **Control (mean value, n=6)** | 0.130 | 1570 |
| +HgCl$_2$ | 0.016 | 1370 |
| +H$_2$O$_2$ | 0.091 | 1380 |
| +CuSO$_4$ | 0.029 | 1350 |
| *Laboratory Experiments* | | |
| Abiotic Muds - untreated* | 0.021 | 1525 |
| Abiotic Muds+HgCl$_2$ | 0.020 | -- |
| Abiotic Muds+H$_2$O$_2$ | 0.018 | -- |
| Abiotic Muds+CuSO$_4$ | 0.023 | -- |

*Not treated with biocides but organic-free with saline porewater chemical environment restored.

Additional laboratory erosion experiments using each of the biocides on reformed, organic-free muds were conducted in order to establish whether the observed reduction in CES in field experiments upon biocide application may simply be due to interference to the electrochemical grain-grain cohesive bonds. However, very similar values for the CES between the untreated abiotic sediment (which is simply an organic-free sample of estuary mud reformed to approximately the *in situ* density and with a saline porewater chemical environment), and the same sediment with each of the biocide treatments (Table I) indicates that the biocides do not measurably influence electrochemical sediment cohesion (although they can in cases interfere with the chemical assays used to measure sediment biological properties).One might expect, for instance, a lowering of the CES were the electrochemical cohesion chemically diminished by the biocides, leading perhaps to the generation of a very fluid, weakly structured bed.

The apparent increase in erosion upon application of biocides has been observed elsewhere (e.g. de Boer, 1981; Daborn et al., 1993; Underwood and Paterson, 1993a). The reduction in CES values from the Carew estuary are similar to comparable studies reported in the literature (see Paterson and Daborn, 1991), although percentage increases in $\varepsilon_p$ (600-700%) are considerably higher than previous studies (e.g. Parchure, 1980; Paulic et al., 1986). It is perhaps dangerous to describe a hierarchy of biocide effectiveness with respect to the CES data, since differences in effectiveness could be due to heterogeneity between experimental areas, which is unknown. However, clearly copper sulphate is seen to be the most effective in increasing sediment erosion, followed by mercuric chloride and then hydrogen peroxide. The limited effect of hydrogen peroxide, in spite of its known strong oxidising properties, is ascribed to significant consumption by detrital and macroscopic organic particles distributed over the mudflat surface. Little peroxide is then be available to oxidise the important (to cohesive strength) organic-mineral bonds.

A physical explanation for the observed decreases in erosion resistance is perhaps possible. Direct micro-scale measurements on the surface millimetre layer reveal a small reduction in the wet bulk density for those sites treated with biocides (Table I). This would contribute to both a lowering of the CES and an increase in $\varepsilon_p$ (Thorn and Parsons, 1980). However, whilst this mechanism may account for the smaller reduction in the CES for mud treated with hydrogen peroxide, it is not regarded to be of sufficient magnitude to produce the greater erosion recorded for beds exposed to copper sulphate and mercuric chloride. Supplementary winter-time experiments in the Carew reveal that reductions in bulk density of this magnitude would reduce the CES by less than ca. 20% (Andrews, 1990). Measurements of particle size reveal the most dramatic change in the physical structure of the sediment. These may explain the observations.

Particle size analysis of treated muds reveal that the surficial flocs have undergone gross disaggregation (Table II). The non-dispersed median grain diameters of biocide sites are only slightly greater than that of experimentally produced organic-free muds from the estuary (12.11 μm).

Table II. Biological and physical properties of the surficial sediment layer (top-most millimetre only): comparison of control, biocide and laboratory sediments.

| | Grain Size Percentiles (μm)[*] | | | Chl. a $\mu gg^{-1}$ | Colloid CHO[**] $\mu gGEg^{-1}$ | ATP $\mu gg^{-1}$ |
|---|---|---|---|---|---|---|
| | $\varphi_{10}$ | $\varphi_{50}$ | $\varphi_{90}$ | | | |
| *Field Experiments* | | | | | | |
| **Control #1** | **8.92** | **131.02** | **390.62** | **128.6** | **353.53** | **4.77** |
| +HgCl$_2$ | 1.96 | 15.71 | 56.80 | 6.08 | 00.00 | 0.24 |
| +H$_2$O$_2$ | 1.75 | 31.77 | 192.11 | 44.50 | 27.20 | 0.76 |
| +CuSO$_4$ | 4.99 | 19.45 | 80.71 | 8.01 | 00.00 | 0.09 |
| *Laboratory Experiment* | | | | | | |
| Abiotic Muds | 3.84 | 12.11 | 72.31 | 0.00 | 0.00 | 0.00 |

*Measurements on non-dispersed samples; **Colloid carbohydrate

A preliminary interpretation of this disaggregation might simply be that each of the treatments exert their effect by directly oxidising the organic-mineral bonds. However, whilst this may in theory be the case for hydrogen peroxide (notwithstanding the earlier observation regarding peroxide consumption by surficial detritus), it is unlikely that mercuric chloride and the copper sulphate achieve their effects through this mechanism. These chemicals are primarily metabolic poisons, and their effectiveness in the doses prescribed in this study has been firmly established in sediment trap studies (e.g. Lee *et al.*, 1992). The laboratory experiments conducted here have shown the biocides themselves do not increase sediment erosion through interference with electro-chemical cohesion (Table I). It is suggested, therefore, that this disaggregation can only have been produced by direct lethal action of the biocides on the sediment micro-biota. There is a marked reduction in the levels of chlorophyll *a*, colloid carbohydrate and ATP within the treated areas (Table II), which suggests that this has indeed happened. Upon their death, therefore, the binding influence of the micro-organisms is eliminated, the floc structure falls apart, and the resistance to erosion decreases substantially. This conclusion suggests that metabolically active micro-organisms are essential to biological stabilisation of estuarine muds i.e. it is not merely a function of bridging by organic mucus. The data support the findings of Daborn *et al.*, (1993) who reported lower rates of production of bio-cohesion upon treatment of muds with a photosynthetic inhibitor (DCMU) during the Canadian LISP program. These results demonstrate that in the absence of the binding influence of the micro-biota, these sediments may be washed away by even moderate tidal currents.

Erosion resistance for Carew muds is therefore envisaged to be a combination of primarily biological and secondarily physical factors, in which the aggregating influence of benthic micro-organisms is central to the ability of the mudflat to withstand shear stresses imposed by tidal flows. Obviously there will be periods when intense physical disturbance e.g. winter storms, during which even the strongest biological binding will be obliterated. Biogenic stabilisation is thus likely to be seasonally variable.

## Biological Indicators of Erosion Resistance

In an effort to understand whether natural variations in the biological component of naturally occurring cohesive muds exert any control over the erosion resistance, measurements of sediment ATP, chlorophyll $a$, colloid carbohydrate and bulk organic content from the surface millimetre layer were plotted against the CES as the dependant variable. Although researchers commonly use such an approach (e.g. Grant and Gust, 1987) it is not strictly valid as there is no control over the independent variable, as there is in laboratory experiments. Nevertheless, some interesting trends emerge.

Figure 4a-d shows the relationship of each of the four biological indices to the CES.

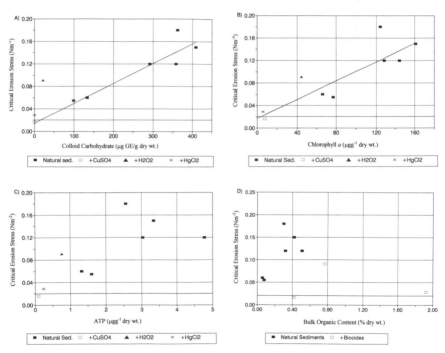

**Figure 4** Dependence of selected biological sedimentary variables on the critical erosion threshold (CES) a) colloid carbohydrate ($r^2$=0.91, P<0.001) b) chlorophyll $a$ ($r^2$=0.72, P<0.05) c) ATP ($r^2$=0.432) d) bulk organic content. The solid horizontal line represents the CES measured in a laboratory experiment performed on remoulded, organic-free muds (abiotic control). The CES for abiotic quartz sediments of approximately the same median diameter as Carew muds established by Mantz (1977) is 0.025 Nm$^{-2}$.

Significant linear best fit equations may be applied to the colloid carbohydrate chlorophyll *a* (Fig. 4a and b, ). Values of the correlation coefficient ($r^2$) are 0.91 and 0.72, respectively. Those muds with little or no carbohydrate and chlorophyll *a* have depressed CES values, which are close to that of the abiotic laboratory experiment (0.021 $Nm^{-2}$, Table I).

The correlation of the CES with sedimentary chlorophyll is intriguing and implies that the cohesive strength of surficial floc aggregates may in some way be dependant on the presence, or absence, of microscopic microalgae (diatoms, euglenoids, cyanobacteria). Similar conclusions have been reported from other field and laboratory studies (e.g. Montague *et al.*, 1992; Madsen *et al.*, 1993; Underwood and Paterson, 1993b; Paterson *et al.*, 1994) in which the same chemical assays but different methods of measuring erosion resistance have been used. These studies have pointed to the importance of diatoms, in particular. The confirmation of a statistically significant monoparameter correlation with chlorophyll from the natural mudflats of the Carew indicates that at least one of these micro-organism classes may be a predominating influence on the cohesive strength of the bed. It is not unreasonable to presume that the mineral particles alone in the Carew give rise to a sediment which is only weakly cohesive, since CES values for organic-free muds are very close to the expected value for non-cohesive particles of similar diameter (Table I). Perhaps, then, the observed relatively high cohesion in the natural mudflats of the Carew is due, in large part, to the presence of micro-orgaisms. Such a situation, in fact, exists on the high shore region in the Bay of Fundy, wherein Paterson (1991) demonstrated quite clearly how removal of the indigenous micro-biota leaves essentially a rippled, non-cohesive silt-flat.

The significant correlation also with colloidal carbohydrate, which is a rudimentary indicator of microbial extracellular polymeric substances (EPS, see Paterson, this volume), lends further support to the suggestion that benthic micro-organisms are contributing in some fashion to the cohesive strength of the sediments in the Carew. It is becoming more widely appreciated that micro-organisms, particularly on tidal flats during mid-day low tides, can influence the cohesiveness, and hence the stability, of a variety of natural sediments through secretion of adhesive EPS (Paterson *et al.*, 1990; Paterson and Daborn, 1991). Indeed, as noted by Hobbie and Lee (1980), the extracellular mucopolysaccharides of microbes are frequently more abundant than the microbes themselves.

Preliminary visual evidence from the Carew, using a specialised low-temperature scanning electron microscope (see Paterson, 1986), reveals that many of the sediment particles throughout the upper millimetres of the sediment are indeed encapsulated and interconnected by extensive sheets and threads of EPS, and that the EPS is secreted in the main by diatoms (Black, 1991, p.142). The relationship suggested here between sediment-inhabiting diatoms and EPS is further substantiated by the strong co-variance measured between chlorophyll and EPS ($r^2=0.95$) in 35 randomly selected surface samples from the mudflat (Black, 1993, Fig. 5). Thus, the binding influence of benthic diatoms in particular, and the validity of using chlorophyll *a* and colloid carbohydrate as indices of sediment erosion resistance, is becoming more firmly established.

ATP (adenosine triphosphate) is an indicator of the total living biomass in the sediments. Natural levels of sub-0.5mm ATP display a positive but non-significant relation to the CES ($r^2$=0.432; Fig. 4c). The ratio of chlorophyll $a$ to total ATP is variable (between 25 and 88) and it is possible (in spite of the effort to exclude meiofauna) that non-microbial ATP was measured by the assay. This need not necessarily bear a positive relationship to sediment strength, although some faunal groups are known to secrete a binding polysaccharide mucus (e.g. nematodes; Krumbein et al., 1993).

Controlled laboratory experiments by Rhoads et al., (1978) have demonstrated the utility of ATP as a biological indicator of microbial binding. However, the results obtained from the Carew estuary (i.e. for entirely undisturbed, naturally deposited muds) are less conclusive, and ATP is not considered as reliable an indicator of sediment erosion resistance as either chlorophyll $a$ or colloid carbohydrate. The contrast here with the work of Rhoads et al. (op. cit.) is likely associated with differences in microbial homogeneity of Carew (natural) muds in comparison to Rhoads et al's artificial microbe cultures and uniform glass bead sediments. Further, consistently higher values of the coefficient of variation (100*standard deviation/mean value) for ATP at each deployment of the flume (55-81% versus <45% for chlorophyll $a$ and colloid carbohydrate) indicate that total microbial biomass may be more patchily distributed than either chlorophyll or colloid carbohydrate on the mudflat surface (Black, 1991). In order to use ATP in this manner, therefore, a more extensive sampling strategy is required. Paulic et al., (1986) reported similar problems to those encountered in the Carew estuary when working with natural muds, highlighting patchiness of total biomass as a confounding influence during their experiments.

It is interesting that the bulk organic content also correlates poorly with the CES (Fig. 4d), particularly since bulk organic content has been reported to co-vary in a positive fashion with the CES by some previous researchers (e.g. Harrison and Owen, 1971; Young and Southard, 1978). However, it appears from the work here that specific indicators relating to the primary mucus (EPS) producing groups rather than blanket indicators of organic mass like ATP and bulk organic content are better suited as predictors of erosion resistance. The recent work of Dade et al., (1990) addressed this area, but their experiments would undoubtedly be very difficult to perform under in situ conditions.

## CONCLUSIONS

The results presented document a reasonably strong case for the phenomenon of biogenic adhesion within natural estuarine muds as a predominating influence over classical electro-chemical particle-particle interactions. Micro-organisms, in particular benthic diatoms in the Carew estuary, contribute to bed stability by increasing the strength of inter-floc bonds through the secretion of an adhesive, carbohydrate-rich mucus. Addition of chemical treatments reverses these effects causing dissociation of surficial floc aggregates with ensuing increases in erosion rate and concomitant decreases in the fluid stress necessary for sediment entrainment. It is implied herein that the effect of the chemical treatments is via a lethal effect on the benthic micro-organisms rather that via direct oxidation of organic-mineral bonds. The structural integrity of floc aggregates, and the natural *in situ* microbial characteristics, are thus fundamentally important to the erosion resistance, and hence continued existence, of naturally formed mudflats.

## ACKNOWLEDGEMENTS

This work was conducted while the author was in receipt of a University of Wales postgraduate studentship. The support of Dr. Adrian Cramp and John McEvoy during this period is gratefully acknowledged. The author would like to thank the reviewers anonymously for their helpful criticism.

## REFERENCES

Amos, C.L., Daborn, G.R., Christian, H.A., Atkinson, A., and Robertson, A., 1992 *In situ* erosion measurements on fine-grained sediments from the Bay of Fundy. Marine Geology 108:175-196.

Andrews, S., 1990 On the stability of cohesive sediments in the Carew estuary, S.W. Wales. Unpublished Bsc. Thesis, University of Wales, Swansea, 34pp.

Black, K.S., 1991 The erosion characteristics of cohesive estuarine sediments: some in situ experiments and observations. Unpub. PhD Thesis, University of Wales.

Black, K.S., 1993 The turbulent resuspension of cohesive inter-tidal muds. In (Sterr, H., Hofstede, J., and Plag, H-P., Eds.) Proceedings of the First International Coastal Congress, Kiel, Germany, p.223-239.

Black, K.S., and Cramp, A.C., 1995 A device to examine the *in situ* response of intertidal cohesive sediment deposits to fluid shear. Continental Shelf Research 15(15):1945-1954.

Bancroft, K., Paul, E.A., Wiebe, W.J., 1976 The extraction and measurement of ATP from marine sediments. Limnology and Oceanography 21:473-480.

Bulleid, N.C., 1976 An improved method for the extraction of ATP from marine sediment and water. Limnology and Oceanography 23:174-178.

Daborn, G.R., Amos, C.L., Brylinsky, M., Christian, H., Drapeau, G., Faas, R.W., Grant, J., Long, B., Paterson, D.M., Perillo, G.M.E., and Piccolo, M.C., 1993 An ecological cascade effect: migratory birds influence stability of intertidal sediments. Limnology and Oceanography **38**:225-231.

Dade, W.B., Davis, J.D., Nichols, P.D., Nowell, A.R.M., Thistle, D., Trexler, M.B., and White, D.C., 1990 Effects of bacterial exopolymer adhesion on the entrainment of sand. Geomicrobiological Journal **8**: 1-16.

De Boer, P.L., 1981 Mechanical efects of micro-organisms on intertidal bedform migration. Sedimentology **28**:129-132.

Downing, J., 1983 An optical instrument for monitoring suspended particulates in the ocean and laboratory. In Proceedings of 'Oceans 1983', pp.199-202.

Grant, J., Bathmann, U.V., arid Mills, E.L., 1986 The interaction between benthic diatom films and sediment transport. Estuarine Coastal Shelf Science **23**:225-238.

Grant, J., and Gust, G., 1987 Prediction of coastal sediment stability from photopigment content of mats of purple sulphur bacteria. Nature **330**:244-246.

Harrison, A.J.M., and Owen, M.W., 1971 Siltation of fine sediments in estuaries. Proc. 14 Congress of IAHR, Vol. 4, Paper D1.

Hobbie, J.E., and Lee, C., 1980 Microbial production of extracellular material: importance in benthic ecology. In: (Tenore, K.R., and Coull, B.C., Eds.) Marine Benthic Dynamics, pp.341-346, Univ. South Carolina Press, Columbia.

Krumbein, W.E, Paterson, D.M., Stal, L.J., and Wippermann, T., 1993 Microbially mediated processes in tide-influenced deposits and their importance in stabilisation and diagenesis of sediments. In (Barthel *et al.*, Eds.) MAST Days and EUROMAR MArket, 15-17 March, Project Reports, Volume 1, pp.242-259.

Lee, C., Hedges, J.I., Wakeham, S.G., and Zhu, N., 1992 Effectiveness of various treatments in retarding microbial activity in sediment trap material. Limnology and Oceanography **37**:117-127.

Madsen, N.P., Nillson, P., and Sundbacl, K., 1993 The influence of benthic microalgae and the stabilisation of a subtidal sediment. Journal of Experimental Marine Biology and Ecology **170**:159-178.

Mantz, P.A., 1977 Incipient transport of fine grains and flakes by fluids - an extended Shields diagram. ASCE Journal of the Hydraulics Division **103**:601-615.

Mehta, A.J., and Partheniades, E. 1982 Resuspension of deposited cohesive sediment beds. 18th. Conference Coastal Engineering, pp.:1569-1588.

Montague, C.L., 1986 Influence of biota on the erodibility of sediments. In (Mehta, A.J., Ed.) Estuarine Cohesive Sediment Dynamics, Springer-Verlag, Berlin, pp.251-268.

Montague, C.L., Parchure, T., and Paulic, M.J., 1992 The stability of sediments containing microbial communities: initial experiments with varying light intensity. In (Mehta, A.J., Ed.) Nearshore and Estuarine Cohesive Sediment Transport (Springer-Verlag, New York.

Paterson, D.M., 1986 The migratory behaviour of diatom assemblages in a laboratory tidal micro-ecosystem examined by low-temperature scanning electron microscopy. Diatom Res. 1:227-239.

Paterson, D.M.,1989 Short term changes in the erodibility of intertidal cohesive sediments related to the migratory behaviour of epipelic diatoms. Limnology Oceanography 34: 223-234.

Paterson, D.M., 1991 S3.8b biological effects on surface cohesion. In: (Daborn et al., Eds.) Littoral Investigation of Sediment Properties, Minas Basin, Bay Of Fundy, compiled Final Report. Acadia Centre for Estuarine Research, Publ. No. 17, p.183-184.

Paterson, D.M., 1995 Biogenic structure of early sediment fabric visualized by low-temperature scanning electron microscopy. J. Geological Society London 152:131-140.

Paterson, D.M., and Daborn, G.R., 1991 Sediment stabilisation by biological action: significance for coastal engineering. In (Peregrinc, D.H., and Lovelace, J.H., Eds.) Developments in Coastal Engineering. University of Bristol Press, pp111-119.

Paterson, D.M., Crawford, R.M., and Little, C., 1990 Subaerial exposure and changes in the stability of intertidal estuarine sediments. Estuarine Coastal Shelf Science 30:541-556.

Paterson, D.M., Yallop, M.L., and George, C., 1994 Spatial variability in sediment erodibility on the island of Texel. In (Krumbein, W.E., Paterson, D.M., and Stal, L.J., Eds.) Biostabilization of Sediments Verlag/Vertrieb, Oldenburg BIS, pp107-120.

Paulic, M.J., Montague, C.L., and Mehta, A.J., 1986 The influence of light on sediment erodibility. In (Shen, H.W., Ed.) 3rd International Symposium on River Sedimentation, University of Mississippi, pp1758-1764.

Raudkivi, A.J., 1991 Loose boundary Hydraulics, Third Edition, Pergamon.

Rhoads, D.C., Yingst, J.Y., and Ullman, W., 1978 Seafloor stability in central Long Island Sound. In (Wiley, M.L., Ed.) Estuarine Interactions, Academic Press.

Thorn, M.F.C., and Parsons, J. G., 1980 Erosion of cohesive sediments in estuaries: an engineering guide. Proc. 3rd International Symposium on Dredging Technology, Paper F1, Bordeaux, pp349-358.

Underwood, G.J.C., and Paterson, D.M., 1993a Recovery of intertidal benthic diatoms after biocide treatments and associated sediment dynamics. Journal of the Marine Biological Association, U.K., **73**:24-45.

Underwood, G.J.C., and Paterson, D.M., 1993b Seasonal changes in diatom biomass, sediment stability and biogenic stabilisation in the Severn Estuary. Journal of the Marine Biological Association, U.K., **73**:871-887.

Young, R.A., and Southard, J.B., 1978 Erosion of fine-grained marine sediments: seafloor and laboratory experiments. Geological Society America Bulletin **89**:663-67.

# 16 Erosion of mixed cohesive/non-cohesive sediments in uniform flow

**H. TORFS**

*Hydraulics Laboratory, Katholieke Universiteit Leuven, Belgium*

## INTRODUCTION

Natural muds are generally a mixture of sand (>63 μm) and fines (<63 μm). Depending on the mixture composition and the bed structure the sand fraction can play an important role in the erosional properties of the mud. So far investigations on coastal sediments have concentrated on cohesive sediments and sand separately because of their different physical properties. But, since the bottom sediments in estuaries are usually a mixture of different size fractions, it is important to understand the behaviour of those mixed sediments.

Therefore, in the framework of both a Belgian interuniversity project and a MAST2 project a laboratory investigation was started. The aim of the research was to study the effect of different parameters such as sand content, type of material, bulk density, bed formation, etc. on the erosional behaviour of mixed sediments. Hence, using different types of cohesive material (clay and natural mud) and a uniform fine sand, different mixtures at different densities were prepared as placed or deposited beds and they were eroded under uniform flow conditions.

## EXPERIMENTS

### Experimental flume

The experiments were carried out in a 9 m long straight, recirculating flume with a rectangular cross section of 40 by 40 cm. The 4 m long inflow section with a fixed bottom is followed by a 3 m measuring section containing the 8 cm thick sediment bed. Further downstream a sediment trap collects the bed load and a 1.5 m outflow section with fixed bed ensures an undisturbed flow. Immediately upstream and downstream of the measuring section the water level is measured with pressure gauges, and suspended load samples can be taken. Discharge is measured with an electromagnetic flow meter on the inflow pipe and velocities are monitored with a pitot tube connected to a differential pressure transducer. The weight of the sediment trap is continuously recorded by a load cell.

*Cohesive Sediments.*
Edited by Neville Burt, Reg Parker and Jacqueline Watts
©1997 John Wiley & Sons Ltd.

## Uniform placed beds

For the experiments with uniform mixed sediments the necessary amounts of sand, cohesive material and water are mechanically mixed for sufficiently long time so that a uniform mixture results. This mixture is placed in the flume, levelled with a steel plate and left to consolidate over night under a water layer. After the experiments a sediment sample is taken to determine the exact mixture composition, the bulk density and the water content.

## Deposited beds

For this set of experiments a 1 m high settling tank was constructed, which could be mounted on top of the flume in the measuring section. Again sand, cohesive material and water are mechanically mixed and this slurry is slowly pumped into the settling tank, which was previously filled with salt water (3 g/l). Simultaneously a proportional amount of slurry is poured into a settling column (1 m high, 10 cm diameter) also filled with salt water. This filling procedure is repeated twice a day (tidal situation) until a sediment depth of ± 8 cm is reached. In the perspex settling column the layer thickness can be measured and using a gamma-densimeter also the density profile is recorded.

## Sediments used

The sand used in all experiments is a uniform fine sand, $d_{50} = 0.23$ mm. For the cohesive fraction 4 different materials are used: a pottery clay, which is mainly montmorillonite, kaolinite (China Clay) and two types of mud from the river Scheldt at Antwerp, mud1 from below the low water level and mud2 from the intertidal zone, which contained a lot of organic matter including worms.

## Experimental procedure

During an erosion experiment step by step the discharge is increased until the erosion starts. Then the discharge is kept constant for a longer period. Continuously a computer measures water levels, point velocity, discharge and the weight of the sediment trap. Suspended load samples are taken every 15 minutes and are filtered afterwards to determine the concentration. This procedure is repeated for a few (higher) discharges. The experiment is stopped when the erosion becomes too massive and the destruction of the bed causes the flow to be no longer uniform.

From the difference in suspended load concentrations upstream and downstream and from the slope of the curve of the weight of the sediment trap versus time erosion rates will be calculated. The bed shear stress is calculated from the slope of the energy line, using a side wall correction method. More information on the experimental procedure and the set up can be found in Torfs (1995).

## DISCUSSION ON INCIPIENT MOTION AND EROSION RATES

From an extensive set of experiments the influence of certain parameters on the erosional properties of mixed sediments becomes clear.

### Mixture composition

Mixture composition includes both the amount and type of cohesive material. The amount of cohesive material in the mixture will be expressed as % fines, i.e. the percentage by weight smaller than 63 mm.

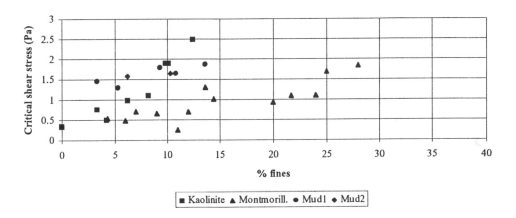

**Figure 1:** Incipient motion for different types of cohesive material.

Increasing the amount of cohesive material in the mixture increases the critical shear stress for erosion of the bed (see figure 1). A small percentage of cohesive material can enhance greatly the resistance to erosion of the sediment (Collins, 1989). In some cases however - e.g. between 7 and 13 % fines for montmorillonite in figure 1 - there is a transition zone in which the increase is less clear. This transition zone indicates a change in the mode of erosion from cohesionless at low % fines - with ripple and dune formation - to cohesive erosion at a higher clay content. For small amounts of cohesive material in the mixture, the fines will be washed out of the top layer at low discharges, the sand is transported in the bed forms near the bed. At higher % fines the sediment becomes cohesive and bonds between all particles exist. The clay particles adhere to the sand grains and the pore spaces in the sand are filled with the finer, adhering particles.

Depending on the type of cohesive material, for the same mixture composition, a difference in critical shear stress for erosion can exist. The kaolinite and the mud mixtures have critical shear stresses that are more or less the double of the values for montmorillonite (figure 1). The transition from cohesionless to cohesive behaviour also depends on the type of cohesive material. For the kaolinite mixtures only for the 3 % mixtures a cohesionless transport behaviour is noticed. This is illustrated in figure 2, for the same excess shear stress (= bed shear stress minus critical shear stress) the erosion rates for the 3 % mixture are an order of magnitude higher than the others. In the flume ripples and dunes were formed. For higher clay contents a totally different mode of erosion was found, which created a wavy surface.

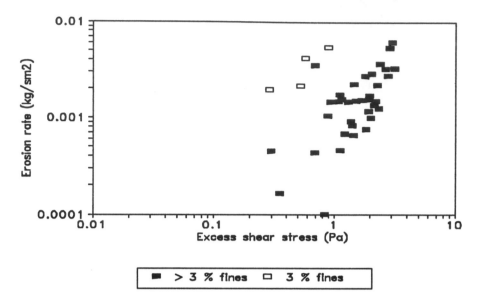

**Figure 2** : Erosion rates for the different kaolinite/sand mixtures.

The mode of cohesive erosion is also strongly dependent on the type of cohesive material used. For the kaolinite and the mud mixtures a surface erosion phenomenon takes place. Individual grains or flocs erode, the fines go into suspension and the sand fraction is mainly transported as bed load. At the same bed density from a sediment bed, which is a mixture of sand and montmorillonite, whole crumbs of material are being removed and transported as bed load (bulk erosion).

A possible explanation for the different modes of erosion could be armouring. For some mixtures or sediment types a form of armouring of the bed by the sand grains would limit the occurrence of surface erosion. One of the reasons for this difference could be the grain size distribution. The montmorillonite in itself already contains a sand fraction (> 63 μm), mixing with some extra sand results in a gradually varying sieve analysis curve. For the kaolinite mixtures the mixture contains in fact two separate fractions.

Bed density

Higher bed densities will increase the erosion resistance of the bed significantly. But also the mode of erosion will change. The low density mud slurries are more easily suspended and the relative importance of bed load will decrease.

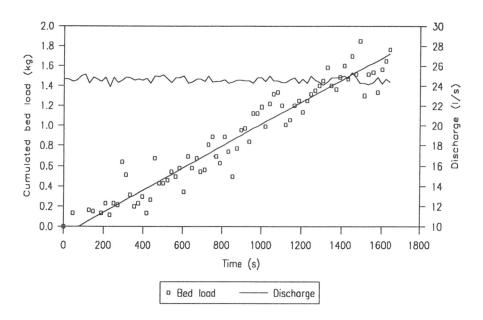

**Figure 3:** Erosion of a uniform mud1 mixture

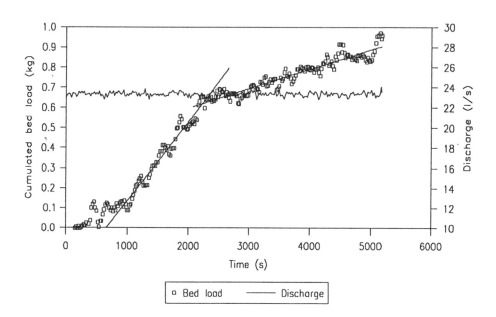

**Figure 4:** Erosion of a layered mud1 mixture.

249

Also the density gradients in the bed (density profile) have an important effect on the erosion. For the uniform beds at a constant discharge (constant bed shear stress higher than the critical value) the erosion rate will be constant as well. When there is an increase in bed density with depth, the erosion rate, at a constant bed shear stress, will decrease in time until a layer is reached of such high density that the applied shear stress is lower than the critical shear stress for that density; hence, the erosion will stop. This is illustrated in figures 3 and 4. Figure 3 gives the weight of the sediment trap as a function of time for an erosion experiment on a uniform mixture. The slope of that curve gives the bed load transport, which is seen to be constant. In figure 4 the results of an erosion experiment on a layered bed is given. It is clear that in this case the bed load transport is decreasing in time.

## Bed structure

Figure 5 gives the bed density profile, as it was measured in the settling column, before the experiment plotted in figure 4. The layered structure, which is the result of the consecutive inputs of the mixture, is clear. The high density peaks at the bottom of each layer indicate that the sand has settled more quickly than the mud fraction and that practically all the sand is found at the bottom of a layer. The previous layer already developed enough structural strength during the few hours in between two fillings to carry the next layer.

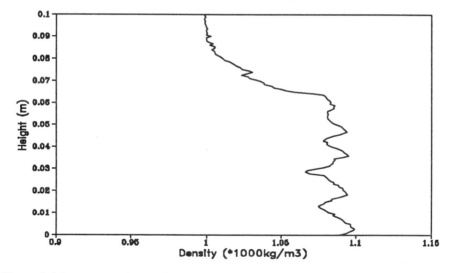

**Figure 5:** Measured density profile.

The presence of a high density sand layer depends on different factors: the percentage of sand in the mixture (a slurry can only contain a certain amount of sand within its matrix), the density of the mixture (for higher initial densities hindered settling occurs preventing the sand of falling through the mud), type of cohesive sediment, supply rate (see also Williamson et al 1992).

The sand content also determines the layer thickness, the peak density and the mean density of the bed (Torfs 1994). For one type of mud, the layer thickness decreases with increasing sand content, the peaks become higher and the mean density (after a fixed consolidation period) increases as well.

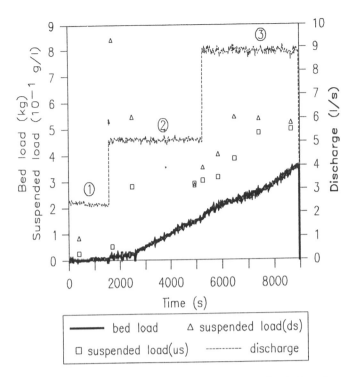

**Figure 6:** Overview of an erosion experiment for a layered sand/mud deposit.

During the erosion of a layered bed there is a continuous sequence of bed and suspended load erosion (see figure 6). At low discharges (1, figure 6) the soft top layer will be eroded and goes into suspension. No bed load is measured. The discharge is increased (2) and even more material is eroded and transported in suspension, i.e. there are large differences between the suspension concentrations measured upstream (us) and downstream (ds) of the sediment bed. After a while, at first upstream, a sand layer becomes available which will be eroded as bed load mainly. Suspended load erosion decreases. Increasing the discharge again (3), gradually the sand transport diminishes as the sand no longer covers the whole length of the bed. Again a soft mud layer is reached, which is suspended, ...

## CONCLUSIONS

The bottom sediments in estuaries are usually a mixture of mud and sand and it is that mixture that will be eroded during periods of high flow. Understanding the erosional properties of such mixtures is thus important. Many parameters influence the behaviour of mixed sediments. The clay content as well as the type of cohesive material determine the erosion resistance and the mode of transport. The bulk density of the bed has an effect on erosion resistance. Also, for high density materials the relative importance of suspended load transport versus bed load transport decreases. When the density increases with depth, erosion rates at constant flow conditions decrease in time. For certain combinations of suspension concentration, sand content, type of mud, etc. a layered bed will result after deposition at quiescent flow conditions. The erosion of these bottoms is a continuous sequence of bed load and suspended load erosion.

## ACKNOWLEDGEMENTS

The erosional behaviour of partly cohesive, partly non cohesive sediments in uniform flow is studied in the framework of an interuniversity research project, financed by the Belgian Fund for Collective Fundamental Research. The work was also a part of the G8 Coastal Morphodynamics programme and was partly funded by the Commission of the European Communities, Directorate General for Science, Research and Development, under contract no. MAS2-CT-92-0027.

## REFERENCES

Collins M. (1989) The behaviour of cohesive and non-cohesive sediments. Proceedings of the International Seminar on the Environmental Aspects of Dredging Activities, 27 November - 1 December, Nantes.

Torfs, H. (1994) Erosion of layered sand-mud beds in uniform flow. Proceedings of the 24th International Conference on Coastal Engineering, 23-28 October, Kobe, Japan, pp. 3360-3368.

Torfs, H. (1995) Erosion of mud/sand mixtures. PhD thesis, Katholieke Universiteit Leuven, Department of Civil Engineering.

Williamson, H.J. and Ockenden, M.C. (1992) Tidal transport of mud/sand mixtures, laboratory tests. HR Wallingford, Report SR 257.

# 17 Size-dependent erosion of two silty-clay sediment mixtures

**A. M. TEETER, T. M. PARCHURE and W. H. McANALLY JR**
*Waterways Experiment Station, Vicksburg, MS, USA*

## INTRODUCTION

Cohesive sediment erodibility varies widely from site to site which makes characterization and erosion testing a frequent requirement for practical studies of estuarine shoaling, dredged material dispersion from disposal sites, contaminated sediment mobility, etc. Characterization of test material allows correlation of results to other situations in some cases and can be helpful in qualitatively predicting erodibility. However, insufficient characterization data have been reported in earlier studies which often resulted in providing empirical correlations. Theoretical developments have not been advanced that show important causal relationships between characterization parameters and erodibility.

Most cohesive sediments in U.S. coastal waters are in the clay and very fine silt size range, often mixed with coarser silt. Previous erosion tests (Teeter and Pankow, 1989) have identified erodibility differences between grain size classes, but lacked detail on particle size changes. In this study, more thorough particle size analysis of the eroded suspensions was done to explore erosion characteristics of silty clay mixtures.

Standardized techniques have not been developed for conducting laboratory erosion tests on cohesive sediment. Results reported in literature often lack corresponding data on sediment characterization and/or an evaluation of the erosion test device. In addition to the erosion device used, variation in results could also occur due to procedure used for preparation of bed, method used for measuring erosion, spatial distribution of shear stress over the bed, and interpretation of experimental results. Comparative tests using different erosion devices and identical sediments as well as test procedure are not readily available, makes comparison between erosion test results quite difficult.

The present erosion tests were to establish an independent calibration for one of the test devices, test the possible dependence of calibration on sediment characteristics, determine test repeatability, and refine procedures for testing and analysis. The tests compared the erosion of identical sediments in two test devices and examined the dependence of erosion on particle size. Two erosion devices were used, a Particle Entrainment Simulator (*PES*) and a Vertical Loop Sediment Tunnel (*VOST*).

*Cohesive Sediments.*
Edited by Neville Burt, Reg Parker and Jacqueline Watts
©1997 John Wiley & Sons Ltd.

## TEST MATERIALS

A mixture of kaolinite and silt and a natural marine sediment were used. The commercial kaolinite had a Cation Exchange Capacity (CEC) of 14.4 meq/100g, and a median dispersed particle size of 0.0017 mm. The kaolinite tests used tap water (pH 8.5) which had 19.2 ppm C1, 8.6 ppm Ca, 18.1 ppm Nà, and 10.3 ppm Mg. It was noted that within the first few days after mixing the kaolinite with tap water, an appreciable fraction was filterable through a 0.0004 mm pore-size filter while after aging over six months the material became completely non-filterable. Prior to testing, the kaolinite was placed under tap water and periodically stirred for about three years to fully equilibrate it with tap water.

The silt (median size of 0.035 mm) in the sediment mix was separated from a loess soil local to the Mississippi area. The material was first passed through a 0.074 mm mesh sieve, mixed and washed with tap water, and then settled to remove clays. Sediment particle size distributions were determined using an electronic particle analyzer. The test mixture was prepared with 37 percent silt and 63 percent kaolinite by dry weight. The wet bulk density of the mixture was 1.36 g/cu cm. The marine mud was obtained from Tiger Pass, Louisiana, collected about 1.6 km from the open coast. The bulk wet density of the material was 1.20 g/cu cm. The CEC was 45 meq/100g.

For these tests, slurries of test sediments were prepared, mixed thoroughly, and poured into the sediment bed chambers. The beds were tilted to even out sediment thicknesses, but not trawled or physically molded. They were then sprayed with water to cover their surfaces and allowed to stand between 1 and 24 hours to settle and gel.

## *PES* TEST PROCEDURES

*PES* at the U.S. Army Engineer Waterways Experiment Station (WES) is a variation of the device described by Tsai and Lick (1986) and used extensively for testing undisturbed core sections (Lavelle and Davis 1987, Davis 1993). It has an erosion chamber geometry identical to that used by several other investigators and is intended primarily as a portable field tool for a quick assessment of erodibility of undisturbed 11.75 cm diameter cohesive sediment core sections. The mechanical aspects of the WES *PES* were slightly changed, but the dimensions of the erosion chamber and oscillating disk were true to the original. The *PES* test chamber is a vertical cylinder of 11.75 cm inside diameter and 25 cm height with a water column of 12.7 cm height over the sediment water interface. A perforated oscillating grid inside the vertical cylinder generates turbulence above the sediment to simulate shear stress. The oscillating disk is located at a minimum distance of 5.1 cm from the sediment bed and has an excursion of 2.54 cm. Additional details may be found in Tsai and Lick (1986). The mean bed shear stress, which normally correlatse erosion with a flow-induced bed shear stress, is not available as a directly measurable parameter in *PES*; therefore *PES* must be calibrated using another erosion device. Originally *PES* was calibrated by comparing test suspension concentrations to those from an annular flume. The original calibration was used to correlate oscillation rate to an equivalent shear stress.

A flow through nephelometer continuously monitored suspension concentrations. The nephelometer tubing increased the volume of the system to 1.4 L. Samples of 30-70 ml were withdrawn for filtration tests and electronic particle size analyses. During sampling, particle free water was introduced to keep the water level constant in the *PES* erosion chamber. Necessary corrections were made in the calculations to account for this dilution.

*PES* tests started with an initial 2 min period at 100 rpm to suspend any very loose sediment resulting from the bed preparation. This initial step was equivalent to about 0.1 Pa and was meant only to prepare the sediment bed. Equivalent shear stresses were increased step-wise during erosion tests and the concentration and particle size distributions of suspensions were monitored with time. Each shear stress equivalent to 0.2, 0.3, 0.4, and 0.5 Pa was maintained over a duration of 15 to 30 min.

## *VOST* TEST PROCEDURES

*VOST* is a smaller, higher-flow version of a sediment water tunnel consisting of two rectangular horizontal and two circular vertical sections arranged in a vertical plane (Teeter and Pankow(1989). Flow in the *VOST* is driven by a propeller pump in one of the two 15.24 cm diameter circular sections and the horizontal tunnel sections are 7.6 cm high by 24.1 cm wide. Test material is placed in a sample tray that has an area of 403 sq cm. The flow cross-sectional area averages 183 sq cm, the flow length around the *VOST* is 3.5 m and the volume of the system is 64 L. The propeller pump is 2.6 m upstream from the bed sediment sample tray. Flows in the *VOST* can reach up to 1.54 m/sec, generating a maximum average shear stress of almost 3 Pa.

The *VOST* shear stresses were calibrated by correlating propeller pump motor voltage and flow speed to measured shear stress. Calibration shear stresses were measured with a hot film sensor at nine locations over a clear acrylic plate positioned at the normal sediment bed level. The shear stress sensor was calibrated in a laminar flow duct at known flow rates and pressure drops. The spatial variation of shear stress observed over the sediment bed was such that standard deviations divided by the mean shear stresses were 27 percent for shear stresses less than 1.1 Pa and decreased to 17 percent at the maximum shear stress of about 3 Pa.

*VOST* tests were performed on placed sediment beds. The shear stresses were applied in steps, maintaining each stress for a duration of 15 to 30 minutes. Sediment suspensions samples were analyzed for suspended sediment concentration and particle size.

## SAMPLE ANALYSIS

Samples of sediment suspensions were analyzed to determine the mass of eroded sediment and the particle sizes eroded from bed surface. Samples were filtered through 0.45 micron polycarbonate Nuclepore© filters to determine total suspended material. Because the suspension volume of the *PES* is relatively small, the *PES* suspension concentrations were corrected for the mass withdrawn during sampling. During sampling, particle-free water was added to keep the water surface at a constant level. For a perfectly well-mixed chamber of constant volume:

$$V \frac{dC}{dt} = -C \, Q \tag{1}$$

$$C = Co \; \exp \; [-\frac{Vs}{V}] \tag{2}$$

where $C$ is suspension concentration, $t$ is time, $V$ is the chamber volume, and $Q$ is the sampling and replenishment rate. For an initial concentration $Co$, and sampling volume $Vs = Q \, t$: a correction equal to the calculated decrease in concentration due to the sampling and chamber replenishment was added to the results of subsequent samples.

Particle size information was obtained by analyzing samples with a model Elzone 80XY instrument manufactured by Particle Data, Inc. The instrument measures the electrical displacement as particles pass through 0.095 or 0.240 mm orifices, counts the particles, and resolves them into 128 logarithmically-spaced size classes. The minimum particle size sensed is 0.0016 mm when using the 0.095 mm orifice and 0.004 mm when using a 0.240 mm orifice. Analytic procedures are given by Teeter (1993).

Summary statistics for particle size mean, sorting (standard deviation), and skewness were computed after conversion of equivalent particle diameters to phi units. Phi diameters are the negative base 2 logarithm of the diameter in millimeters. Eleven quantiles ranging from 5 to 95 percent of the cumulative distributions were used in the calculation of summary statistics (Teeter 1993).

## EROSION RATES RESULTS

The *PES* tests displayed initial high erosion rates that decreased in time and became about constant by the middle portion of the test steps. The eroded suspension concentrations did not reach a clear steady-state value during the test steps;therefore, erosion rates did not decrease to zero. The *VOST* test steps displayed more uniform erosion rates.

Figure 1 shows three replicate *PES* tests (P4, P5, and P7) on identically prepared kaolinite and silt mixure with the concentrations adjusted for sample withdrawal. Two tests yielded suspension concentration time histories in close agreement. The third test (P7) started with a temporary over ranging of oscillation rate at the 0.2 Pa step which appeared to cause concentrations to rise above the other tests. Test P7's higher concentrations were maintained throughout the test or possibly increased slightly.

Two duplicate *VOST* tests (V17 and V19) were performed on the kaolinite and silt mixure. A suspension time history example is shown in Figure 2 for the lower shear stress range.

Erosion rates at 0.5 Pa shear stress (or equivalent shear stress) over 30 min periods averaged 0.29 g/sq m/min for the three *PES* tests and 0.45 g/sq m/min for the two *VOST* tests. The erosion rates were calculated by differencing eroded masses per unit bed area over the same time periods.

Figure 3 shows the 0.5 Pa test step for P5 (shown in Figure 1). Data were collected at 0.5-min intervals and converted from nephelometric to sediment concentration values. At the end of 30 minute time step, one liter of suspension was drawn off while simultaneously replacing it with sediment free water, which resulted in a reduction of suspension concentration to about 58 percent of its original value. The erosion test continued for about 10 more minutes. It was noticed that concentration continued to

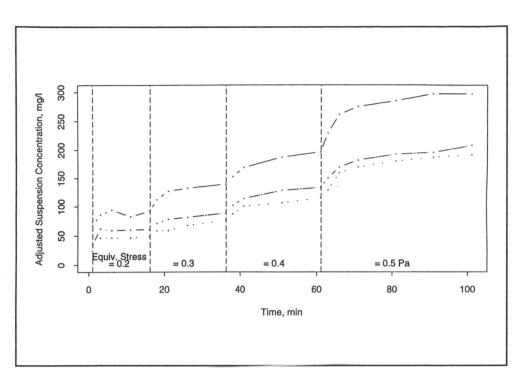

**Figure 1.** Suspension concentrations for *PES* tests P4, P5, P7 for kaolinite silt mixture

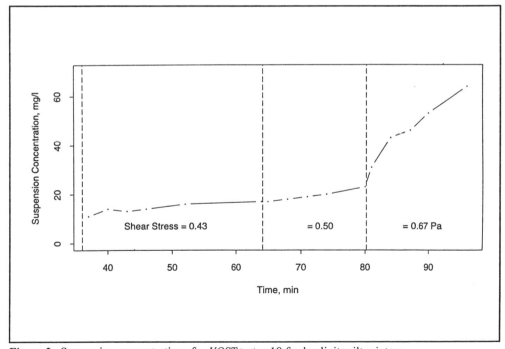

**Figure 2.** Susenpsion concentrations for *VOST* tests v19 for kaolinite silt mixture

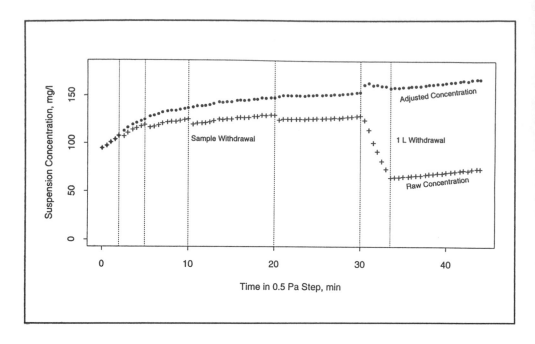

**Figure 3.** Measured and corrected suspension concentrations for *PES* test P5 after multiple withdrawals for 0.5 Pa step

increase at a rate about equal to that before the withdrawal. The original data (lower line in Figure 3) show the decreases in suspension concentration associated with sampling. The corrected data (higher line) are also shown in Figure 3. The erosion rate reached a constant value after about 15 min of this test step.

The marine mud was tested twice in the *PES* and once in the *VOST*. The erosion rates for the first 10 min of the 0.5 Pa steps were 0.41 g/sq m/min for the *PES* and 0.32 g/sq m/min for the *VOST*.

## RESULTS OF PARTICLE SIZE CHANGES

The eroded suspensions were analyzed for particle size distribution using the same orifices as the corresponding test bed material. Since there are variabilities in the size distributions due to sampling and analytic errors, particle size of eroded suspensions changed during tests. In general, however, the suspension particle size distributions were finer than bed sediment sizes in early portions of tests and approached those of the bed sediments by the end of the tests. The summary statistics calculated from the distributions of the test bed material are shown in Table 1.

Figure 4 shows the phi mean, sorting, and skewness plotted versus total suspended material (corrected) for test P5. There were visual signs that silts were left behind by the erosion process in some *VOST* tests of kaolinite and silt mixtures, as eroded mixtures displayed a darker surface color characteristic of the silt compared to the lighter color underneath. It was noted that the mean suspended particle sizes increased with time

Table 1: Summary Stastics of Particle Size in Test Bed Material

| | Orifice Size mm | Summary Statistics, Phi units | | |
| --- | --- | --- | --- | --- |
| | | Mean | Sorting | Skewness |
| Kaolinite | 0.095 | 8.17 | 0.81 | -0.16 |
| Marine Mud | 0.095 | 7.55 | 1.26 | -0.18 |
| Kaolinite & Silt | 0.240 | 6.23 | 0.91 | -0.17 |

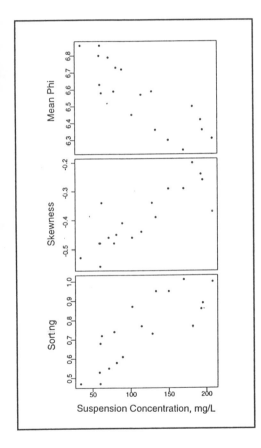

**Figure 4.** Scatter plots of summary statistics and suspension concentration for P5

during *PES* and *VOST* tests for both sediment materials. Furthermore, particle size distributions were significantly more graded and more positively skewed as the test progressed.

Figure 5 shows mean suspended sediment size versus eroded mass for *PES* tests P4, P5, and P7. Individual tests show good linear relationships between mean eroded sediment size and eroded mass.

## DISCUSSION

Both sediments, kaolinite/silt mixture and marine mud, displayed cohesion and showed similarity in erosion. Kaolinite is known to flocculate in freshwater (Lau and Krishnappan, 1994). The kaolinite/silt mixture had a much higher density which appeared to offset the erodibility effect of its lower cohesion (as indicated by lower CEC).

When averaged over time, the erosion rates found in the *PES* and *VOST* were similar at intermediate shear stresses where both devices were within their optimum shear stress ranges. However, erosion rates in the *PES* typically decreased to much lower values than the *VOST* during test steps. The *VOST* test steps had more constant erosion rates than the *PES* steps and suspension concentrations climbed linearly with time.

*VOST* tests had greater maximum shear stresses than the *PES*. In the shear stress range above that of the *PES*, a break point occurred when *VOST* erosion rates increased markedly for both test sediments.

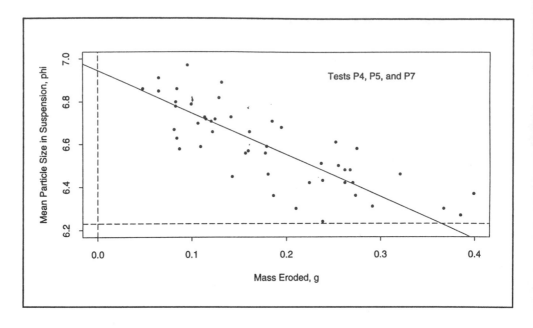

**Figure 5.** Mean particle size for P4, P5, and P7

The one liter withdrawal test, for example Figure 2, at the end of the highest *PES* shear stress confirmed that deposition did not occur simultaneous with erosion in either the kaolinite/silt or the marine mud tests. Deposition is concentration dependent and the change in chamber concentration associated with the withdrawal would have changed the apparent erosion rate had deposition been occurring. A similar result was found for kaolinite by Lau and Krishnappan (1994) and Parchure (1984). The changes in erosion rate during the *PES* test steps must be explained by some other process such as vertical variations in bed shear strength or particle availability as discussed below.

A trend analysis, which uses a combination of mean, sorting, and skewness (Teeter 1993, McLaren and Bowles 1985) was used to identify size distribution changes during tests. A Z score was calculated for each of the possible eight cases involving increases or decreases in mean, sorting, and skewness. All tests with sufficient sample numbers showed highly significant Z scores for the case of coarsening, more graded, and more skewed toward the fine end of size distributions. Though the laboratory erosion test data showed very obvious trends in the summary statistics of the eroded suspensions, field suspended samples are more variable and difficult to interpret. The Z score is useful in determining whether valid trends exist in the particle size distribution data.

Trends similar to the eroded suspensions in this study have been identified by the first author in field sample sequences in strong estuarine tidal flow. Mean suspended particle size at a station near the Golden Gate Bridge at the mouth of San Francisco Bay increased with tidal flow at near-bottom sampling stations. The corresponding Z scores were significant ($p < 0.05$). Samples from a total of three each near-bottom, mid-depth, and near-surface stations were analyzed. While the near-bottom stations showed coarsening, more graded, and fine-skewed trends with increased tidal discharge, the mid-depth stations did not display significant trends. Most of the near-surface stations displayed other trends possibly related to mixing lags and circulation effects. Such factors complicate suspension size composition and the analysis of estuarine sediment dynamics.

Fine particles eroded first from natural mud and laboratory clay/silt mixtures indicating that coarser bed material particles were left behind on the bed surface. Erodibility is not equal for all particles and is statistically distributed among particle sizes; therefore, the higher the equivalent shear stress, the greater the presence of silts in suspension.

The linearity of the increase in mean eroded particle size suggests that a fraction of erodible bed material increases with shear stress and/or eroded mass. The dashed line at the bottom of Figure 5 indicates the mean size of the bed or erosion source. The intercept of eroded suspension line with this line indicates the eroded mass at which all bed particles are being eroded. The fraction ($f$) of bed particles eroded can be related to the mean size of eroded suspension ($u$) using the mean bed size ($ub$) and the intercept ($uo$) of the regression line with the zero mass eroded axis as mixing end members.

$$f = \frac{uo - u}{uo - ub} \tag{3}$$

The active erosion layer thickness ($L$) is defined as that surface bed thickness contributing eroded particles to the suspension. If ($A$) is the bed area,

$$L = \frac{M}{A f Cs} \tag{4}$$

where $M$ is the eroded mass and $Cs$ is the solids content of the bed. Test results indicated that $L$ was fairly constant and had a value on the order of the largest bed material particle diameter taken at about the 90 percentile level (D90). The datasets do not explain whether or not equilibrium suspensions of constant particle size distribution or concentration would eventually form, had the test steps been continued. It is not clear whether coarser particles were less erodible or whether they merely eroded more slowly.

## SUMMARY AND CONCLUSIONS

The two test sediments eroded similarly in the same test device. Although test sediment erosion rates for the *PES* and *VOST* were similar by some measures, no conclusions could be drawn as to the comparability of these erosion test devices. Additional work is underway to improve *PES* repeatability and to decrease shear stress variation across the *VOST* test sample tray. These two devices are probably most useful in different ranges of shear stress.

Erosion rates in *PES* for a given equivalent shear stress generally decreased in time but did not reach zero during the time step used. Simultaneous erosion and deposition did not occur in *PES* tests. *VOST* test steps had more constant erosion rates. A more significant mode of erosion was found in *VOST* tests at shear stresses above those of the *PES*.

Both erosion test devices showed an increase in the mean particle size of suspended sediment with eroded mass and/or shear stress. Fine particles were eroded first, leaving coarser particles behind on the bed surface. Erodibility was not equal for all particles and was statistically distributed among particle sizes. At shear stresses of about 0.5 Pa and greater, all or nearly all bed sediment particle sizes were eroded. For these tests, the thickness of the active erosion layer was approximately the D90 of the source sediment (about 0.04 - 0.05 mm).

In addition to increases in mean particle size of suspension, particle grading also increased and skewness of particle size distributions became more positive toward the fine sizes as shear stress and eroded mass increased.

## ACKNOWLEDGEMENTS

Research presented here was funded by the U.S. Army Engineer Dredging Research Program. Dr. Nicholas C. Kraus was Manager of Technical Area 1 during the period when much of the research was conducted. Mr. E. Clark McNair, Jr. was Program Manager at WES. Thanks to Ms. Clara Coleman, Mr. Thad Pratt, and Mr. Joe Parman who performed laboratory tests and analyses. Permission to publish this paper was granted by the Chief of Engineers.

## REFERENCES

Davis, W. R., 1993, The role of bioturbation in sediment resuspension and its interaction with physical shearing, J. Exp. Mar. Bio Ecol., 171, Elseveir, pp. 187-200.

Lau, Y. L., and Krishnappan, B. G., 1994, Does rentrainment occur during cohesive sediment settling?, J. Hydraulic Engineering, Vol. 120, No. 2., ASCE, pp. 236-244.

Lavelle, J. W. and Davis, W. R. 1987. Measurements of benthic sediment erodibility in Puget Sound, Washington, NOAA Tech. Memorandum ERL PMEL-72, Pacific Marine Environmental Laboratory, Seattle, WA.

McLaren, P. and Bowles, D., 1985, The effects of sediment transport on grain-size distributions, J. of Sedimentary Petrology, Vol. 55, No. 4, pp. 457-470.

Parchure T. M., 1984, Erosional behavior of deposited cohesive sediments, Ph. D. Thesis, Univ. of Florida, Gainesville, FL, USA, Coastal and Oceanographic Engineering Department, No. TR/071.

Teeter, A. M. and Pankow, W., 1989, Deposition and erosion testing on the composite dredged material sediment sample from New Bedford harbor, Massachusetts, Technical Report HL-89-11, U.S. Army Engineer Waterways Experiment Station, Vicksburg, MS, USA.

Teeter, A. M., 1993, Suspended transport and sediment-size transport effects in a well-mixed, meso-tidal estuary, in Nearshore and Estuarine Cohesive Sediment Transport, Ed. A. J. Mehta, Coastal and Estuarine Studies, 42, American Geophysical Union, Washington, DC, pp. 411-429.

Tsai, C. H. and Lick, W., 1986, A portable device for measuring sediment resuspension, J. Great Lakes Res., 12(4) pp. 314-321.m

# 18  Erosion of fluid mud by entrainment

**J. C. WINTERWERP[1] and C. KRANENBURG[2]**

[1]*Delft Hydraulics, Marine, Coastal and Estuarine Management,*
*The Netherlands*
[2]*Delft University of Technology, Department of Civil Engineering,*
*The Netherlands*

## INTRODUCTION

Fluid mud is a highly concentrated near-bed suspension of mud, either mobile or stationary. It is generated by liquefaction of muddy beds by waves, or during rapid deposition, when the deposition rate exceeds the consolidation rate. Fluid mud is commonly observed in many estuaries, navigational channels, harbour basins and along some coasts (Kirby, 1988). Van Leussen & Van Velzen (1993) presented a summary of fluid mud occurrences in The Netherlands.

Fluid mud appearances are a nuisance to many authorities, as these can cause large siltation rates. Therefore, a strong need exists to model the generation, transport and subsequent fate of these fluid mud layers. Ross & Mehta (1988) developed a one-dimensional vertical model, mainly meant to analyze the relevant processes. Le Hir (1994) used a two-dimensional vertical model to analyze the fluid mud behaviour in the Loire Estuary in France. More common is a two-layer approach where the fluid mud layer is conceived as the lower layer and the overflowing water as the upper layer. Such a model was applied by Odd & Cooper (1989), for instance, to simulate the fluid mud behaviour in the Severn Estuary in Great Britain, and by Kusuda & Futawatari (1992) to simulate the sediment transport in the Rokkaku River in Japan.

A major problem with these models, however, is the lack of physically accurate and correct mathematical formulae describing the relevant processes, such as the rheological properties related to the history and the water content of the mud, the mobility of the mud, the re-entrainment behaviour, liquefaction and compaction processes, damping of turbulence and internal waves.

Mehta & Srinivas (1993) used the turbulent kinetic energy (TKE) equation of the upper (mixed) layer to analyze the entrainment process. By using similarity consider-ations and dimensional analyses they obtained an integral entrainment model. The coefficients in this model were obtained from experiments in a so-called race track flume with kaolinite, a mixture of kaolinite and bentonite and with lake mud.

Winterwerp et al. (1993) showed experimentally that the fluid mud layer can be (partly) eroded by an entrainment process that closely resembles well-known entrainment processes observed in two-layer fresh/saline water systems. Experiments were carried out in an annular flume under flow conditions in which the velocity varied sinusoidally

*Cohesive Sediments.*
Edited by Neville Burt, Reg Parker and Jacqueline Watts
©1997 John Wiley & Sons Ltd.

with time. During decelerating flow and slack tide, a fluid mud layer formed on the bed. This layer was eroded during accelerating flow, first rapidly due to the entrainment process, then more gently as floc erosion became dominant.

The present paper focuses on the re-entrainment process of fluid mud layers. An entrainment model is derived herein, also using the TKE equation. However, this equation now is integrated formally across the mixed layer. The resulting entrainment equation is applied in conjunction with the momentum and mass balance equations and an empirical relation for the concentration distribution in the fluid mud layer. The empirical model constants involved are obtained from experiments reported in the literature and from two series of experiments in an annular flume, one with fresh and saline water and the other with a kaolinite suspension, described in this paper.

## THEORY - AN INTEGRAL ENTRAINMENT MODEL

Sherman et al. (1978) and Imberger & Hamblin (1982), amongst others, reviewed entrainment models for mixed-layer deepening in lakes and reservoirs, which are based on an integral form of the TKE equation. This approach is adopted herein to model the entrainment of a fluid-mud layer by turbulent flow. A more detailed analysis is given by Kranenburg (1994).

Assuming uniform flow, no longitudinal concentration gradients and adopting the Boussinesq approximation, the integral TKE equation for the upper water (mixed) layer can be written as

$$\frac{d}{dt}\int_0^H \overline{q}\,dz + \int_0^H \overline{u'w'}\frac{\partial \overline{u}}{\partial z}dz - 2\frac{\lambda}{W}U^3H - \Delta\frac{g}{\rho_w}\int_0^H \overline{w's'}\,dz + \int_0^H \epsilon\,dz = 0 \qquad (1)$$

$$\quad I \qquad\qquad II \qquad\qquad III \qquad\qquad IV \qquad\qquad V$$

where $t$ is time, $H$ is the water depth, $\overline{q}$ the TKE, $z$ the vertical coordinate ($z = 0$ at free surface or screen and is positive downwards), $W$ the width of the flume or channel, $\lambda$ a sidewall friction coefficient, $U$ the mean water velocity, $\Delta = (\rho_s-\rho_w)/\rho_s$ is the fractional density difference of sediment and water, $s$ the dry density, $g$ the acceleration of gravity, $\epsilon$ the dissipation rate, and $u$ and $w$ horizontal (streamwise) and vertical velocity components. A sketch is given in Figure 1. The terms I - V represent storage of TKE, shear production by the vertical velocity gradient, shear production by sidewall shear stress, buoyancy destruction and dissipation rate, respectively.

The width-averaged balance equations for momentum and mass are invoked to eliminate the turbulent transport terms $\overline{u'w'}$ and $\overline{w's'}$ in (1). These equations read

$$\frac{\partial}{\partial z}\overline{u'w'} = -\frac{\partial \overline{u}}{\partial t} - \frac{1}{\rho_w}\frac{\partial \overline{p}}{\partial x} - 2\lambda\frac{U^2}{W} \qquad (2)$$

$$\frac{\partial}{\partial z}\overline{w's'} = -\frac{\partial \overline{s}}{\partial t} - \frac{\partial}{\partial z}\overline{w_s\,s} \qquad (3)$$

where $p$ is the pressure and $w_s$ the setting velocity. Using boundary conditions at $z = 0$ and at $z = H^+$ (at the base of the mixed layer), integral forms of (2) and (3) become

$$\frac{d}{dt}UH \approx u_*^2 - u_H^2 + U_H\frac{dH}{dt} - \frac{H}{\rho_w}\frac{\partial \overline{p}}{\partial x} - 2\frac{\lambda}{H}U^2 H \qquad (4)$$

$$\int_0^H \overline{s}\,dz = H\tilde{s} = H_0 s_0 + \int_{H_0}^H s_b(z)\,dz \qquad (5)$$

where $u_*$ and $u_H$ are friction velocities at $z = 0$ and $z = H^+$, $U_H$ is a possible velocity (near the lutocline) of the mud layer, $s_b(z)$ the dry density in the bed, $\tilde{s}$ the mean dry density in the water layer, and the subscript $0$ refers to the initial situation ($t = 0$).

As in any turbulence model, approximations are needed to obtain a closed set of equations. Approximations in this case are concerned with the various terms in (1). As for the storage term I it is assumed that $\overline{q} \propto u_+^2$, where $u_+ = \max [u_*, u_H]$ so that

$$\text{term I} \propto \frac{d}{dt}u_+^2 H \qquad (6)$$

Substituting from (2) and another approximation give for the second term

$$\text{term II} \approx -\left[u_H^2(U-U_H) + u_*^2(U-U_H) + \frac{1}{2}(U-U_H)^2\frac{dH}{dt}\right] \qquad (7)$$

Multiplying (3) by $z$ and integrating from $z = 0$ to $z = H^+$ and approximating gives for the fourth term

$$\text{term IV} \approx -\Delta\frac{g}{\rho_w}\left[\frac{1}{2}H(s_H - \tilde{s})\frac{dH}{dt} + H\tilde{w}_s\tilde{s}\right] \qquad (8)$$

where $s_H = s_b(H^+)$. The dissipation term V is accounted for by assuming that the production terms II and III are reduced by empirical factors. In addition the mechanical work done to loosen and disperse entrained material from the bed is included (the term containing the empirical friction velocity $u_m$ - see (9)). A modelled form of (1) may then be written as

$$c_q \frac{d}{dt}\left(u_*^2 H\right) - 2c_H u_H^2 (U - U_H) - c_* u_*^2 (U - U_H) - c_H (U - U_H)^2 \frac{dH}{dt}$$

$$-2c_w \lambda \frac{H}{W} U^3 + u_m^2 \frac{dH}{dt} + B\frac{dH}{dt} + 2\frac{\Delta \bar{s}}{\rho_w} g\bar{w}_s H = 0 \quad \text{for} \quad \frac{dH}{dt} > 0$$

(9)

where $B = \Delta g H(s_H - \bar{s})/\rho_w$ is the total buoyancy, and $c_q$, $c_H$, $c_*$ and $c_w$ are empirical coefficients. Empirical evidence (Kantha et al, 1977 - see Figure 2 -, Price 1979, Kranenburg 1987, amongst others) suggest $c_* + c_H = 0.5 \pm 0.1$, $c_H/c_* = 1.3 \pm 0.3$, $c_q = 6.4 \pm 1.3$ and $c_w \approx 0.085$. For strongly stratified flows the precise value of $c_q$ is irrelevant, as is the value of $c_w$ for small aspect rations $H/W$. A stirring term proportional to $u_*^3$ has been neglected in (9).

The value of $u_m$, representing the mechanical work done in the fluid mud layer, is unknown and is to be determined from experiments as described herein, for example. It is likely that $u_m$ increases with $s_H$.

Finally, an empirical relation is used to describe the sediment concentration in the fluid mud layer:

$$s_b(z) = s(H_0) + \left[\frac{z - H_0}{H_0}\right]^\alpha (s(h) - s(H_0))$$

(10)

where $H_0$ is the initial depth of the mixed layer, $h$ is the total depth of the flume and $\alpha$ is an empirical coefficient.


## EXPERIMENTAL SET-UP AND PROCEDURES

Two series of experiments were carried out in the annular flume of Delft University of Technology with an outer diameter of 4.0 m, an inner diameter of 3.4 m and a maximal height of 0.4 m. The fluid within the flume is driven by a screen on top of the flume; the flume itself can rotate in the opposite direction to minimize secondary currents near the bed. For fresh water, the three-dimensional flow and shear stress distribution in the flume were measured in detail with a Laser-Doppler Anemometer (LDA) system. These measurements, and further details on the flume are reported by Booij et al. (1993) and by Visser et al. (1994).

During the present tests the height of the flume was set at 27.3 cm and the ratio of rotational speeds of screen and flume was set at -1.88, which is the optimal value for this water depth.

In the first series, entrainment experiments with saline and fresh water were carried out. First salt water was let into the flume. On top of this saline layer, fresh water was added through a diffusor. Thus a two-layer system was obtained with an initial interface thickness of about 1 cm.

Prior to an entrainment experiment, both flume and screen were gradually accelerated to the rotational speed selected for the flume. In this way it was achieved that at the start and during most part of the experiment the lower, saline layer remained stagnant with respect to the flume, so that no additional turbulence was generated in this layer. One experiment with a non-rotating flume was carried out for comparison. An experiment was started by smoothly accelerating the screen till the selected speed, which took about 10 seconds.

During the experiments the rotational speeds of the screen and the flume were monitored, and the tangential shear stress distribution close to the screen was measured with an LDA-system. Comparison with fresh water measurements showed that the shear stress at the screen was not affected by stratification effects.

The entrainment rate was obtained from the decreasing height of the interface, which was monitored visually and with video cameras at the inner and outer walls of the flume, using a shadowgraph technique. The detailed experimental programme is given in the next section.

In the second series, fluid mud experiments were carried out with kaolinite. Saline water at a salinity of 5 ppt was let into the flume. Then a dense suspension of kaolinite was added to obtain a concentration of about 60 g/l. This suspension was rotated and mixed in the flume for a week. Prior to and after the entrainment experiments, samples were taken to carry out additional consolidation experiments in separate columns.

At the start of an experiment, flume and screen were set at the selected speed for the flume as with the saline/fresh water experiments. Due to the high viscosity of the suspension, the fluid became stagnant relative to the flume within a short time and the kaolinite started to settle. After the selected settling/consolidation period an entrainment experiment was started by accelerating the screen till the selected speed in the opposite direction.

During such an experiment, the following parameters were measured:
- the rotational speeds of flume and screen,
- the suspended kaolinite concentration in the centre line of the flume at three heights (10, 16, 24 cm from the bottom of the flume) with turbidity sensors (OSLIM),
- at regular time intervals samples were taken to calibrate the OSLIM's,
- the height of the interface was measured visually and from video-recordings at the inner and outer wall of the flume, and
- during two tests, the vertical and tangential velocity in the centre line of the flume was measured at 10 cm height with an immersible electro-magnetic current meter (EMS); data were sampled at a frequency of 20 Hz.

During a separate consolidation experiment in still water the vertical distribution of the kaolinite concentrations in the fluid mud layer was measured for various consolidation periods with an electrical conductivity probe (ECP).

The detailed experimental programme is given in the second-next section.

## FRESH/SALINE WATER EXPERIMENTS

The entrainment experiments with fresh and saline water were carried out for comparison with the fluid-mud experiments. The programme of these experiments is given in
Table 1. The first experiment aimed at reproducing results obtained by Kantha et al.
(1977, their Figure 5) and the second to detect a possible effect of the rotation of the
flume. The remaining six experiments were similar to those carried out with kaolinite:
the depths of the upper layer and the densities of the lower layer were comparable to
those in the fluid-mud experiments.

Experimental results are shown in Table 1 and in Figure 2. Here $w_e = dH/dt$ is the
entrainment velocity, $E_* = w_e/u_*$ is the nondimensional entrainment rate and $Ri_* = (\Delta gH/u_*^2$ is an overall Richardson number. In two-layer experiments $Ri_*$ is a constant.
$E_*$ for experiment 1 is a bit large. This is caused by the relatively low shear stress
measured at the screen, when compared with experiment 2, for instance. For the six
remaining experiments, the shadowgraph images indicated that a part of the lower layer
(sublayer) was dragged along by the surface layer. The theoretical thickness of this
viscous layer is about $v_s/w_e$, where $v_s$ is the kinematic viscosity of the lower layer.
Calculated values of $v_s/w_e$ amount to 5 mm for experiment 3 and 15 to 40 mm for
experiments 4-8.

| | Experimental programme | | | | Results | |
|---|---|---|---|---|---|---|
| Run No | Density of lower layer $\rho_s$ [kg/m³] | Speed $U_s$ of screen [m/s] | Speed $U_f$ of flume [m/s] | Initial depth of lower layer [m] | Entrainment velocity $w_e$ [m/s] | Friction veloc- ity at screen $u_*$ [m/s] |
| 1 | 1100 | 1.39 | 0 | 0.22 | (5.1 ± 1.0) 10⁻³ | 0.030 ± 0.004 |
| 2 | 1100 | 0.90 | -0.48 | 0.22 | (4.4 ± 0.6) 10⁻³ | 0.037 ± 0.005 |
| 3 | 1100 | 0.91 | -0.48 | 0.05 | (2.8 ± 0.3) 10⁻⁴ | 0.035 ± 0.005 |
| 4 | 1200 | 0.84 | -0.44 | 0.05 | (5.8 ± 0.4) 10⁻⁵ | 0.032 ± 0.004 |
| 5 | 1100 | 0.59 | -0.32 | 0.05 | (3.7 ± 0.4) 10⁻⁵ | 0.024 ± 0.003 |
| 6 | 1100 | 0.68 | -0.36 | 0.05 | (5.7 ± 0.5) 10⁻⁵ | 0.028 ± 0.004 |
| 7 | 1100 | 0.79 | -0.42 | 0.05 | (1.0 ± 0.2) 10⁻⁴ | 0.033 ± 0.005 |
| 8 | 1050 | 0.57 | -0.30 | 0.05 | (5.4 ± 0.2) 10⁻⁵ | 0.023 ± 0.003 |

*Table 1: Fresh/saline water experiments.*

The time scale of the formation of the viscous boundary layer is about $0.2v_s/w_e^2$,
which amounts to 4 s for experiment 3 and to 30 - 230 s for experiments 4-8. These
time scales are well below the duration of the entrainment experiments so that sufficient

time was available for the viscous layer to develop. Because observed vertical turbulence scales were comparable to experiment 3, or less than the calculated viscous-layer thickness, it is concluded that experiments 3-8 were affected by viscous effects, i.e. part of the lower layer was dragged along by the surface layer.

The entrainment diagram of Figure 2 also shows the theoretical results for (i) the case where sidewall friction is absent and $u_H = 0$ and (ii) the case where sidewall friction is present and $U_H = U$ (see also equ's (13) and (14)). These results indicate that experiments 1 and 2 were hardly influenced by sidewall friction and viscous drag, whereas experiments 3-8 were dominated by these effects. The theoretical result for experiments 4-8 even overestimates the observed entrainment rates by about 40%. This discrepancy may have been caused by the large aspect ratio ($H/W \approx 0.8$) in these experiments, while the entrainment model (i.e. the empirical coefficients) was tuned to experiments with smaller aspect ratio ($H/W \approx 0.25$). In addition, the proximity of the bottom of the flume may have suppressed the turbulent fluctuations.

## FLUID MUD EXPERIMENTS

The artificial clay used for the fluid-mud experiments consisted of kaolinite (85 percent by weight) and mica. Particles with sizes less then 2 $\mu$m (10 $\mu$m) formed 31 (84) % of the total weight, and all particles were smaller than 40 $\mu$m. The density was 2620 kg/m$^3$, the specific surface area 24 m$^2$/g, and the Cation Exchange Capacity 3.3 meq/100g. The viscosity and Bingham yield strength were obtained from the flow curves for clay suspensions measured at various concentration using a Haake Rotoviscometer CV100 with measuring system ZA30 (concentric cylinders). The measuring period for each flow curve was 6 min. The results are shown in Table 2.

| Concentration [g/l]: | 58 | 113 | 172 | 237 | 275 |
|---|---|---|---|---|---|
| Viscosity $\mu$ [mPas] | 1 | 2 | 3 | 4 | 5 |
| Bingham strength $\tau_B$ [Pa] | - | 0.1 | 0.4 | 1.0 | 1.5 |

*Table 2: Viscosity and Bingham strength of kaolinite suspension.*

A typical profile of the sediment concentration in the fluid mud layer after a consolidation time of 120 minutes is shown in Figure 3, together with the power law fit (10), using $\alpha = 5$ and $H_0 = 0.083$ m.

The experimental programme of the fluid-mud experiments is shown in Table 3. Settling still proceeded when experiments K1 en K2 were started, whereas all material had been deposited on the bed when experiments K3-K8 commenced.

The shear stresses at the screen are assumed to be equal to those in the fresh/saline water experiments, because in these experiments no influence of stratification on these stresses could be measured.

The entrainment velocity was assessed from four different techniques: a direct visual observation of the height of the interface at the outer glass wall of the flume, from the video recordings of the position of the interface at the inner and outer glass wall, and indirectly from the measured concentration distributions in the water column and fluid mud layer. From these observations it was concluded that the interface was tilted by about 1 to 2 cm due to centrifugal forces. However, the entrainment rate itself was almost the same for all methods.

| Run No | Experimental programme | | | Results for initial phase | |
|---|---|---|---|---|---|
| | Settling/ cons. period [min] | Speed of screen $U_s$ [m/s] | Speed of flume $U_f$ [m/s] | Initial entr. velocity $w_{e,i}$ [m/s] | Mean conc. top of bed [g/l] |
| K1 | 33 | 1.39 | 0 | $(7.0 \pm 1.0)\ 10^{-3}$ | 60 |
| K2 | 28 | 0.91 | -0.48 | $(2.5 \pm 1.0)\ 10^{-3}$ | 60 |
| K3 | 133 | 0.91 | -0.48 | $(3.4 \pm 0.1)\ 10^{-4}$ | 175 |
| K4 | 129 | 0.84 | -0.44 | $(2.4 \pm 0.1)\ 10^{-4}$ | 175 |
| K5 | 136 | 0.59 | -0.32 | $(8.5 \pm 0.3)\ 10^{-5}$ | 175 |
| K6 | 205 | 0.91 | -0.48 | $(2.5 \pm 0.1)\ 10^{-4}$ | 180 |
| K7 | 180 | 0.84 | -0.44 | $(2.1 \pm 0.1)\ 10^{-4}$ | 180 |
| K8 | 180 | 0.59 | -0.32 | $(7.1 \pm 0.3)\ 10^{-5}$ | 180 |

*Table 3: Fluid mud experiments and initial entrainment velocity $w_{e,i}$.*

The concentration distribution in the upper part of the fluid mud layer (Figure 3) show only little variation. Hence, the entrainment rates at which this upper part is eroded, can be compared directly with those of the fresh/saline water experiments. The initial entrainment velocities $w_{e,i}$ are listed in Table 3 together with the concentrations (accuracy about $\pm$ 5 g/l) in the upper part of the bed. Figure 2 shows that the *initial* entrainment rates of fluid mud at large $Ri_*$ are somewhat, but consistently larger than those for the fresh/saline water experiments. This may have been caused by the initial thickness of the lower layers: this initial thickness for the fresh/saline water experiments was about 5 cm, whereas for the fluid mud experiments it was about 8 cm. Hence the influence of the flume bottom may have been larger during the fresh/saline water experiments.

The argument concerning viscous effects on the entrainment process presented in the previous section, also applies to the fluid mud experiments, because the viscosity of the fluid mud bed was considerably larger than of the saline water used (Table 2). The theoretical thickness of the viscous layer for the experiments K1 and K2 is estimated at 0.1 to 0.8 mm, and for the experiments K3 to K8 at 1 to 4 cm.

Summarizing, it may be concluded that, for the present experiments with settling/consolidation times up to 205 minutes, the upper part of the bed behaved as a viscous fluid.

Analysis of the later phases of the experiments requires numerical integration of Equations 4, 5 and 9. This was done with a simple predictor-corrector scheme. Numerical model simulations were carried out for the experiments K3 through K8. As viscous effects are important, it was assumed that $U_H = U$, hence the value of the coefficients $c_H$ and $c_*$ are not relevant. The other coefficients are set at: $c_w = 0.085$, $c_q = 6.4$, $\lambda = 0.004$ (Blasius' law), $u_*^2 = 0.0018 \, (U_s - U)^2$ and $w_s = 0$ or 0.05 mm/s (the latter value was obtained from the consolidation experiments). At $t = 0$ the simulation is started with $U$ and $\tilde{s} = 0$ and $U_s$ according to Table 3.

For experiment K3, $w_s = 0$ mm/s the computed entrainment velocity $w_e$, the mean flow velocity $U$ in the surface layer, and the suspended sediment concentration $\tilde{s}$ are drawn in Figure 4 as a function of time, together with the measured suspended sediment concentration. Also given is $\tilde{s}$ for $w_s = 0.05$ mm/s. It is shown that $U$ first increases with time up to about 0.5 m/s, but for $t > 100$ s $U$ starts to decrease. This initial increase is caused by the acceleration processes of the fluid, the subsequent decrease is induced by side wall friction and entrainment, and by the fact that the thickness of the mixed layer increases.

The entrainment velocity $w_e$ first decreases, again due to the acceleration of the flow. At $t = 0$, $U = 0$ and $U_s - U$, hence $u_*$ is maximal. After about 40 s, $w_e$ increases again caused by the production of TKE by side wall friction and entrainment. The subsequent decrease is caused by a continuing increase in bed density, hence an increase in $Ri_*$. The effect of $w_s$ apparently is small for these test conditions.

Comparison with the experimental data shows a fair overall agreement. The deviations at small $t$ are due to inaccuracies in the experimental data and spin-up effects (for instance it takes time to have a fully developed turbulent boundary layer) that are not taken into account in the model. For $t > 350$ s, the model overestimates the actual $\tilde{s}$, possibly for the same reasons as in the case of the fresh/saline water experiments. Furthermore, the work done in the fluid mud layer may also play a role. This latter effect is not yet elaborated.

Figure 5 shows the results for experiment K8. In a qualitative sense, the picture is similar to that for experiment K3, though the entrainment process itself is much slower, as a result of a smaller $u_*$. This also explains why the influence of $w_s$ is much larger: it will take the same amount of turbulent energy to keep the particles in suspension; however, relatively less energy is available in K8 test conditions. Beyond $t = 800$ s, the model starts to overestimate the measured $\tilde{s}$ considerably. This is caused by the gradual change in the erosion process, as the work done within the bed becomes progressively more important, a process not (yet) taken into account in the model. This last phase of the entrainment process is often referred at as surface erosion. At the end of experiment K8 a layer of kaolinite remained uneroded in the flume with a thickness of about 1 to 2 cm.

The EMS functioned properly in experiments K2 and K4 only. Time histories of mean velocity $\overline{u}$, turbulent shear stress $\overline{u'w'}$ and turbulence intensities were obtained using a moving-average technique (the averaging period was 50 s). Estimates of $\overline{u}$ and $\overline{u'w'}$ were used to calculate flux Richardson numbers. The flux Richardson number $R_f$ is defined as the ratio of buoyancy destruction to shear production:

$$R_f = -\frac{\Delta g \overline{w's'}}{\rho \overline{u'w'} \dfrac{\partial \overline{u}}{\partial z}} \tag{11}$$

Assuming, during entrainment, a linear velocity profile near the mud-water interface (Kranenburg, 1984) gives $\partial \overline{u}/\partial z \approx \overline{u}/\delta z$, where $\delta z$ is the elevation of the current meter above the bed. Assuming furthermore that $\overline{w's'}$ varies linearly with depth, we find:

$$\overline{w's'} = \frac{(s_H - \tilde{s})w_e(H - \delta z)}{H} \tag{12}$$

Values of $R_f$ thus calculated for experiment K2 increased up to 0.05 - 0.06 at $t = 80$ to 100 s and then decreased again, because $w_e$ decreased. These values of $R_f$ indicate that during experiment K2 stratification effects were moderate at most (e.g. Turner, 1973). This result is related to the low value of $Ri_*$ ($\approx 18$) for experiment K2.

For experiment K4 the flux Richardson number gradually increased from $R_f \approx 0.13$ at $t = 100$ s to $R_f \approx 0.16$ at $t = 300$ s and then rapidly decreased to $R_f \approx 0.04$ at $t = 400$ s and even lower values thereafter. The high values of $R_f$ for $t = 100$ to 300 s indicate near-collapsing turbulence near the bed, the maximum obtainable value of $R_f$ being about 0.15 (Turner, 1973). This result agrees with observations at large $Ri_*$ in a fresh/saline water system (Kranenburg, 1984).

The influence of stratification is also indicated by the pronounced gradient in the initial vertical concentration distribution observed during experiment K4, for instance (see Figure 6).

## DISCUSSION OF RESULTS

The present entrainment model can be used for further analysis. First the classical fresh/saline water entrainment process is studied, i.e. shear driven entrainment of a heavier fluid. It is assumed that $u_H$, $u_m$, $\partial p/\partial x$, $w_s$ and $H/W = 0$, and that $u_*$ and $\rho_b$ are constant. For equilibrium situations (at large $t$: $dU/dt \approx 0$), the solution of (9), using (4), becomes:

$$\frac{1}{u_*} \frac{dH}{dt} = \left[ \frac{c_* + c_H}{c_q + Ri_*} \right]^{1/2} \tag{13}$$

This is the well known empirical formula relating $E_*$ with $Ri_*^{-\frac{1}{2}}$. It is also drawn in Figure 2 for small $Ri_*$.

At large $Ri_*$ viscous effects become important, so that the lower layer will be dragged along with the upper layer ($U_H \approx U$), the variation in momentum $UH$ becomes negligible, and turbulence production is mainly due to side wall friction:

$$\frac{1}{u_*} \frac{dH}{dt} = \frac{c_w}{\left[ 2\lambda \frac{H}{W} \right]^{1/2} (c_q + Ri_*)} \tag{14}$$

This shows that for large Richardson numbers $E_*$ scales with $Ri_*^{-1}$, a relation often used in literature, and with $(H/W)^{-\frac{1}{2}}$. For large $Ri_*$ this formula is also presented in Figure 2.

The simulation of experiment K8 (Figure 5) clearly shows that work has to be done within the mud layer to erode the bed. This process can be accounted for by the sixth term in (9). However, this is subject to further study. Related to this, the (gradual) change from erosion by entrainment to the so-called "surface erosion" should be modelled as well. This would require a consolidation model and a rheological model relating bed density and bed strength, including the effects of time (thixotropy).

Another important next step is the application of the present results to full scale, prototype conditions. In principle, this can be done using the present integral entrainment model. However, in laboratory studies the influence of screen and/or side wall are dominant, whereas in nature turbulence is mainly produced at the interface.

Moreover, additional experiments with natural sediment will be carried out in a next phase of this study.

## ACKNOWLEDGEMENTS

The experiments were performed in the annular flume at the Hydromechanics Laboratory of the Civil Engineering Department of Delft University of Technology, The Netherlands. The skilful preparation and execution of the experiments by Mr. A. den Toom and Mr. H. Tas and Ms. M. Moot, members of the laboratory's staff and the help with the data processing by Mr. J. Reuber of the Technical University of Aachen are gratefully acknowledged. This study was carried out in the framework of The Netherlands Centre for Coastal Research and was co-sponsored by the Commission of the European Communities, Directorate General for Science, Research and Development under MAST contract No 0035-C.

# REFERENCES

Booij, R., Visser, P.J. & Melis, H., 1993, "Laser Doppler measurements in a rotating annular flume", Proceedings of the Fifth International Conference on Laser Anemometry Advances and Applications, August 1993, Veldhoven, The Netherlands, pp 409-416.

Imberger, J. & Hamblin, P.F., 1982, "Dynamics of lakes, reservoirs and cooling ponds", Annual Review of Fluid Mechanics, Vol 14, pp 153-187.

Kantha, L.H., Phillips, O.M. & Azad, R.S., 1977, "On turbulent entrainment at a stable density interface", Journal of Fluid Mechanics, Vol 37, pp 643-655.

Kirby, R., 1988, "High concentration suspension (fluid mud) layers in estuaries", in: Physical Processes in Estuaries, eds. J. Dronkers and W. van Leussen, Springer Verlag, pp 463 - 487.

Kranenburg, C., 1984, "Wind-induced entrainment in a stably stratified fluid", Journal of Fluid Mechanics, Vol 145, pp 253-273.

Kranenburg, C., 1987, "Boundary induced entrainment in two-layer stratified flows", Journal of Geophysical Research, Vol 92C, pp 5417-5425.

Kranenburg, C., 1994, "An entrainment model for fluid mud", Report 93-10, Delft University of Technology, Department of Civil Engineering.

Kusuda, T. & Futawatari, T., 1992, "Simulation of suspended sediment transport in a tidal river", Water Science Technology, Vol 26, No 5, pp 1421-1430.

Leussen, W. van & Velzen, E. van, 1989, "High concentration suspensions: their origin and importance in Dutch estuaries and coastal waters", Journal of Coastal Research, Special Issue No 5, Summer 1989, pp 1-22.

Le Hir, P., 1994, "Fluid and sediment integrated modelling: basis of a 2DV code and application to fluid mud flows in a macrotidal estuary", submitted to the Proceedings of INTERCOH'94, Oxford, July 1994.

Mehta, A.J. & Srinivas, R., 1993, "Observations on the entrainment of fluid mud by shear flow", in: Coastal and Estuarine Studies, Vol 42, ed. A.J. Mehta, American Geophysical Union, Washington D.C., pp 224-246.

Narimousa, S. & Fernando, H.J.S., 1987, "On the sheared density interface of an entraining stratified fluid", Journal of Fluid Mechanics, Vol 174, pp 1-22.

Price, J.F., 1979, "On the scaling of stress-driven entrainment experiments", Journal of Fluid Mchanics, Vol 90, pp 509-529.

Ross, M.A. & Mehta, A.J., 1989, "On the mechanics of lutoclines and fluid mud", Journal of Coastal Research, Special Issue No 5, Summer 1989, pp 51-62.

Odd, N.V.M. & Cooper, A.J., 1989, "A two-dimensional model of the movement of fluid mud in a high energy turbid estuary", Journal of Coastal Research, Special Issue No 5, Summer 1989, pp 185-194.

Sherman, F.S., Imberger, J. & Corcos, G.M., 1978, "Turbulence and mixing in stably stratified waters", Annual Review of Fluid Mechanics, Vol 10, pp 267-288.

Turner, J.S., 1973, "Buoyancy effects in fluids", Cambridge University Press, p. 160.

Visser, P.J., Booij, R. & Melis, H., 1994, "Flow field observations in a rotating annular flume", Proceedings of Conference on the Remediation of Sediments, November 1992, Rutgers University, New Brunswick, New Jersey, USA, to appear.

Winterwerp, J.C., Cornelisse, J.M. & Kuijper, C., 1993, "A laboratory study on the behaviour of mud from the Western Scheldt under tidal conditions", in: Coastal and Estuarine Studies, Vol 42, ed. A.J. Mehta, American Geophysical Union, Washington D.C., pp 295-314.

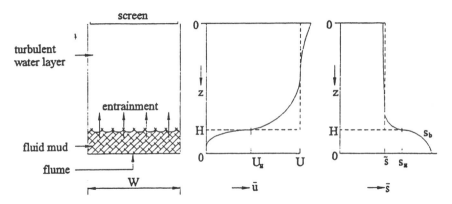

Fig.1 Definition sketch

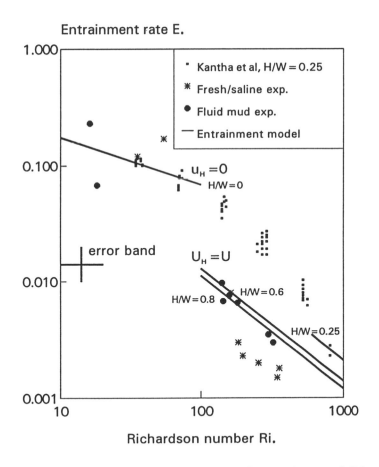

Fig.2 Entrainment rate as a function of Ri.

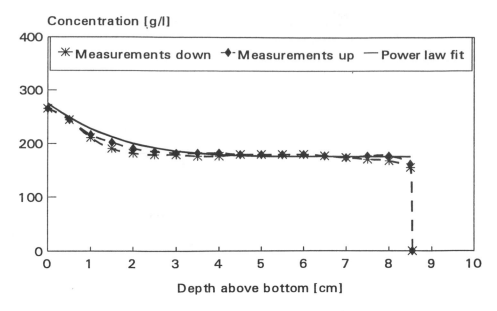

Fig. 3 Bed concentration after
120 minutes consolidation time

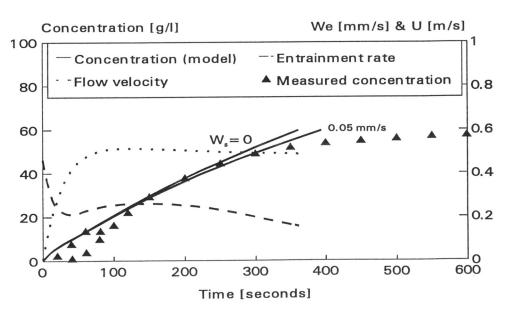

Fig. 4 Results for experiment K3
Comparison model with measurments

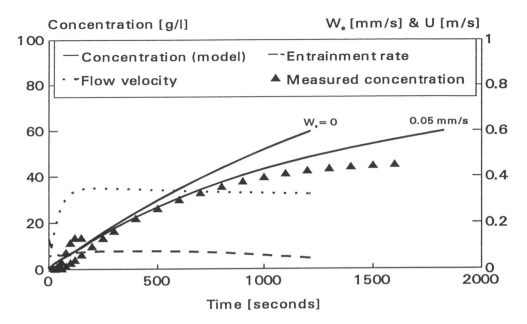

Fig. 5 Results for experiment K8
Comparison model and measurements

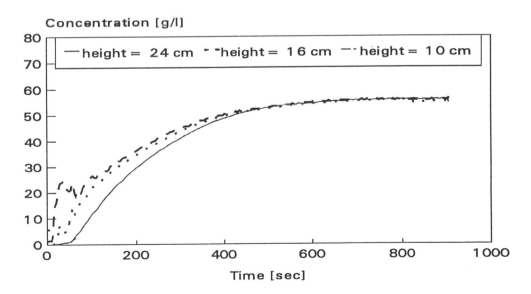

Fig. 6 Suspended sediment concentration
as a function of time; experiment K4

# 19 Critical shear stresses for erosion and deposition of fine suspended sediments of the Fraser river

**B. G. KRISHNAPPAN and P. ENGEL**
*Aquatic Ecosystem Protection Branch,*
*National Water Research Institute, Canada*

## INTRODUCTION

Models of cohesive sediment transport require parameters describing the erosion and deposition rate functions which, at present, can only be obtained by measurements in laboratory flumes using site specific sediments. Conventional straight flumes are not suitable for cohesive sediment studies because the transport processes are time dependent with time scales ranging from hours to days and would require excessively long flumes. Previous studies have shown that an alternative approach is to use circular flumes and to generate the flow by moving the flow boundaries rather than the fluid (Partheniades and Kennedy, 1967; Mehta and Partheniades, 1973; Kuijper et al, 1989; Petersen and Krishnappan, 1994; Krishnappan et al, 1994b). Such a circular flume was designed and built at the National Water Research Institute in Burlington, Ontario, Canada (Krishnappan, 1993) and was used to study the transport processes of sediment from the Fraser River and its tributary the Nechako River near their confluence. A brief description of the flume and the details of the transport study are presented in this paper.

## EXPERIMENTAL EQUIPMENT AND PROCEDURE

### The Flume

The flume is 5.0 m in mean diameter, 0.30 m in width, 0.30 m in depth and rests on a rotating platform. A counter rotating top cover, called the ring, fits inside the flume and makes contact with the surface of the sediment-water mixture in the flume. By rotating the platform and the ring assembly in opposite directions, it is possible to generate turbulent shear flows and to study the behaviour of fine sediments under different flow conditions.

*Cohesive Sediments.*
Edited by Neville Burt, Reg Parker and Jacqueline Watts
©1997 John Wiley & Sons Ltd.

The flume and the ring can each be rotated up to a maximum rate of three revolutions per minute. A sectional view of the flume is shown in Figure 1. The flow characteristics of rotating circular flumes have been studied by a number of investigators (Partheniades and Kennedy, 1967; Mehta and Partheniades, 1973; Krishnappan, 1993; Peterson and Krishnappan, 1994; Krishnappan et al, 1994b; Spork et al, 1994) and it was demonstrated that the rotation-induced secondary currents can be minimized by optimizing the rotational speeds of the flume and the ring.

## Instrumentation

The flume is instrumented with a Laser Doppler Anemometer (LDA) to measure the flow field, a Preston tube to measure the bed shear stress, a Malvern Particle Size Analyzer (MPSA) to measure the in situ particle size distribution in the flow and an optical turbidity sensor (OSLIM) from Delft Hydraulics to measure sediment concentration in the flow. The instrumentation activated for the present experiments were the Preston tube, the MPSA and the OSLIM sensor.

The Preston tube entered the flume at one location through a water tight sleeve attached to the outside flume wall. The difference between the dynamic and static pressures was measured with a Valedyne model DP-45 pressure transducer with a diaphragm having a pressure capacity equivalent to 25.4 millimeters of water. The calibration proposed by Patel (1965) was adopted to compute the shear velocities from the pressure difference measured with the pressure transducer.

The MPSA operates on the light diffraction principle (Fraunhoffer diffraction). Complete details of the method can be found in Weiner (1984). Continuous in situ measurements of the floc sizes were made by mounting the MPSA on the rotating platform below the flume so that the flow-through sensor was located directly below the centre-line of the flume cross-section. The sediment suspension was drawn continuously from the flume by gravity through a 5 mm tube, located on the centre-line of the cross-section, having its intake above the bed at approximately one half of the flow depth, through the sensor and into a reservoir from where it was pumped back into the flume. The end of the tube was bent at a right angle, similar to a Pitot tube, so that the intake would face directly into the flow on the centerline of the cross-section. The length of the withdrawal tube was kept to a minimum to avoid flocs disruption and the withdrawal rate was just large enough to avoid deposition of sediment in the tubes.

The measuring principle of the OSLIM sensor is based on the attenuation of a light beam, caused by light absorption and reflection by the sediment particles. The light source used by this instrument is an infra red light emitting diode (LED). The light absorption is dependent on the particle size distribution but by pumping the suspension through a short, small diameter tube just ahead of the sensor and allowing the shear in the tube to break up the flocs, the error caused by varying size distributions was kept to a minimum. The sediment suspension was pumped continuously from the flume through a 5 mm tube which entered the flume horizontally through the flume wall at approximately a height of about half the flow depth above the flume bed. Once again the tube was bent at a right angle so that the intake would face directly into the flow. The suspension was pumped from the withdrawal tube through the OSLIM sensor and back into the flume downstream of the withdrawal point at a rate just large enough to prevent deposition of sediment in the pumping circuit.

## Measurement of Shear Stress

The Preston tube cannot be used in flows with suspended sediment because the presence of the sediment particles inhibits its functioning. Therefore, a relationship between the bed shear stress and the rotational speed of the flume was established using clear, distilled water prior to the tests with suspended sediment. For the chosen water depth of 12 cm, the ratio of rotational ring speed to rotational flume speed was determined to be 1.167. This gave the best two dimensional cross-sectional shear stress distribution for this flow depth. Details of the shear stress experiments are described in Krishnappan et al (1994b). For a given linear relative flow velocity, computed from the rates of rotation of the flume and ring, the bed shear stress can be obtained from the curve in Figure 2.

## Collection and Preparation of River Water Sample

River water samples were collected on the Fraser and Nechako rivers upstream of two pulp mills. A pumping system, similar to that used for cleaning swimming pools, was used to vacuum the sediment deposited on the gravel bed and pump river water into 100 litre plastic containers which could be sealed tightly for shipping. Eight such containers were filled at each river site for a total of 800 litres. Care was taken to obtain representative samples of the entire cross-section at each site. The effect of the effluent from the pulp mills was also investigated in this study. For this, a 50 litres of effluent sample was collected from the discharge wells of the pulp mills and put into a suitable container for shipping.

All containers were kept in refrigerated storage for a sufficient length of time to allow the sediment to settle. The supernatant was then drawn off the sediment and 500 litres were placed in the flume and the remainder put into separate containers and returned to refrigerated storage. The remaining sediment slurry was combined into one container. The concentration of the slurry was determined so it could be added to the water in the flume resulting in water-sediment mixtures of known initial concentrations.

## Deposition and Erosion Tests

Before beginning a deposition test, the water-sediment suspension was thoroughly mixed in the flume with a mechanical mixer to break up existing flocs. The ring was then lowered to the desired position resulting in a water depth below the ring surface of 12 cm. The ring penetrated the water surface by about 3 mm to ensure proper contact between the ring and the water surface. The MPSA and the OSLIM sensor were then checked to ensure that they were operating properly. Having completed all preparations, the flume and ring were set in motion to run at $N_t = 2$ rpm and $N_r = 2.5$ rpm respectively to obtain a suspended sediment concentration in excess of that sustainable by the flow at the chosen test speeds. After twenty minutes, the flume and ring were slowed down to their respective test speeds in accordance with the established ratio $N_t/N_r = 1.167$ for the water depth of 12 cm. Samples were withdrawn from the flume at intervals of 5 minutes during the first hour of the test and every ten minutes thereafter until completion of the test. Each time a sample was drawn, the volume removed was replaced by clear river water. A test was considered to be complete after the suspended sediment concentration remained virtually constant for at least 1 hour. The samples were taken at mid-depth as tests had shown that the concentration was virtually uniform over the depth. The concentration of each sample was determined by filtration, drying and weighing. Simultaneously, with the manual sampling, sediment concentration was determined with the OSLIM sensor to evaluate its performance under operating conditions. Its main contribution was the ability to obtain real time concentrations which permitted accurate observations of the rates of change in concentration as deposition took place. The size distribution of the suspension was measured at regular intervals with the MPSA and recorded on a disc file and computer print-out. This way the formation of flocs and changes in the size distribution of the flocs with time could be monitored. Once a test was completed, the procedure was repeated for other flume speeds.

Before beginning an erosion test, the water-sediment mixture in the flume was left undisturbed for a chosen length of time to allow the sediment to settle and consolidate on the flume bed. Care was taken to ensure that the water depth was 12 cm, that the ring penetrated the water surface sufficiently and that the MPSA and the OSLIM instruments were operating properly. Having completed all preparations, the flume and ring were set in motion beginning with the lowest flume speed. As before, samples were withdrawn from the flume and the change in concentration monitored with the OSLIM at intervals of 5 to 10 minutes until the sediment concentration became virtually constant. When this stage was reached, the flume and ring were set to the next speed. This sequence was repeated until the largest desired flume speed was reached. The size distributions of the suspended sediment were again measured at regular intervals with the MPSA and recorded on a disc file and computer print-out.

## RESULTS AND DISCUSSION

Altogether, sixteen tests were carried out for the Fraser River sediment. A summary of the experimental conditions is given in Table 1.

## Table 1. Summary of experimental conditions

| Test No. | Experiment Type | Shear Stress N/m² | Initial Conc. Mg/l | Age of deposit in hr | Effluent conc in %by vol. |
|----------|-----------------|-------------------|--------------------|----------------------|----------------------------|
| 1 | Deposition | 0.056 | 200 | n/a | 0 |
| 2 | Deposition | 0.121 | 200 | n/a | 0 |
| 3 | Deposition | 0.169 | 200 | n/a | 0 |
| 4 | Deposition | 0.213 | 200 | n/a | 0 |
| 5 | Deposition | 0.056 | 250 | n/a | 0 |
| 6 | Deposition | 0.121 | 250 | n/a | 0 |
| 7 | Deposition | 0.169 | 250 | n/a | 0 |
| 8 | Deposition | 0.213 | 250 | n/a | 0 |
| 9 | Deposition | 0.056 | 250 | n/a | 3 |
| 10 | Deposition | 0.121 | 250 | n/a | 3 |
| 11 | Deposition | 0.169 | 250 | n/a | 3 |
| 12 | Deposition | 0.213 | 250 | n/a | 3 |
| 13 | Erosion | n/a | n/a | 114 | 0 |
| 14 | Erosion | n/a | n/a | 164 | 0 |
| 15 | Erosion | n/a | n/a | 40 | 2 |
| 16 | Erosion | n/a | n/a | 114 | 2 |

## Deposition Tests

Deposition experiments were carried out for two different initial concentrations and four different bed shear stresses. The influence of the effluent was examined in the tests with the higher initial concentration. The time variation of concentration is shown in Figure 3 for the first four tests. In this figure, the 20 minutes period during which the flume and the ring were rotated at high speeds for initial mixing was also included. It can be seen from this figure that, after the initial period, the concentration decreases gradually and reaches a steady state value for all the runs. For the lowest bed shear stress (i.e. 0.056 N/m²), the drop in the concentration is the biggest; the steady state concentration is only about 1/20th of the initial concentration. A slightly lower shear stress would have produced a nil concentration of sediment in suspension and would correspond to the critical shear stress for deposition for this sediment. As the bed shear stress increases, more and more sediment stays in suspension during the steady state.

Figure 4 shows the depositional characteristics of the sediment under constant bed shear stress but with different initial concentrations. It can be seen that the steady state concentration is a function of the initial concentration and the ratio of the two is a fixed value for a particular bed shear stress. For these tests the ratio is about 0.50. Such behaviour is typical of cohesive sediments and conforms to earlier studies by a number of investigators such as Partheniades and Kennedy (1967), Mehta and Partheniades (1973) and Lick (1982). In a recent study, Lau and Krishnappan (1994) had reconfirmed the earlier notion put forth by Partheniades and Kennedy (1967) and Mehta and Partheniades (1973) that during cohesive sediment settling, there is no simultaneous erosion and settling of the sediment near the bed. This provided further support for the concept that the sediment settles in a flocculated form and only those flocs which are strong enough to settle through the region of high shear near the bed can deposit on the bed. Weaker flocs are broken up at the region of high shear and are brought back into suspension. As there can only be a certain fraction of sediment in the mixture that can form stronger flocs, the amount remaining in suspension becomes a function of the amount in the initial suspension.

Flocculation of the sediment during deposition can be inferred from the size distribution data shown in Figure 5. In this figure, the size distribution of the sediment suspension as measured with the MPSA at different times during deposition for Test No. 4 are shown. During the initial mixing period, the weaker flocs are broken up and the suspension contains sediment flocs with a median size of about 74 microns. As the deposition begins, weaker flocs are reformed and the median size of the flocs increases and attains a steady state value of around 110 microns. This trend is common to all of the deposition tests except the ones with the lowest bed shear stress. For these tests, the deposition of sediment continues without the formation of larger sediment flocs. This may be due to lower concentration of sediment in suspension and decreased turbulence level. The size distribution pattern for Test No. 1 is shown in Figure 6. Similar distributions were obtained for all the tests with the lowest bed shear stress of 0.056 N/m$^2$, which is closer to the critical shear stress for deposition.

When extrapolating the laboratory data on deposition of sediment to field conditions in the Fraser River, it is important to consider the similarity in velocity gradient rather than the similarity in bed shear stress. In laboratory channels with smooth bed, where the viscous sublayer thickness could be several times larger than the particle size, the sediment flocs could be subjected to much larger velocity gradients within the viscous sublayer in comparison to a flow in the field where the bed roughness elements are likely to protrude through the viscous sublayer. One possible solution to the problem could be to derive a scale relationship for the shear stress by assuming similarity in the shear rate between the flows in the flume and in the river. This would require a knowledge of the roughness characteristics of the river flow.

The effect of pulp mill effluent on the depositional characteristics of the sediment is shown in Figures 7 and 8. From these figures, it can be seen that the pulp mill effluent does have an effect on the depositional behaviour of the sediment and it increases the rate of deposition. A possible explanation for this behaviour is that the pulp mill effluent has enhanced the flocculation of the sediment and thereby increased the settling velocity and the deposition rate. A field study carried out in the Athabasca River downstream of a pulp mill at Hinton, Alberta, Canada (Krishnappan et al, 1994a) also showed evidence of increased deposition rate of the ambient sediment.

## Erosion Tests

The experimental conditions for the erosion tests are given in Table 1. Two different ages of sediment deposit and the influence of the pulp mill effluent on the erosional behaviour were tested. A typical result of an erosion test is shown in Figure 9 for Test No. 14 corresponding to a deposit age of 164 hours. In this figure, the shear stress steps and the corresponding concentration profiles are shown. The sediment deposit is fully stable until the bed shear stress step of 0.121 $N/m^2$ is established. After initiation, the sediment concentration in suspension gradually increases and attains a steady state value for each shear stress step. For the maximum shear stress (0.462 $N/m^2$) tested, not all the deposited sediment was re-suspended. The maximum concentration reached was only about sixty percent of the total concentration that would have resulted from complete re-suspension.

The effect of the age of the deposit and the pulp mill effluent are shown in Figures 10 and 11 respectively. In these figures, the steady state concentration for each shear stress step is plotted against the shear stress. It can be seen from Figure 10 that the age of deposit does have an effect at least during the initial stages of erosion. When the age of deposit is higher, the erosion resistance of the sediment is higher and hence less sediment is re-suspended. From Figure 11, it can be seen that the influence of the pulp mill effluent on the erosion process is not as significant as it is for the deposition process shown in Figure 7 & 8.

The size distribution of the re-suspended sediment measured during the erosion tests with the MPSA sheds some further light on the erosion process. From the size distribution data shown in Figure 12, which corresponds to Test No. 14, it appears that the sediment bed is peeled off during the erosion process and the re-suspension contains a large percentage of larger sized flocs. (solid line-0.121 $N/m^2$). As the bed shear stress is increased (0.169 $N/m^2$-0.213 $N/m^2$-0.259 $N/m^2$), these larger flocs break up and become finer and finer and attain a distribution similar to the ones obtained during the initial stages of the deposition experiments.

## SUMMARY AND CONCLUSIONS

Depositional and erosional characteristics of the Fraser River sediment were studied in a rotating circular flume. The influence of the pulp mill effluent on these processes was also examined. The laboratory measurements show that the sediment exhibits transport characteristics peculiar to cohesive sediments.The depositional process is dominated by the flocculation of the sediment and the pulp mill effluent further enhanced the flocculation mechanism. The erosion process of the deposited sediment is characterized by a peeling off of the top layer of the sediment bed rather than by the mobilization of individual particles normally encountered incohesionless sediment.

## ACKNOWLEDGMENT

The writers gratefully acknowledge the contribution of R. Stephens in setting up and conducting the experiments and J.Marsalek for his review of the manuscript.

# REFERENCES

Kuijper, C., J.M. Cornelisse and J.C. Winterwerp, 1989. Research on Erosive Properties of Cohesive Sediments. J. Geophysical Res. 95(C10), 14341-14350.

Krishnappan, B.G., 1993. Rotating Circular Flume. J. Hydraul. Engrg. ASCE 119(6), 758-767.

Krishnappan, B.G., R. Stephens, J.A. Kraft and B.H. Moore, 1994a. Size Distribution and Transport of Suspended Particles in the Athabasca River Near Hinton. NWRI Contribution 94-112, National Water Research Institute, Burlington, Ontario, Canada, 33 pp.

Krishnappan, B.G., P. Engel and R. Stephens, 1994b. Shear Velocity Distribution in a Rotating Flume. NWRI Contribution 94-102, National Water Research Institute, Burlington, Ontario, Canada, 23 pp.

Lau, Y.L. and B.G. Krishnappan, 1994. Does Reentrainment Occur During Cohesive Sediment Settling ?. J. Hydraul. Engrg. ASCE 120(2), 236-244.

Lick, W., 1982. Entrainment, Deposition and Transport of Fine Grained Sediments in Lakes. Hydrobiologia (91), 31-40.

Mehta, A.J. and E. Partheniades, 1973. Depositional Behaviour of Cohesive Sediments. UFL/COEL-TR/016, Coastal and Oceanographic Engineering Department, University of Florida, Gainesville, Florida, 297 pp.

Partheniades, E. and J.F. Kennedy, 1967. Depositional Behaviour of Fine Sediment in a Turbulent Fluid Motion.Proc. 10th Conf. on Coastal Engineering, Tokyo, Japan 2, 707-729.

Patel, V.C., 1965. Calibration of the Preston Tube and Limitations on its Use in Pressure Gradients. J. Fluid Mech. 23(1), 185-208.

Petersen, O. and B.G. Krishnappan, 1994. Measurement and Analysis of Flow Characteristics in a Rotating Circular Flume. J. Hydraul. Res., IAHR, Vol. 32(4). pp. 483-494.

Spork, V., P. Ruland, B. Schneider and G. Rouve, 1994. A New Rotating Annular Flume for Investigations on Sediment Transport. International Journal of Sediment Research. Vol. 9 Special Issue. pp. 141-147.

Weiner, B.B., 1984. Particle and Droplet Sizing Using Fraunhofer Diffraction. Modern Methods of Particle Size Analysis, H.G. Barth, ed.,John Wiley and Sons, New York, N.Y., 135-172.

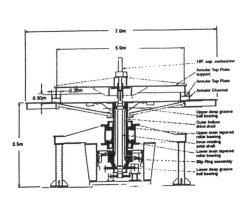

Figure 1. Sectional view of the rotating flume assembly

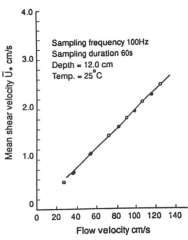

Figure 2. Two dimensional flow shear velocity as a function of flow velocity.

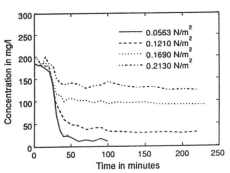

Figure 3. Variation of concentration for different shear stress

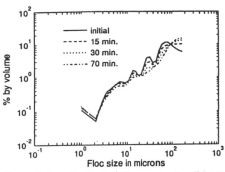

Figure 5. Size distribution of suspended sediment in Test No. 4

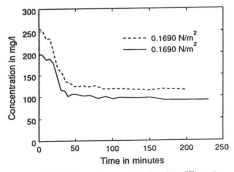

Figure 4. Variation of concentration for different initial concentrations

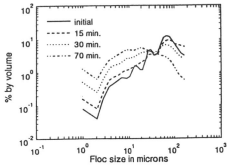

Figure 6. Size distribution of suspended sediment in Test No. 1

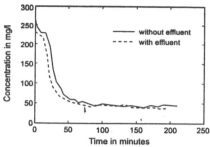

Figure 7. Effect of pulp mill effluent during deposition - shear stress = 0.121N/m²

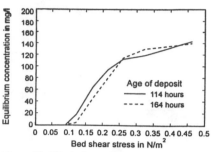

Figure 10. Effect of age of deposition on erosion

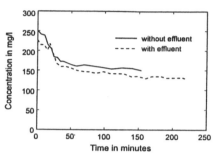

Figure 8. Effect of pulp mill effluent during deposition - shear stress = 0.213N/m²

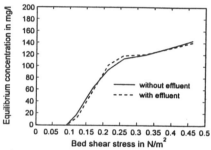

Figure 11. Effect of pulp mill effluent on erosion

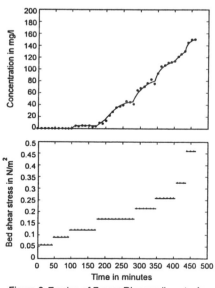

Figure 9. Erosion of Fraser River sediment - Age of deposit: 164 hours

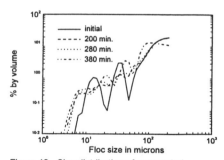

Figure 12. Size distribution of suspended sediment in Test No. 14

# 20 Interpreting observations of near-bed sediment concentration and estimation of 'pick-up' function constants

**J. N. ALDRIDGE and J. M. REES**
*MAFF Fisheries Laboratory, Lowestoft, UK*

## 1. Introduction

The physics of the interaction of an erodable bed, particularly a cohesive one, with the turbulent flow above it is sufficiently complex that, at present, the only hope of incorporating it into large scale sediment transport models is via gross parameterisations involving empirically derived 'constants'.

Although laboratory work is invaluable in trying to understand the physical processes, it has been recognised, particularly for cohesive sediment, that observations taken in the field are necessary to find out what actually happens in the natural environment. Thus, a number of instrument packages, e.g. Butman & Folger (1979), Sternberg et al. (1991), Wright et al. (1992), Diserens (1993), Humphrey & Moores (1994), have been developed to monitor the near bed flow and sediment concentrations under field conditions. One such set of instruments are the Tetrapod (Green *et. al.* 1992) , Quadrapod and 'Minipods', developed by the COhesive SEdiments Dynamics Study (COSEDS) group, a collection of workers based at a number of UK universities and research institutes. This paper briefly describes the instruments, shows data from two long deployments in the eastern Irish Sea, and outlines preliminary work in using the data to obtain values of empirical constants for parameterisations of bed erosion.

## 2. Instrumentation

Three types of bottom landing instrument packages have been developed. All are based on a central logging system that stores data from a configurable set of sensors onto a hard disc. The central logger can control each sensor, including the ability to dynamically switch it on and off and vary the sampling rate. Typically, foreground and background modes are defined with two different levels of data collection. Transition to foreground mode, where all instruments are sampling, can be based on significant wave height if storm events, for example, are deemed to be of particular interest. The instrument packages are described in more detail below.

*Cohesive Sediments.*
Edited by Neville Burt, Reg Parker and Jacqueline Watts
© Crown copyright, 1996. Published 1997 by John Wiley & Sons Ltd.

**Tetrapod** Designed to be hydrodynamicaly 'clean', in its normal configuration it has an array of five electromagnetic current meters (ECMs) ranged above the bed from 50 to 120cm, a pressure sensor for estimation of wave activity, and three devices for measuring sediment concentrations:

1. two transmissometers,
2. four mini optical backscatter (MOBS) probes,
3. a three frequency acoustic backscatter (ABS) probe.

An acoustic travel time sensor (BASS) is presently being tested for three component, high frequency measurement of the turbulent flow. Due to the flexibility of the logging system, additional sensors or different configurations of the present ones can be quite easily accommodated.

**Quadrapod** Less hydrodynamicaly clean than the Tetrapod. Mounted on it are MOBS sensors; water samples can be taken at pre-programmed times for MOBS calibration. Presently, a settling box for measuring fall velocity *in-situ*, and a sideways looking acoustic device for assessing bed roughness are being developed. Also being developed is a sensor to measure bed porosity also *in-situ*.

**Minipods** These are cut down versions of the Tetrapod, enabling several to be built and deployed, and giving information on spatial variability at a site. A single ECM, two MOBS, a pressure sensor and a two frequency ABS are mounted.

3.       *Comparison of Morecambe bay and St. Bees deployments.*

The data from two Tetrapod deployments, both in the Eastern Irish sea, are examined here in some detail. Deployment 86 was outside Morecambe bay in 20m of water and on a mixed substrate of gravel, sand and mud. Deployment 77 was off the Cumbrian coast at St. Bees over a muddy sand bed with a water depth of 21m. In both cases the equipment operated successfully for the full duration of the deployment (about a month) and captured data for a variety of wave and tide conditions.

It is interesting to compare and contrast, the two data sets by looking at plots of current speed, wave activity and MOBS over the whole deployment. Figures 1a,1b,1c are plots of near-bed current, uncalibrated MOBS signal, and near-bed wave orbital velocity respectively, for the Morecambe bay deployment. The velocities are dominated by a fairly strong semi-diurnal tide ($U_{max}$ ~0.5 m/s, figure 1a) with a clear Spring Neap variation cycle. There are also peaks corresponding to wind events. We see the expected correlation between wave activity, figure 1c, and MOBS activity, figure 1b. The relatively strong tides also impose a signal, however. The middle spring tide regime, which coincides with a period of very little wave activity, is clearly discernible in the MOBS record.

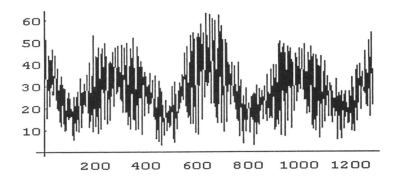

Figure 1a: Current meter record. (cm/s). Morecambe bay deployment. Horizontal scale is hours.

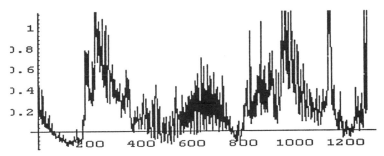

Figure 1b: MOBS record (uncalibrated). Morecambe bay deployment. Horizontal scale is hours.

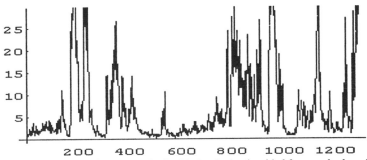

Figure 1c: Significant near-bed orbital velocity (cm/s). Morecambe bay deployment. Horizontal scale is hours.

By contrast, the St Bees deployment, in a much weaker tidal regime (U < 0.3 m/s), shows no significant resuspension without wave activity (figures 2a,2b,2c). This supports the view that only during storm events is there likely to be significant resuspension and hence transport of material off the Sellafield mud patch.

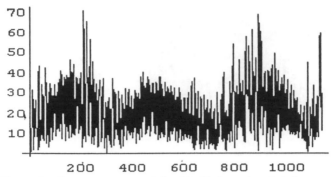

**Figure 2a**: Current meter record. (cm/s). Horizontal scale is hours. St Bees bay deployment.

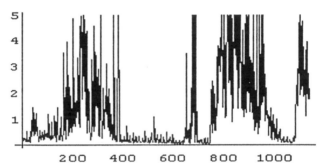

**Figure 2b** : MOBS record (uncalibrated). Horizontal scale is hours. St Bees bay deployment.

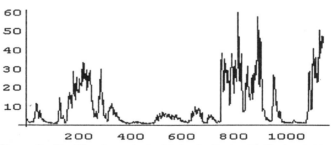

**Figure 2c** :Significant near-bed orbital velocity (cm/s). Horizontal scale is hours. St Bees deployment.

Coming to a more detailed understanding of what is happening on a tidal cycle by tidal cycle basis is more difficult. At these time scales the relation between current velocity, wave activity and the MOBS signal is not always easy to interpret. Some examples of records from the Morecambe and St Bees deployments are now given, together with interpretations regarding the physical processes that might be responsible for the observed time series.

## 3.1    Deployment 86 - Outside Morecambe bay

Figure 3 shows a record of current velocity and MOBS signal (uncalibrated at present) both measured at ~1m from the bed and both representing hourly 10 minute averages. The vertical axis scale is the current velocity and the horizontal scale is the time in hours since the start of the deployment

We see a characteristic `twin peaked' signal that appears frequently in the Tetrapod data at this site. This sort of signal is associated with the advection of a concentration gradient past the sensor, together with local settling and pickup as the tide turns - see for example Jago *et. al.* (1994) or Weeks et. el. (1993). There seems to be a lag here between minimum current and minimum MOBS signal, usually the two coincide. Such lags occur occasionally in the Tetrapod records with no obvious explanation.  Here it may possibly be due to fine material falling out of suspension only slowly. At the time of the deployment the instruments did not have the ability to take water samples for an analysis of suspended particle size.

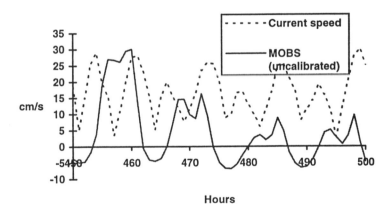

**Figure 3**: 'Twin peaked' advective signal measured at 1m from the bed

Figure 4 shows an interesting sequence during a period of minimal wave activity. A 'twin peak' advective signal (records 500-540) gives way a regime in which there is a direct relation between the current speed (i.e. bed stress) and the amount of material in suspension (560-600).  The latter is characteristic of local erosion and deposition of material with no supply limitation.

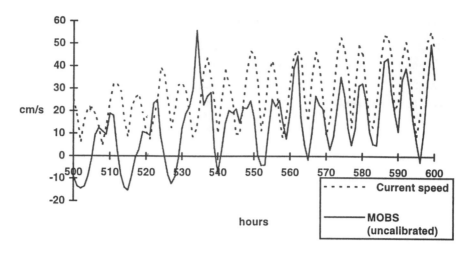

**Figure 4:** Transition from advection to local pickup measured at 1m from the bed

It is difficult to suggest a mechanism to account for this transition. It occurs as we move into a spring tide regime; thus it is possible that the higher stress is eroding bed material that could not be eroded during the neaps. However, there is no evidence of a critical erosion stress, material seems to be coming off the bed immediately the flow accelerates and therefore at stresses that occurred in the previous neap regime. It appears more as if a local supply of material became available at around record 550 but there is no obvious indication that a large supply was advected in just prior to this. See Aldridge (1996) for a possible explanation of this section of the time series.

Figure 5b shows the significant wave orbital velocity from another period in the deployment. The orbital velocity is not measure directly but is calculated from the pressure record. Pressure variations are converted to a corresponding wave height, assuming linear water wave theory. A significant wave height is calculated and a corresponding orbital velocity just outside the wave boundary layer is calculated again using linear wave theory.

There are a large number of other records that are not easy to understand, for example figures 5a,b. Here there are drops in concentration not obviously connected with the tidal signal or changes in wave activity. Some parts of the record could be material being advected out, and other parts are possibly local pickup from the bed. Further complications occur due to a possible mixture of particle sizes in suspension. In general, this section of the time series is complex and not easy to interpret. When a storm event occurs it is often very difficult to tell if the corresponding peak in concentration is material coming off the bed or if it is being advected in. Advection is generally only distinguishable when a relatively stable concentration gradient is advected by tidal flows.

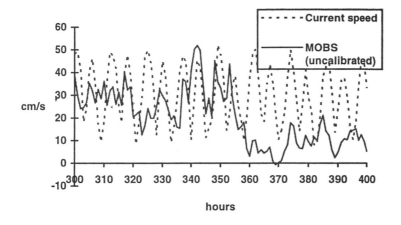

**Figure 5a**: MOBS and current meter record 1m from the bed.

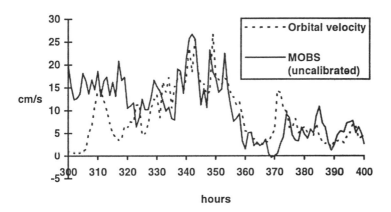

**Figure 5b**: MOBS and significant wave orbital velocity just outside wave boundary layer.

In summary, having examined the record from the whole deployment, this site appears to be primarily an advective, supply limited region. This does in fact correspond with what is known about the site. A study of the sediment transport regime in Morecambe bay (McLaren 1989), was unable to obtain grab samples in the region in which the Tetrapod was deployed. Of four grab samples taken on recovery of the Tetrapod, two brought nothing the other two brought up a sand/gravel mixture and a sand over mud mixture. Because of the mixture of material on the bed, the predominance of advective transport and the inability to take water samples, it is not possible to determine the size of the particles being transported. This may be done indirectly by fitting models to the data and seeing which values of fall velocity give similar behaviour to that observed.

Nearly all significant MOBS signal for this deployment is connected with wind wave activity. Thus in figure 6a,b we see virtually no activity in the MOBS record until the wind wave activity becomes significant after record 750 (fig 6b). Although the wave activity affects the amount of material in suspension, the tidal signal still controls the temporal variation of the signal (fig 6a). Thus in this record for example, there is a close relation between the MOBS signal and the peaks and troughs of the tidal current, indicating local pickup of sediment. Drops in concentration at slack water occur, even though the wave induced bed stress is high, because there is no current boundary layer to diffuse the material to the level of the MOBS sensor at ~1m from the bed. We would expect material to still be in suspension within the wave boundary layer (i.e. within say ~ 0.1m of the bed).

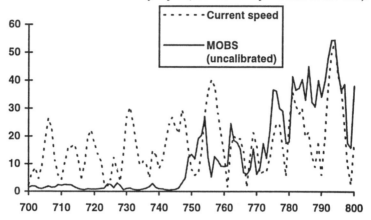

**Figure 6a**: Ecm and MOBS records 1m from the bed.

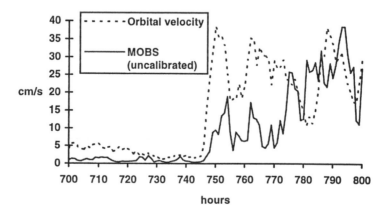

**Figure 6b**: MOBS and significant wave orbital velocity just outside wave boundary layer.

Other records at the same site indicate an advective signal during periods of wave activity, e.g. figure 7a, b. This record also shows some puzzling behaviour. Where there is no wave activity, before record 120, there are small "kickups" of material on one side of the tide only; this occurs for extended periods during tidal dominated sections at this site. It is not clear why there is activity only when the current is flowing in one direction. As the wave activity increases we begin to see 'twin peak' signals indicating advection of material in and out of the area.

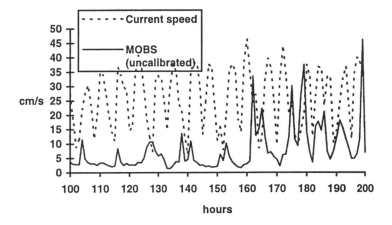

**Figure 7a**: MOBS and ecm record showing 'twin peaks' type advective signal

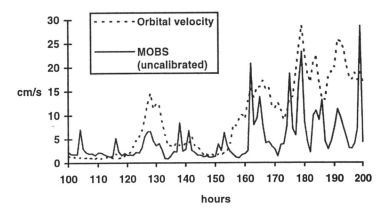

**Figure 7b**: MOBS and wave activity record just outside wave boundary layer.

As with the Morecambe bay site, there are records that appear to show a transition from an advective to local pickup of material, although at the St. Bees site wave activity was required to get significant material into suspension.

The fairly rapid fallout of material at times of slack water indicates that the particles have a relatively large fall velocity. In the MOBS record at least, there does not appear to be evidence of large quantities of fine material in suspension even during storm events, although it is possible that the material is flocculating at slack water. An added complication is the instrument sensitivity to different particle sizes. Instruments currently being developed to take water samples and to make in *situ* measurements of fall velocity should be available for future deployments and will be invaluable in aiding the interpretation of the data.

In summary, for the St Bees deployment, only wave activity can get large amounts of material into suspension; however, once this occurs the tides play a major role in determining the temporal variations in concentration. The relatively rapid response to changes in current speed seems to indicate relatively large fall velocities. This is surprising for a site where the bed material has a large mud content. Again there are sections of the MOBS time series that are very difficult to interpret.

### 3.3    Evidence of cohesive behaviour

It is not clear how cohesive effects in the bed and in the suspended material would be evident in records of concentration of the sort shown above. Certainly deployment 77 over muddy sand should contain periods when muddy material was in suspension, but there is no obvious evidence that this is having a significant effect on the observed behaviour (for example, the existance of a critical deposition stress). Measurements of fall velocity distribution *in-situ* and water sample collection will be essential requirements in interpreting the MOBS data. The other key component is knowledge of the condition of the bed.

### 4.    Extracting quantitative information

Although it is extremely valuable to inspect records to try to understand, in a general way, what processes are operating at a particular site, it is desirable to try and extract quantitative information from the records. In particular, information on critical shear stress, and erosion rate constants are required in parameterisations of sediment erosion and deposition in large scale models of sediment transport.

In order to obtain values of parameters that cannot be measured directly, it is necessary to set up a model of the system and make parameter estimations by fitting the model to the observations. In our case the observations are time series of concentrations measured at a number of points above the bed. In previous examples of this approach e.g. Lavelle & et al. (1984), Luettich *et. al.* (1990), it was possible to use a model based solely on local pickup and neglect both advective processes and supply limitation of available bed material. For the Irish sea tetrapod observations described above, neither of these assumptions is valid (a similar conclusion was made by Jago *et.al.* 1994 for their North sea deployments).

The following set of conservation equations describe the relation between the quantity of material in the water column, $C$ (units kg m$^{-2}$), and the amount on the bed, $b$,

$$\frac{db}{dt} = -f_{-h} \tag{1}$$

$$\frac{dc}{dt} = f_{-h} + f_a \tag{2}$$

The flux from the bed is parameterised in the form $f_{-h} = -w_f(\beta c - c_r)$, where $w_f > 0$ is the fall velocity. This form of boundary condition causes the amount of material in suspension to relax toward a steady state equilibrium. The quantity $c_r = c_r(\tau)$ is the steady state reference concentration at the bed, and is taken to be a function of the instantaneous bed stress $\tau$. As an example, assuming no critical erosion stress, we might have $c_r = \gamma (\tau / \tau_{ref})^n$ where $\gamma$ is to be estimated by fitting the model to the observations. In the absence of advection, the concentration at the bed can be related to the amount of material in the water column by assuming (for example) a Rouse type steady state distribution of concentration and multiplying by a factor $\beta$ (say). This factor $\beta$ is a function of the ratio of fall velocity to the friction velocity and also depends on the water depth.

The value of the fall velocity is not known and cannot be easily determined from the field data. In principle it should be possible to estimate a value from the data itself by fitting the model with $w_f$ as one of the parameters to be extracted. Unfortunately the Kalman filter approach requires a linear problem and the fall velocity appears non-linearly in $\beta$. We therefore specified this parameter 'by hand'; for the example shown below it was set to 0.001 m s$^{-1}$

The advective flux is $f_a = u(t)a(t)$, where $u(t)$ is the measured local flow velocity and $a(t)$ is an unknown concentration gradient that has to be determined from the observations to quantify the advective input.

Bed stress is estimated from the measured current velocity using a quadratic law. The value of the drag coefficient assumes a particular value of the bed roughness, which can, in principle be obtained from the data, either by fitting log profiles or using the inertial dissipation method, (Green 1992). For this very preliminary study no attempt was made to determine a local roughness value and a standard drag coefficient of 0.005 was used. It will be vital to include wave induced stress and wave current interaction into any realistic attempt to relate bed stress to sediment pickup. This can be done using a wave current interaction model or directly from the COSEDS data via the inertial dissipation method.

The above model is obviously very crude, a major simplification is that we deal only with the total quantity of material in suspension, rather than with the depth varying concentration profile. This is to keep the preliminary model as simple as possible. However, the use of $\beta$ derived from a simple assumption of sediment distribution with depth, assuming a steady state and zero advection, is clearly a deficiency. At present, the measured signal from the MOBS is simply assumed to be proportional to the amount of material in suspension and a nominal conversion applied. Sections of the data are used simply as example data to test methods of extracting information by model fitting, as we now discuss. When only local pickup is assumed, a small number of parameters connected with the pickup function have to be estimated. This has generally been done by systematically changing the values until reasonable agreement with the observations is found. Introducing a time varying advection parameter increases the number of unknowns to a point where more sophisticated methods seem to be necessary. It could be argued that if the main aim is to obtain information on erosion constants, then it is only worth trying to fit data where local pickup is clearly occurring and where advective effects are not dominant. However, there is information on critical deposition stress and possibly particle fall velocity, even in the advective 'twin peaks' type profiles, when material settles around slack water. It is also more satisfactory to have a unified model that can fit observations for a range of conditions and so increase our confidence that we are correctly describing the physical processes. Two techniques for carrying out the parameter estimation have been investigated.

### 4.1  Kalman filter algorithm

The Kalman filter provides an algorithm for updating estimates of the model state variables in the light of observed data. In our case, discrete versions of equations (1) and (2) are supplemented with a stochastic relation for the unknown concentration gradient and a relation for $\gamma$ which says essentially that it is a constant. Thus we put

$$a_n = \alpha a_{n-1} + q \tag{3}$$

$$\gamma_n = \gamma_{n-1} \tag{4}$$

where q is a random variable. The equations are then time stepped starting from some guessed initial state; at each stage an observation on the state variable $C$ is assimilated into the model to yield an estimate of the current state vector. A correction is applied to ensure that the material on the bed remains non-negative. Strictly, this alters the model so that it is no longer linear, as assumed in the Kalman filter algorithm. However, this does not seem to cause problems in tests with simulated data, nor when applied to the 'real' data. A further non-linearity will arise if the pickup function does not depend on the constant to be fitted in a linear way.

Typical results are shown in figure 8a,b. Here estimates of the state of the system, which include the quantity of material in suspension, the quantity on the bed, the advective flux, and the pickup parameter $\gamma$, are fitted for every observed value. The estimated value of $\gamma$ is shown in figure 8b and settles down to a reasonably constant value. Again we emphasise that at present the observed data being assimilated into the model is purely fictitious in that it is an uncalibrated MOBS signal, arbitrarily scaled to represent the total material in suspension.

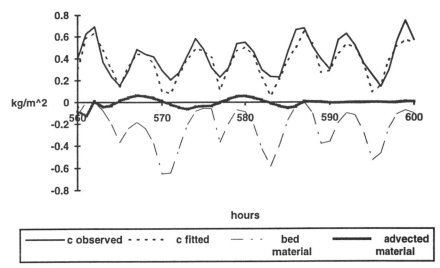

**Figure 8a**: Kalman filter fit to observations.

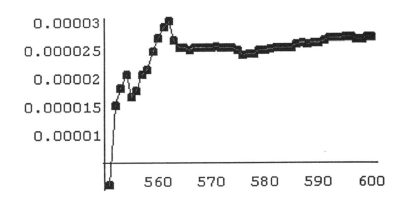

**Figure 8b**: Kalman filter estimate of the pickup constant $\gamma$.

The advantage of the Kalman filter is that it can be programmed easily and naturally with the form of model we are using. There is however a major drawback to the method as presently implemented. The equations are based on conservation of material; however the Kalman correction will not preserve this and the corrected state will not balance with regard to mass conservation. It may be possible to reformulate the model, or to apply a modified Kalman smoothing algorithm to remedy the discrepancy. This is being investigated.

### 4.2    General least squares minimisation

Because of the problem of mass conservation with the Kalman filter, a different approach has been implemented which always satisfies the conservation equations. It is possible to formulate the model as a least squares problem, with $\gamma$, $w_f$ ,and the advective flux at each tidal cycle (say), as unknown parameters. A recurrence relation for the gradient of the cost function with respect to the estimation parameters can be derived. This information is then incorporated into a standard numerical minimisation algorithm to determine the parameter values. The algorithm minimises, in a least squares sense, the difference between the observed and modelled concentrations. As well as satisfying the conservation equations , a further advantage of this approach over the Kalman filter is that the estimation parameters do not need to appear linearly in the model. Further details are given in Aldrige (1995).

5.      *Summary and conclusions*

The instrument packages developed within the COSEDS program have been briefly described and include the Tetrapod, Quadrapod and Minipod landing devices.

Data from two contrasting Tetrapod deployments have been examined in detail. Both show enhanced quantities of material in suspension when wind wave activity is significant. However, the Morecambe bay site also showed resuspension of material by tides alone. In both cases, once material was in suspension, tidal effects were evident, with concentrations at the sensor dropping at slack water, and evidence of material suspended by wave activity being advected into the deployment site by tidal currents. We could summarise this be saying that the wave activity seems to control the amount of material in suspension whilst the temporal variation is controlled by the tides.

In order to obtain quantitative information for parametarisations of sediment resuspension, a very crude model has been set up and combined with uncalibrated MOBS data via a Kalman filter algorithm to demonstrate the feasibility of obtaining such information. Other ways of combining the model and data are being investigated to avoid shortcomings of this technique with respect to mass conservation.

# References

**Aldridge J.N.** (1996) ,"Optimal fitting of a model to observations of sediment concentration in the Irish sea'', *Estuarine and Coastal modelling*, proc. 4th International Conference, San Diago USA, pp 417-428

**Butman B. & Folger D.W.,**(1979), "An instrument system for long term sediment transport studies on continental shelf", *J. Geophys. Res.*, 89,p 8197- 8203.

**Diserens A.P, Ockenden M.C. Delo E.A.** (1993), "Application of a mathematical model to investigate sedimentation at Eastham dock , Mersey estuary", In A.J. Mehta (ed), Near-shore and Estuarine Sediment Transport . Coastal and Estuarine studies No. 42, American Geophys, Union, pp 486-503.

**M.O. Green,**(1992), "Spectral estimates of Bed Shear Stress at Subcritical Reynold Numbers in a Tidal boundary layer", *J. Physical Ocean.*, 22(8),p 903- 917

**M.O. Green, N. Pearson N. D. Thomas, C.D. Rees, J.M. Rees, T.R.E. Owen,** (1992), "Design of a data logger and instrument-mounting platform for seabed sediment transport research", Contin. Shelf. Res. ,**12**,p 543-562.

**J.D. Humphrey, S.P. Moores** (1994) "STABLE II - an improved Benthic lander for the study of turbulent wave-current interactions and associated sediment transport", procedings, *Electronic Engineering in Oceanography,* held at Churchill College, Cambridge, UK, publication 394, IEE, London,.

**C. Jago,A.J. Bale,M.O. Green,S. Jones,M.J. .M.J.Howarth, I.N. McCave,G. E.Millward,A.W. Morris,A.A. Rowden,J.J. Williams,**(1994), "Resuspension processes and seston dynamics, southern North Sea", in *Understanding the North Sea System,* pp 97-114., ed Charnock, Dyer, Huthnance, Liss, Simpson, Tett, pub Chapman & Hall

**Lavelle J.W. & Mofjeld H.O** (1984) "An in situ erosion rate for a fine grained marine sediment.", *J. Geophys Res,* **89**(C4), pp 6543-6552.

**Luettich R.A Jr, Harleman D.R.F., & Somlyody L.,**(1990), "Dynamic behaviour of suspended sediment concentrations in a shallow lake perturbed by episodic wind events",*Limnol. Oceanogr.*, **35**(5),pp 1050---1067.

**McLaren P.** (1989), *pers. comm.*

**Sternberg,-R.W.; Kineke,-G.C.; Johnson,-R.,** (1991), "An instrument system for profiling suspended sediment, fluid, and flow conditions in shallow marine environments.", *Continental Shelf Res.* **11**(2), pp. 109-122

**Weeks A.R., Simpson J.H., Bowers D.,**(1993), "The relationship between concentrations of suspended material and tidal processes in the Irish Sea",*Continental Shelf Res.*, **13**(12),pp 1325-1993.

**Wright L.D. Boon J.D., Xu J.P.** and **S.C.Kim,**(1992), "The bottom boundary layer of the Bay Stem Plains environment of the Lower Cheaspeke Bay" *Estuarine, Coastal and Shelf Sci.* **35**,17-36.

303

# 21 Experiments on erosion of mud from the Danish Wadden Sea

C. JOHANSEN, T. LARSEN and O. PETERSEN

*Department of Civil Engineering, Aalborg University, Denmark*

## INTRODUCTION

Estuarine fine sediment beds which are controlled by the tides, such as in the Wadden Sea Estuary, occur in different stages of consolidation. Suspended sediment deposited at low flow velocities, are soft with a high water content and low aggregate shear strength. The main process by which sediment is brought back into suspension is erosion of cohesive particulate aggregates due to current, or a combination of current and waves. Erosion occurs when the bed shear stress ($\tau_b$) exceeds a critical shear stress ($\tau_s$), that depends on the bed material characteristics (sediment composition and texture), pore water character, eroding fluid character and bed structure (Nichols, 1986).

There are three modes of erosion: surface erosion which is a removal of particles and/or aggregates, mass erosion which is generated by a geotechnical failure within the bed such that all the material above a plane is almost instantly brought into suspension, and re-entrainment of a high density suspension where the bed first is fluidized, and flow-induced destabilisation of the fluid mud-water interface thus formed causes interfacial entrainment and mixing (Mehta, 1991; Van Rijn, 1989).

In laboratory studies where the bed is hydraulically smooth, the fluid shear stress at the bed is a measure of the entrainment force (Nichols, 1986). The interparticle electrochemical bonds must be broken before resuspension occurs and the critical shear stress be exceeded before surface erosion can begin. The complex interaction between the bed shear stress and the critical shear stress means that there is no general formula for the erosion rate.

The consensus on the critical shear stress is that it is mainly related to the sediment concentration (dry density), e.g. Parchure and Mehta, (1985), Van Rijn, (1989) and Mehta, (1986). The increase in the critical shear stress with depth or consolidation results in a decrease in erosion rate because of the decrease in the excess shear stress $\tau_b$ - $\tau_s$. The purpose of this work is to evaluate and if necessary extend this hypothesis especially in respect to the influence of the consolidation time.

*Cohesive Sediments.*
Edited by Neville Burt, Reg Parker and Jacqueline Watts
©1997 John Wiley & Sons Ltd.

## CHARACTERISTICS OF THE SEDIMENT

The material used in the erosion experiments was collected in Esbjerg harbour located in the Danish part of the Wadden Sea (figure 1). The sediment was collected from a small vessel using a Van Veen grab.

The sediment characterisation is presented in table 1, indicating that the suspended sediment is similar to illite with respect to the CEC (Cation Exchange Capacity) and the Attenberg Limit, and that the sediment is a highly plastic material with a large content of clay. The sediment seems to be similar to San Francisco bay mud (Partheniades, 1965).

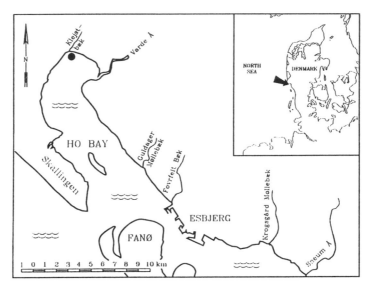

**Figure 1.** Map over the northern part of the Danish Wadden Sea

The settling velocity was determined using an instrument based on the Owen tube principle. The experiments were conducted in saline water with a salinity of 21 o/oo and temperature around 20 °C.

To determine the relation between the suspended sediment concentration and the settling velocity experiments were conducted with concentrations ranging from 1.1 to 3.0 kg/m$^3$. The experiments showed that the relationship between concentration and settling velocity can be expressed by equation (1) (Mehta, 1986; Van Rijn, 1989; Van Leussen, 1988).

$$W_s = k_1 C^m \qquad (1)$$

where $W_s$ is the settling velocity, $k_1$ is an empirical constant, C is the suspended sediment concentration and m is an empirical exponent.

**Table 1.** Characterisation of Esbjerg harbour sediment. (Johansen and Meldgaard, 1993; Møller-Jensen, 1993)

|  | Esbjerg Harbour sediment |
|---|---|
| CEC | 25 meq/100g dry soil |
| Salinity | 23.6 kg/m$^3$ |
| SAR | 2.9 meq/100g dry soil |
| Dry density | 2430 kg/m$^3$ |
| Organics | 8% by weight |
| Liquid limit | 100 % |
| Plasticity limit | 46 % |
| Plasticity index | 55 % |

Figure 2 shows the settling velocity versus the suspended concentration for the Thames (in-situ measurements), the Severn Estuary (in-situ measurements), San Francisco bay (laboratory measurements), Ho bay (measurements in natural sea water and in synthetic sea water with varying salinity) and for Esbjerg harbour (measurements in synthetic sea water)

As suggested by the data in figure 2, the settling velocity depends on the sediment and fluid compositions and on the in-situ environment. The in-situ measurements resulted in significantly higher settling velocities than those measured in the laboratory.

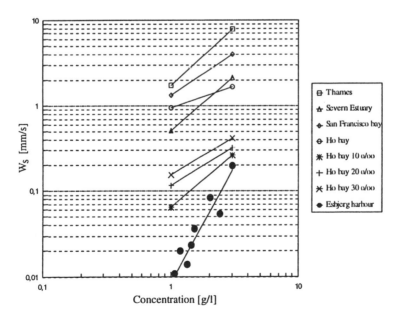

**Figure 2.** Comparisons of settling velocities measured in-situ on different locations and in the laboratory (Burt, 1986; Van Leussen, 1988; Møller-Jensen, 1993; Johansen and Meldgaard, 1993).

307

## EXPERIMENTAL SETUP AND INSTRUMENTATION

The circular flume used in this study is stationary and the flow field is induced by a rotating lid placed at the upper surface boundary. It is 1.9 m in outside diameter, 20 cm wide and 26 cm deep. The lid fits inside the flume with a tolerance of 3.0 mm on both sides. The lid can be rotated with variable speed between 0 and 3 m/s in either direction. The flume has horizontal and vertical ports for the instruments used to measure velocity and turbidity.

The turbidity measurements were carried out using an OSLIM turbidity sensor (Delft, 1991). The measuring method is based on the attenuation of a lightbeam from a LED, caused by light absorption and reflection due to the presence of particles. Light absorption depends in principle on the particle size distribution, but by pumping the suspension through a tube to the sensor the shear in the tube breaks the aggregates up thus limiting the error caused by a varying size distribution. Light absorption is also dependent on the surface of the particles, the content of organic matter, etc. Thus it is necessary to calibrate the instrument with each specific material.

In a stationary circular flume with a rotating lid the velocity increases across the flume from the inner wall to the outer wall due to secondary circulation and decreases from the top to the bottom, hence affecting the vertical distribution of suspended sediment. The secondary circulation can be reduced but not eliminated by making both the upper lid and the flume rotate in opposite directions (Partheniades and Kennedy, 1967; Krishnappan, 1992; Petersen and Krishnappan, 1994). Numerical simulations using the commercial CFD model, CFX-F3D, show that the secondary shear stress in average is less than 20% of the tangential shear stress. The resulting stress vector is twisted 11 degrees from the tangential stress vector. Thereby, the secondary currents only account for 2% of the resulting shear stress, which in relation to the erosion experiments is insignificant.

A relation between the rotating lid velocity and the bed shear stress was determined by measuring the velocity profiles using a micro-propeller and resulted in a mean value of the bed shear stress equal to (Møller-Jensen, 1993):

$$\overline{\tau}_b = kU_L^2 \rho \tag{2}$$

where k is an empirical constant equal to $2.9 \ 10^{-4}$, $U_L$ is the lid velocity and $\rho$ is the water density.

With the experimental setup it was possible to analyse how the critical shear stress and the erosion rate depend upon the bed shear stress and the consolidation time. The experimental procedure included three phases:

1. Mixing, in which the sediment was brought into suspension with a shear stress large enough to prevent the suspended material to deposit.

2. Deposition and consolidation, during which the particles flocculate and settle out on the flume bottom because the bed shear stress is reduced to zero. The sediment starts to consolidate subsequent to deposition.

3. Erosion, in which the sediment is eroded in steps with different bed shear stresses.

A schematic description of the phases is shown in figure 3.

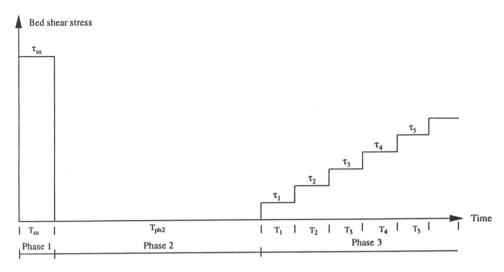

**Figure 3.** The experimental procedure used in the erosion experiments, where $T_1 = T_2 = T_3 = T_4 = T_i$.

The erosion experiments were conducted in saline water with salinity 21 o/oo and temperature from 20 to 23 °C. The concentration of the initially suspended sediment was in all erosion experiments 13 kg/m³. The samples were taken out at a depth of 13 cm. Table 2 shows the data for the experiments with the notation from figure 3.

The bed shear stress was kept constant during each erosion step. After each erosion step the suspended concentration reached a constant concentration, because the bed shear stress eroded the sediment down to the depth where the bed shear stress equaled the shear strength of the sediment.

The consolidation experiments, used to determined the dry density, were conducted using a similar setup than the apparatus described by Mehta, (1986). The experimental setup consisted in a 54 cm high and 20 cm wide column, in which 9 small tubes, ranging in height from 1 to 9 cm, were placed at the bottom of the column. The column was then filled with the sediment suspension and the sediment allowed to settle out to the bottom and into the small tubes. The difference in mass between the different tubes was then transformed to a dry density profile. The experiments were conducted in saline water with salinity 21 o/oo and temperature from 20 to 23 °C. The consolidation period varied from 0.5 to 96 hours and suspended sediment concentration ranging from 48 to 63 kg/m³.

**Table 2.** Data for the erosion experiments (Johansen and Meldgaard, 1993)

| | | | Erosion experiments |
|---|---|---|---|
| Phase 1 | $T_m$ | [hours] | 5.00 |
| | $\tau_m$ | [N/m²] | 1.81 |
| Phase 2 | $T_{ph2}$ | [hours] | 12,24,48 |
| | $\tau_{ph2}$ | [N/m²] | 0.00 |
| Phase 3 | $T_i$ | [hours] | 4.00 |
| | $\tau_i$ | [N/m²] | 0.055, 0.083, 0.117, 0.157, 0.202, 0.254, 0.311, 0.374, 0.442, 0.517 |

## RESULTS

The results from the erosion experiments are shown in figure 4. The erosion rate (E) is found directly from the change of concentration (ΔC) at the beginning of the erosion step. The erosion rate can then be given by

$$E = h \frac{\Delta C}{\Delta t} \tag{3}$$

where h is the depth of the water.

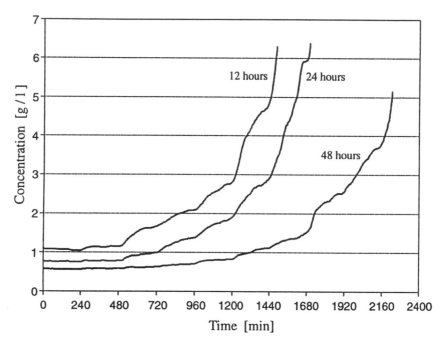

**Figure 4.** Erosion results for 12, 24 and 48 hours consolidation. (Johansen and Meldgaard, 1993)

The initial concentration at the beginning of the experiments differs because the consolidation period was not long enough to deposit the entire suspended material. Further it is seen that the last erosion step did not reach equilibrium because the concentration exceeded the range of measurement (7 g/l).

The shear strength as a function of the eroded depth and the dry density was estimated through consolidation experiments. By assuming that the properties of the sediment applied in the erosion and consolidation experiments are the same, it was possible to approximate the eroded depth and the dry density was thereby found from the equilibrium concentration.

The resulting $\tau_s(z)$ profiles are shown in figure 5 indicating that the shear strength increased with depth and the consolidation period. The increase in the shear strength with depth is due to the flocculated aggregates, formed while the sediment is in suspension, are

310

crushed due to increase in the overburden pressure after they deposit. The first point ($z = 0$) is assumed to be equal to the first applied shear stress ($\tau_b = 0.055$ N/m²).

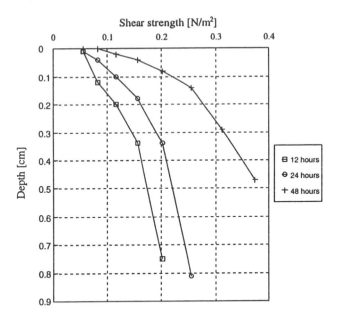

**Figure 5.** The critical shear stress versus depth (Johansen and Meldgaard, 1993).

Analyses of the experimental results yielded the best fit relationship of the form

$$E = k(T_{dc})^\alpha (\tau_b - \tau_s)^\beta \qquad (5)$$

where the constants were found to be: $k = 107,6$, $\alpha = -0,98$ and $\beta = 4,025$, given the consolidation period in hours, the shear stresses in N/m² and the erosion rate in kg/(m² s). Neither in this study has it been possible to obtain a nondimensional expression. The reliability of the regression was found to $R = 0,94$. Equation 5 is plotted in figure 6 along with the erosion rate achieved by the experiments.

The standard deviation ($\bar{e}$) was estimated by (6) and found to 4.4 $10^{-6}$ kg/(m² s)

$$\bar{e} = \sqrt{\frac{\sum\limits_{i=1}^{N}(E_m - E_b)^2}{N-1}} \qquad (6)$$

where N is the number of measurements and calculations, $E_m$ is the measured and $E_b$ the calculated erosion rate. The scatter plot for the calculated versus the measured erosion rate is shown on figure 7.

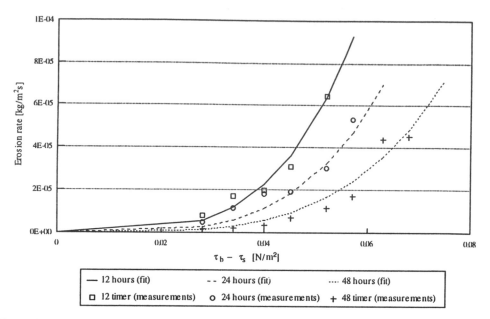

**Figure 6** Erosion rate versus the excess shear stress for 12, 24 and 48 hours consolidation periods (Johansen and Meldgaard, 1993)

## DISCUSSION

Generally, the erosion rate is assumed to increase linearly with the excess shear stress, i.e. Nichols, (1986) and Sheng and Lick, (1979). However, field observations have shown that the erosion rate can be non-linear (Olsen and Kjelds, 1991 and DHI, 1990) and can be expressed mathematically by

$$E = E_0 (\tau_b - \tau_s)^n \quad \text{for } \tau_b > \tau_s$$

where $E_0$ is the erosion rate constant and n is an empirical exponent.

The erosion experiments indicated that the erosion rate is a function of the strength of the bed material. Furthermore, the experiments conducted in this study implied that the erosion rate constant depends on the consolidation time. For consolidated beds the equations found in Mehta et al., (1982), where the erosion rate varies linearly with the excess shear stress, can be used.

## CONCLUSIONS

Results of the laboratory experiments indicate that the shear strength increases within the first one or two centimetres. Underneath this layer the sediment is more consolidated resulting in a more uniform shear strength distribution. The erosion rate was found to increase with the excess shear stress with an erosion rate constant that was a function of the consolidation period. The critical shear stress and the erosion rate were found to depend strongly on the bed consolidation time. Indicating that a non-linear erosion rate description should be used for the deposited bed.

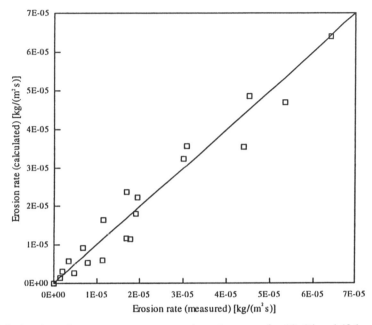

**Figure 7** Calculated erosion rates versus measured erosion rates for 12, 24 and 48 hours consolidation periods (Johansen and Meldgaard, 1993)

## REFERENCES

Burt, T.N., 1986, Field settling velocities of estuary mud, published in *Estuarine Cohesive sediment Dynamics,* Lecture Notes on Coastal and Estuarine Studies, Ed. A. J. Mehta, Vol. 14, pp. 126-150, Springer Verlag

Delft hydraulics, 1991, *Manual for OSLIM, an optical silt measuring instrument*, Delft, The Netherlands

DHI - Danish Hydraulic Institute and Water Quality Institute, 1990, *Newport Borough Council, River Usk Barrage, sediment transport modelling*, Vol. 2. DHI, Hørsholm, Denmark

Johansen, C. and Meldgaard, H., 1993, Transport af kohæsive sedimenter i Grådyb tidevandsområde (Cohesive sediment transport in the Grådyb tidal area). *M. Sc. Thesis.* Department of Civil Engineering. University of Aalborg, Aalborg, Denmark. (in Danish)

Krishnappan, B. G., 1993, Rotating Circular flume, *Journal of Hydraulic Engineering*, ASCE 119(6), 758 - 767

Møller-Jensen, P., 1993, Wadden Sea Mud, Methods for Estimation, Erosion and Consolidation of Marine Cohesive Sediments. *Ph. D. Thesis.* Department of Civil Engineering. University of Aalborg, Aalborg, Denmark.

Mehta, A. J., Parchure, T. M., Dixit, J. G. and Ariathurai, R., 1982, Resuspension potential of deposited cohesive sediment beds, in *Estuarine Comparisons*, Ed. V.S. Kennedy, Academic Press.

Mehta, A. J., 1986 On Estuarine Cohesive Sediment Suspension behaviour,, *Journal of Geophysical research,* Vol. 94, No. c10, pp. 14.303 - 14.314, Springer Verlag

Mehta, A. J., 1988 Laboratory studies on cohesive sediments deposition and erosion, in *Physical Processes in Estuaries*, Eds. J. Donkers and W. Van Leussen, Springer Verlag

Mehta ,A. J., 1991 Review notes on cohesive sediment erosion, *Proc. Coastal Sediments '91*, ASCE, 1, pp. 40-53.

Nichols, M. M., 1986 Effects of fine sediment resuspension in estuaries, in *Estuarine Cohesive sediment Dynamics*, Lecture Notes on Coastal and Estuarine Studies, Ed. A. J. Mehta, Vol. 14, pp. 5-42, Springer Verlag

Partheniades, E., 1965, Erosion and deposition of cohesive soils, *Journal of the hydraulics division,* ASCE, vol. 91, HY1, pp. 105 -139.

Partheniades, E. and Kennedy, J. F., 1967, Depositional behaviour of fine sediment in a turbulent fluid motion, *Proc. 10th Conf. on Coastal Engineering*, Tokyo, Japan, 2.707-729.

Petersen, O. and Krishnappan, B. G., 1994. Measurement and analysis of flow characteristics in a rotating circular flume, *Journal of Hydraulic Research*, Vol. 32, No. 4, pp. 483 - 494

Sheng Y.P. and Lick. W., 1979, The Transport and Resuspension of Sediments in a Shallow Lake, *Journal of Geophysical Research*, Vol. 84, No. C4, pp. 1809 - 1825

Van Leussen, W., 1988, Aggregation of particles, settling velocity of mud flocs. -a Review, in *Physical Processes in Estuaries*, Eds. J. Donkers and W. Van Leussen, Springer Verlag

Van Rijn, L. C., 1989 *Handbook Sediment Transport by Currents and Waves*, Delft Hydraulics, Report H. 461. pp. 12.1 - 12.24.

# RHEOLOGY AND WAVE EFFECTS

# 22    A review of rheometric methods for use with fine sediments

**T. E. R. JONES**

*Department of Mathematics and Statistics, University of Plymouth, UK*

## INTRODUCTION

Many studies have been carried out on the flow properties of mud and this review is concerned with the art of measuring these properties using various rheometrical techniques. Very many workers have studied the flow properties of suspensions (Barnes[1]). Various research groups have used differing techniques to study the flow properties of mud suspensions. Flow geometries have varied from complicated mixing type impellers to very well defined flow systems such as cone and plate and concentric cylinder flow measurement devices. The disadvantage of the impeller method is that the shear rate is not well defined. The fluid is thus subjected to varying rates of shear which can cause problems, particularly, for materials which exhibit a yield stress or are affected by settling. Most of the rheometry experiments have been carried out using either a cone and plate or a concentric cylinder geometry and it is the effect of using these geometries which we shall now consider. Various studies have been performed on a range of estuarine muds taken from various rivers e.g. Scheldt, Brisbane, Restronguet and Parrett[2].

In 1963 a comprehensive study of the rheological properties of estuarial sediments was carried out between the Hydraulic Engineering Laboratory and the Sanitary Engineering Research Laboratory of the University of California, Berkeley[3]. In this study, using a capillary viscometer, the shear rate varied from zero to a particular value. Therefore if the measured Bingham value is correct then only a part of the sediment is being sheared and plug flow can occur. Also experiments were not carried out at low enough shear rates and therefore the existence of a yield value is difficult to justify.

Experiments were also carried out on a rotating cylinder viscometer. In this part of the work consideration was given to the thixotropic behaviour of the sediments. It was concluded that, as expected, a certain time was required before equilibrium was reached. An analysis was carried out to model the sediment behaviour in terms of either Newtonian or Bingham behaviour but the lack of data at very low shear rates raised a doubt as to the validity of using these particular models. It was concluded that these measurements provided complimentary information to that obtained using the capillary viscometer.

*Cohesive Sediments.*
Edited by Neville Burt, Reg Parker and Jacqueline Watts
©1997 John Wiley & Sons Ltd.

In 1979 Gularte, Kelly and Nacci[4] carried out a study on the rheological methods for predicting cohesive erosion. This research was directed towards determining relationship between hydrodynamic erosional parameters and the rheological properties of cohesive materials. A comparison was carried out between the Bingham parameters measured on both a specially designed microminiature vane viscometer and the conventional rotating cylinder viscometer. It was found that in general the critical shear stress increased with increasing Bingham yield. It appeared that the critical shear stress was less dependent on Bingham yield stress at low salinities but became more dependent as salinity increased. The main problem in this work is the lack of agreement between data obtained on a vane viscometer and a rotational viscometer. This is probably due to difficulties in defining the shear rate regime in a vane viscometer. Calibrations can be carried out with Newtonian fluids systems but if the material under test is highly non-Newtonian then this method can lead to errors.

In 1979 Yong, Sethi, Ludwig and Jorgensen[5] carried out a study of the interparticle action and rheology of dispersive clays. Various tests were used in the identification and characterisation of dispersive clays and these involved the measurement of the force required to detach particles and, the floc size and double layer interactions between various particles and floc.

A rotating cylinder device was used to provide rheological information of relevance to the first parameter. A series of experiments were carried out on a range of clay minerals and by measuring the torque required to maintain a given shear rate, the shear stress was calculated and the flow behaviour of the material was obtained. This paper contains a comprehensive range of shear stress/shear rate rheograms which show differences between dispersed and flocculated clay systems. A Bingham model was used to analyse the data but as most of the data was taken at fairly high shear rates (>10(1/sec)) obviously the magnitude of the yield value cannot be relied upon.

A range of experiments was carried out by Bryant, James and Williams[6] (1980) using a coni-cylindrical cell on a Weissenberg Rheogoniometer. This geometry was chosen so that the shear rate was constant throughout the material during the experiment. For the range of concentrations considered there was no evidence for the existence of a yield stress, however at higher concentrations there is a possibility of its existence. It is concluded in this paper that the rheological behaviour of model cohesive sediments is strongly influenced by physico-chemical factors such as pH, salt concentration and mineralogical composition. This influence is largely reflected in the way in which these factors affect the flocculation process. This paper anticipates a similar influence on the rheological behaviour of cohesive sediments in estuarine and coastal waters.

This paper also concludes that mud systems do not exhibit a yield stress if careful experiments are carried out at low shear rates. However, it should be said that the concept of yield stress is obviously dependent on the time of the experiment. Therefore in certain flow situations the existence of a yield stress could certainly prove to be very useful. The work considered in the above paper is continued on a paper published by James and

Williams[7] in 1982. Experiments are carried out using the same geometry but data is obtained for higher concentration suspensions. The results are interpreted in terms of a residual stress arising from the residual effects of inter-particle potential while a slope of the shear stress/shear rate curves at high shear rate gives the plastic viscosity arising from purely hydro-dynamic effects. The increasing pseudo-plastic behaviour with increasing % kaolinite suggests that the development of a flocculated structure associated with interactions between the constituent particles is an important factor influencing the rheological behaviour of these suspensions.

Dzuy & Boger (1983,1985)[8,9], James et. al.(1987)[10], Barnes & Carnali (1990)[11] and Alderman et. al. (1990)[12] have considered the problem of measuring the yield stress of suspension type materials using vane rheometry. Dzuy and Boger have shown that a single point measurement with the vane device is sufficient to determine accurately the yield stress. they also conclude that this method does not rely on any previous shearing of a suspension. James et. al. consider a direct measurement of static yield properties of cohesive sediments based upon the combined use of an applied stress rheometer and an instrument for measuring the propagation velocity of small amplitude torsional shear waves. Results are presented which show that the technique does not result in the excessive disruption of material associated with applied shear rate instruments and wall slip is avoided by the use of the vane geometry. Barnes & Carnali considered the numerical simulation of the vane in cup geometry using a power law fluid and conclude that for a shear thinning fluid of power law index less than 0.5 the fluid within the periphery of the blades essentially trapped there and turns with the blade as a solid body. The question arises is what significance this result has for the use of this geometry with cohesive sediments.

In 1986 O'Brien and Julien[13] considered the Rheology of non-Newtonian fine sediment mixtures using a rotational viscometer. It was shown that the Bingham model adequately represented the shear stress/shear rate relationship. It was also shown that the non-Newtonian fluid properties of the mixtures were largely functions of the concentration and the type of clay in the fluid matrix. The apparent viscosity and Bingham yield stress depends on the shear rate and increases exponentially with sediment concentration in agreement with work published by Jones and Golden[2] (1990). It was also shown that sand in the fluid matrix has no significant effect on the rheological parameters of non-Newtonian mixtures for concentrations less than 20% by volume.

Nguyen and Boger (1987)[14] developed a theoretical analysis of the flow of yield stress materials in a concentric geometry. They show that in cylindrical systems with large radius ratios , as is usually the case with suspensions, the yield stress induces two possible flow regimes in the annulus. Unless the yield stress is exceeded everywhere in the gap only part of the of the fluid can be sheared while the remaining section behaves like a solid plug. However most reported mud systems data is produced with fairly small radius ratios and therefore this effect is most probably not relevant. However spurious data with this geometry could be due to plug flow effects if large gaps are employed.

Mud samples from the harbour of Zeebrugge and from the river Scheldt at Antwerp were used by Verreet et al[15] (1987) to determine the relationship between physico-chemical and rheological properties of fine-grained muds. Rheological properties, in particular yield stress, were correlated with density, consolidation time, granulometry, specific surface and the content and type of organic matter. It has shown that yield stress increased substantially with an increase of the fine fraction ($<\mu$m) and colloidal matter. The low shear data contained in this paper is very sparse which raises certain questions as to the accuracy of the yield stresses measured.

A review of the 'Rheology and non-Newtonian behaviour of sea and estuarine mud' was carried out by Verreet and Berlamont[16] in 1987. This includes a section on the different models used to describe the flow behaviour of mud systems. The effect of time dependency and visco-elasticity is discussed but the latter property suffers from a lack of data. It is concluded that it is important to be able to define exactly the flow regime since approximations in calculations of shear stress and shear rate can lead to significant errors in prediction of low parameters such as apparent viscosity and yield stress. A very interesting section is included on the properties which affect the rheology of the sediment such as flocculation, salinity, pH, solid particle properties, clay minerals and organic matter.

In 1988 O'Brien and Julien[17] published work on the laboratory analysis of mudflow 'low' shear rate properties. A specially designed viscometer was used to obtain data under temperature controlled conditions. This study emphasised the importance of conducting experiments at low shear rates, because these are the conditions found in natural channels and the slippage problems observed at large sediment concentrations can be avoided. The Bingham model was fitted to the data and it was shown that both the yield stress and viscosity increased by three orders of magnitude as the volume concentration of sediments changes from 10-40%. The effect of adding sand particles to certain suspensions is negligible provided the sand concentration is less than 20% by volume. However, the shear rates used were not really low as it is possible to use a controlled stress rheometer to perform experiments at shear rates of the order of 0.00001.

## EXPERIMENTAL APPARATUS

Most conventional viscometers have been designed to characterise materials which obey Newton's law of constant viscosity. The behaviour of such materials is characterized by a constant viscosity coefficient and the design of viscometers to determine this constant is a fairly simple process . Elastico-viscous liquids are not 'Newtonian' in their behaviour and certainly need more than one constant to characterize their behaviour. Some of the conventional viscometers can be modified to measure the apparent viscosity and hence shear stress of elastico-viscous liquids at various shear rates.

It is clear that very versatile instruments are required to characterize non-Newtonian materials in any satisfactory way. Controlled shear rate instruments (e.g. Weissenberg Rheogoniometer) or Controlled Stress rheometers (e.g. Carrimed rheometer) are commercial rheometers which incorporate these features. There are various advantages

d disadvantages in using both types of instruments The controlled stress instrument differs from
ntrolled shear rate instrument in that a shear stress is applied to the sample to be measured. This
ables more meaningful low shear data to be produced since the instrument reacts in 'sympathy' to the
iid behaviour and does not force the material to move as in a controlled shear rate experiment. This
viously has advantages when the 'yield behaviour' of a material is to be studied i.e. the minimum
ess to cause flow of a material may be directly measured. This experiment does not provide a
mplete rheological characterization of complex materials. It is often necessary to examine the
namic behaviour of a fluid. Oscillatory stress experiments can be carried out on a controlled stress
eometer and provides an alternative method for measuring the viscoelastic properties of materials.
ie advantage of this technique is that if experiments are carried out at very low amplitudes and
equency of oscillation then the structure of the material is maintained and not destroyed as may be
e case in certain steady stress experiments. However this aspect will not be considered in this review.

HEORETICAL INTERPRETATION

eady Shear

periments can be carried out using cone and plate, parallel plate, concentric cylinder and double
ncentric cylinder geometries. The relationship between the applied couple shear stress shear rate and
scosity is given by the following expressions.

ne and Plate

$$\dot{\gamma} = \Omega / \theta_0$$

$$\eta(\dot{\gamma}) = \frac{3C}{2\pi a^3 \dot{\gamma}}$$

rallel Plate

$$\dot{\gamma}_a = \frac{\Omega a}{h}$$

$$\eta(\dot{\gamma}_a) = \frac{3C}{2\pi a^3 \dot{\gamma}_a}\left[1 + \frac{1}{3}\frac{d \ln C}{d \ln \dot{\gamma}_a}\right]$$

## Concentric Cylinders

$$\lambda = \frac{r_2 \Omega}{r_1 - r_2}$$

$$\eta(\dot{\gamma}) = \frac{C}{2\pi r_1^2 h_1 \dot{\gamma}}$$

$$\sigma(\dot{\gamma}) = \eta(\dot{\gamma})\dot{\gamma}$$

where

| | | |
|---|---|---|
| $\dot{\gamma}$ | – | shear rate |
| a | – | radius of member |
| $\Omega$ | – | speed of rotation |
| $\theta_0$ | – | cone angle |
| C | – | applied couple |
| $\dot{\gamma}$ a | – | rim shear rate |
| h | – | gap between plates |
| $\eta$ | – | apparent viscosity |
| $r_2$ | – | radius of outer cylinder |
| $r_1$ | – | radius of inner cylinder |
| $h_1$ | – | $r_2 - r_1$ |
| $\sigma$ | – | shear stress |

For the experiments described in this review results were obtained using a double cylinder geometry. The theory for the double cylinder geometry is similar to the above concentric cylinder theory.

There are certain advantages and disadvantages in using these geometries which tend to be a function of the materials to be measured. The cone and plate system has the advantage of the shear rate being constant in the gap between the platens when the cone is less than four degrees. The other advantage, especially for fluid mud systems, is that the fluid is kept fully mixed during the experiment especially at high shear rates. However the disadvantage is that in most instruments the gap between the truncated cone and the bottom plate is small (varies from 20 microns up to 100 microns dependent on the cone angle) which can cause problems when the flow properties of particulate matter is being measured. Particles can be caught in the gap giving rise to spurious experimental data. Therefore as a rule of thumb the truncated gap should be at least ten times the largest particle size in the material These problems can be overcome by using a parallel plate geometry where the gap can be varied. However the gap cannot be too large because of the possibility of materials being extruded from the gap during shear. The disadvantage of this geometry is that the shear rate varies from zero to a maximum value at the edge of the plate which can cause problems when yield stress materials are tested. This could also lead to plug flow which could also lead to errors in the measured shear rates. The concentric cylinder geometry can overcome both the problems of gap size and variable shear rate especially when the radius is large and the gap is small compared to the radius. The gap should not be too small so that particles can be caught in the gap. Gaps up to two thousand microns can be accommodated without invalidating the condition for a fairly constant shear rate across the gap. The major disadvantage is the possibility of settling especially with low concentration fluid samples. Very many studies have been performed using concentric cylinder measurement devices (Barnes)[1]. The advantage, especially for very thin fluids, is that more of the fluid is in contact with the measurement system which can produce more accurate data. Also if the gap between the cylinders is small, compared to the radius of the inner cylinder, then the shear rate is fairly constant. The disadvantage of this geometry is that if settling is a problem then it can be difficult to remix the sample in situ. The particles migrate (due to settling) to the area immediately below the inner cylinder. Therefore under shear, it will not be possible to re-suspend this part of the sample since this flow region is not subjected to shear. This effect obviously changes the concentration of the remainder of the sample which can lead to incorrect flow data.

These problems can be overcome, to a certain extent, by using a double concentric cylinder geometry, which has not previously been used to study the flow properties of mud suspensions. This geometry consists of a circular annular channel into which is inserted a thin walled cylinder. The sample is in contact with both sides of the inner cylinder which has the advantage of being able to produce more accurate data. This geometry also removes the stagnation point associated with the concentric cylinder geometry and the small gap between the inner cylinder and the annular channel aids in keeping the sample in suspension during shear. However settling can occur because the flow is always laminar which can lead to problems.

The apparatus used at University of Plymouth is the Carrimed controlled stress rheometer and the stress is applied to a free moving upper platen by means of an electronically controlled induction motor. The Carrimed instrument has two outputs, one enables the displacement to be measured (used in creep and oscillatory stress experiments). An Opus V microcomputer with a Tecmar IEEE interface card is used to control the rheometer. Special software has been developed which enables automatic analysis of output data from steady stress, oscillatory stress and combined steady and oscillatory stress experiments to be performed.

The controlled stress rheometer has a fixed lower platen while the upper platen is free to rotate under an applied stress. Once the level between the cylinders is set the computer automatically raises the inner cylinder to a predetermined level.. A computer is also used to control the operating temperature which for the experiments described in this review was set at 20 degrees centigrade.

The supplied software with the controlled stress rheometer enables analysis of the flow curves in terms of various models such as Newtonian, Power Law, Bingham, Casson, Herschel Bulkley, and Polynomial form.

## EXPERIMENTAL METHOD

The main problem in performing experiments, with low concentration fluid mud systems, is keeping the sample fully mixed during the experiments. Settling can produce density gradients which will obviously change the flow properties of the mud. Therefore a separate experiment was carried out to determine the applied stress necessary to 'mix' the sample sufficiently such that no settling occurred during the time taken to produce a flow curve from the controlled stress rheometer. Care was also taken so that the stress applied did not, in any way, permanently change the structure of the mud sample. However, accepting the fact that settling is a problem, later experiments will consider the effect of settling on the flow properties.

The following procedure was used to obtain flow curves for each sample. The fluid was placed in the instrument and sufficient time was allowed until the equilibrium temperature was reached. With the low concentration samples (less than 80g/l) flow occurred immediately a stress was applied, therefore it was not necessary to predetermine the peak torque necessary for the material to flow. The optimum ascent and descent time of applied stress was determined from the previous applied stress experiments.

Having completed the flow curve experiments and analysis, with respect to low and Bingham yield values, separate experiments were carried out to determine the optimum time for equilibrium to be achieved and to monitor the effort of settling. This was carried out by applying a range of fixed applied stresses, between the low yield and Bingham yield values, and monitoring the shear rate behaviour with time.

The Carrimed instrument, previously used to carry out these type of experiments, has been modified to include an optical encoding device for the measurement of output displacement. This modification has enabled more accurate data to be obtained, especially at low shear rates. In order to illustrate some of the effects discussed in the previous sections the flow behaviour of mud taken from the river Parrett will be considered.

## EXPERIMENTAL RESULTS

Steady stress experiments were carried out on various concentrations of fluid mud samples taken from the River Parrett, UK (Jones and Golden)[2]. Figure 1 contains typical shear stress/shear rate curves for a 163g/l sample. The curve shows a typical thixotropic behaviour i.e. the down curve lies below the up curve. The data seems to indicate that a certain stress must be exceeded before flow is initiated i.e. the materials exhibit a yield value. However, with the new more accurate measurement available on the controlled stress rheometer it is possible to take data at much lower shear rates than before. Therefore the results seems to indicate that the flow curves will ultimately intercept the shear stress axis at the origin (data taken at lower shear stresses produces lower shear rate values until the limit of measurement is reached) which means that these materials will not exhibit a yield value. However, the results are not conclusive and a apparent yield value can be measured for a range of concentrations. It is certainly true that all the mud concentration samples measured show a very high resistance to flow at very low shear rates (i.e. a high apparent viscosity).

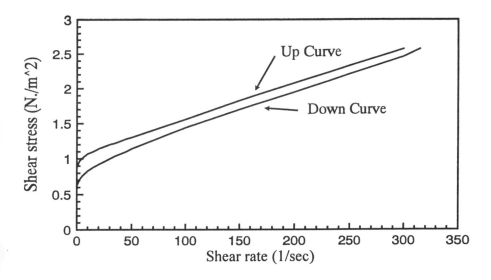

**Figure 1** River Parrett mud flow curve, 163g/l

All the concentration curves also show two distinct regions of flow, especially the high concentration curves. The initial highly shear thinning region is followed by a Bingham type behaviour i.e. a linear variation of shear stress with increase in shear rate. Therefore these curves could have been analysed in terms of a lower yield value (when flow is initiated as indicated by a movement of the optical encoder) and an upper yield value (by using a Bingham model).The shear stress/shear rate comparison (up curves only) between the various concentrations are given in Figure 2 All the data shows that for a particular shear rate the shear stress increases with increase in concentration, as expected.

A separate experiment was carried out to determine the effect of applying a constant stress over a fixed period of time. The original aim was to determine the length of time necessary to achieve equilibrium and its variation with applied stress. However, as the data shows, it is possible to use this experiment to demonstrate the effect of applied stress on the increase in viscosity which could be due to settling . It should also be pointed out that the results were obtained in a small gap double concentric cylinder apparatus and therefore conclusions can only be drawn for this particular geometry. The settling characteristics for other flow situations might not be the same, due to wall effects etc.

A typical stress response of this mud sample is presented in Figure 3. For the range of applied stresses an initial increase in shear rate is followed by a decrease. The initial increase, indicates the effect of thixotropy, followed by the decrease in shear rate indicating a rise in viscosity which could be due to various factors. The effect could be due to an increase in concentration due to settling but could also be due to anti-thixotropy. A detailed analysis of the apparent settling part of the data curves in Figure 3 lies outside the

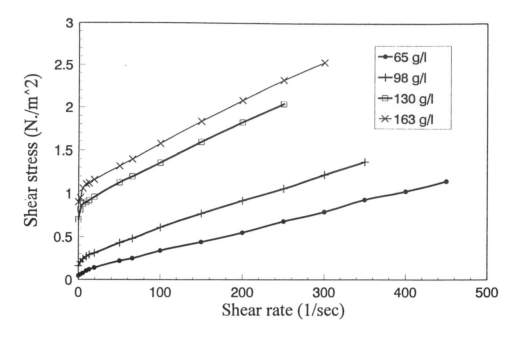

**Figure 2** River Parrett shear stress /shear rate comparison (up curves)

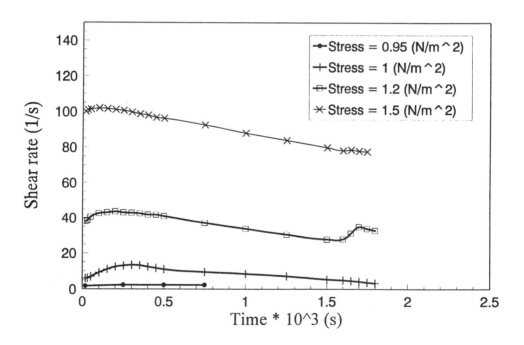

**Figure 3** River Parrett mud-shear rate response, 163g/l

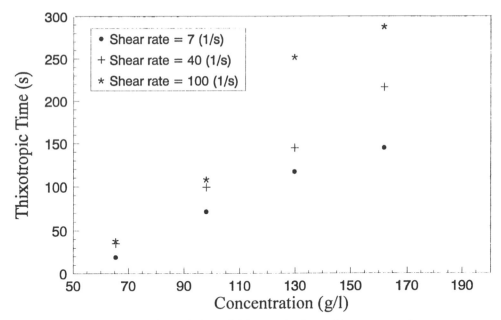

**Figure 4** River Parrett mud- Variation of thixotropic time with concentration

scope of this work but some interesting conclusions can be drawn. The rate of settling increases with increase in applied stress but the percentage change in shear rate and hence in viscosity is greater the lower the applied stress. The data for the other concentrations show essentially the same behaviour with time and stress.

Figure 4 contains the variation in time required for equilibrium (thixotropic time) with concentration, for a range of shear rates. It is concluded that it takes longer to overcome thixotropy, the higher the concentration, and this time also increases with increasing shear rate and applied stress. This result is as expected but it is perhaps surprising that the longest times are of the order of 5 minutes.

## CONCLUSIONS

1.    The type of rheometer employed to carry out rheological experiments should be carefully chosen. The problem with a capillary viscometer is that the shear varies from zero to a particular vale which can cause problems when highly viscous material (at low shear rates) are measured. The shear rate is not well defined in a vane viscometer which can give rise to secondary flow effects especially for low concentration materials. The concentric cylinder geometry is the system most often used, due to the lack of interaction with the particle size but consideration should be given to using the double concentric cylinder referred to by Jones and Golden (1990)[2] in their report on the flow properties of River Parrett mud. It is recommended that a concentric cylinder measurement should not be used to carry

out measurements on low concentration fluid mud systems due to problems which can arise due to settling. This is an effect which is not considered in most of the publications reviewed but obviously could be important when data is taken at low shear rates.

2       The double concentric cylinder geometry, on a controlled stress rheometer, can be used to overcome the migration of particles problem, associated with a normal concentric cylinder. geometry, especially for fluid mud suspensions. Most mud type suspensions exhibit Newtonian type behaviour at low concentrations but show thixotropic pseudo-plastic behaviour at higher concentrations. Flow curve experiments illustrate the differences in flow behaviour between the sample in terms of yield stress, concentration and bulk density. Stress response experiments can be used to quantify the effect of thixotropy and settling.

3.      If the Bingham model is to be used to model the low shear rate properties of mud systems then care should be taken to perform experiments at low shear rates as data taken at shear rates above 1 (1/sec) can lead to errors in prediction of yield stress. It should be pointed out that even if Bingham type behaviour is observed ( plastic viscosity is constant), the apparent viscosity will not be constant and will decrease with increase in shear rate.

4.      In order to obtain relevant experimental data at low shear rates then a rheometer which allows the measurement system to react in sympathy to the movement of the material and not force the material to move, should be used. It is recommended that a controlled stress rheometer should be used instead of a controlled strain rate instrument.

5.      It is generally recognised that if careful experiments are carried out at very low shear rates then whatever the stress applied the material will flow but it is all a matter of time scale. Within the time scale of a particular experiment the concept of a yield stress can be a useful parameter and as shown in this review can provide the experimentalist with useful correlations with various physical properties.

**REFERENCES**

1.      Barnes H A. Dispersion Rheology, Unilever Research (1981).

2       Jones T E R, and Golden K. Various Reports for Hydraulics Research Ltd. 1987-1992.

3.      Krone R B. 'A Study of Rheological Properties of Estuarial Sediments'. Report prepared for the Committee on Tidal Hydraulics under Contract DA-22-079-CIVENG-61-7 with the Waterways Experiment Station, Corps of Engineers, US Army, 1963.

4.      Gularte R C, Kelly W E, and Nacci V A. 'Rheological Methods for Predicting Cohesive Erosion'. Paper presented at the Annual Conference of the Marine Technological Society, New Orleans, USA, 1979.

5.      Yong R N, Sethi A J, Ludwig H P, and Jorgensen M A. 'Inparticle Action and Rheology of Dispersive Clays', Jn Geotech Eng Div, 1193-1209, 1979.

6.      Bryant R, James A E, and Williams D J A. 'Rheology of Cohesive Sediments' Industrialised Embayments and their Environmental Problems (1980).

7.      James A E and Williams D J A. 'Flocculation and Rheology of Kaolinite/Quartz Suspensions' Rheol. Acta 21, 176-183, 1982

8.      Dzuy N Q and Boger D V 'Yield Stress Measurement for Concentrated Suspensions' J.Rheol, 27(4), 321-349, 1983.

9.      Dzuy N Q and Boger D V 'Direct Yield Stress Measurement with the Vane Method' J.Rheol, 29(3), 335-347, 1985.

10.     James A E, Williams D J A.& Williams P R   'Direct Measurement of Static Yield Properties of Cohesive Sediments' Rheol. Acta 26, 437-446, 1987

11      Barnes H A & Carnali J O 'The Vane in-cup as a Novel Rheometer Geometry for Shear Thinning and Thixotropic Materials' J. Rheol., 34(6), 841-866, 1990.

12.     Alderman N J, Meeton G H & Sherwood J D 'Vane Rheometry of Bentonite Gels' J.N.N.F.M., 1990

13.     O'Brien J S, and Julien P Y.   'Rheology of Non-Newtonian Fine Sediment Mixtures', Aerodynamics/Fluid Mechanics/Hydraulics, 988-997. 1986.

14.     Nguyen Q D & Boger D.V. 'Characterization of Yield Stress Fluids with Concentric Cylinder Viscometers' Rheol Acta., 26, 508-515,1987

15.     Verreet G, Van Geothem J, Viaene W, Berlamont J, Houthuys R, and Berleur E. 'Relations between Physico-chemical Properties of Fine-Grained Muds',   Third International Symposium on River Sedimentation, Jackson, Mississippi. 1986.

16.     Verreet G, and Berlamont J. 'Rheology and Non-Newtonian Behaviour of Sea and Estuarine Mud'. Encyclopedia of Fluid Mechanics, V1, 7, Cheremisioff. 1988.

17.     O'Brien J S, and Julien P Y. 'Laboratory Analysis of Mudflow Properties'. Journal of Hydraulic Engineering (American Society of Civil Engineers), vl 114, No 8, 877-887. 1988.

# 23 On the liquefaction and erosion of mud due to waves and current

P. J. DE WIT and C. KRANENBURG
*Department of Civil Engineering, Delft University of Technology,
The Netherlands*

## INTRODUCTION

The interaction between surface waves and a cohesive, soft bed has been studied experimentally and theoretically by several researchers. In some laboratory experiments it was observed that a layer of fluid mud was formed and that waves were damped due to the dissipation of wave energy in the fluid mud (Maa & Mehta, 1990; Lindenberg et al., 1989; Sakakiyama & Bijker, 1989). The fluid mud may be transported easily by a current, resulting in a relatively large transport of sediment near the bed. Consequently, the mechanisms underlying the generation of fluid mud are of great importance to the response of a muddy bed to waves and current, and play an important role in the transport of cohesive sediments.

In preliminary experiments the liquefaction mechanism was studied by Ross (1988) and Feng (1992) using soil mechanics principles. In the experiments made it was found, among other things, that during wave action the pore pressures increased and consequently the effective stresses decreased. However, the mechanism underlying the generation of fluid mud, and the interaction between waves and current on the one hand and a cohesive bed on the other hand are not well understood at present. Therefore, a research project was started on the liquefaction of a mud bed due to surface waves at Delft University of Technology. The objectives of this project were to analyze (1) the liquefaction mechanism, (2) the wave damping over a liquefied bed and (3) the transport and erosion of a liquefied mud layer caused by a current. In the framework of this project experiments were carried out in a wave/current flume using two artificial muds. First results were reported by De Wit & Kranenburg (1993). In addition mathematical models related to the first two items mentioned were elaborated. The experimental results and the results of the models used are presented in this paper.

## SEDIMENTS

Two different artificial clays were used in the flume experiments, namely China Clay and Westwald Clay. The main reason for using these artificial muds was to get reproducible measurements. An additional reason is that the small-scale waves in the experiments require a less cohesive mud, which these clays are.

*Cohesive Sediments.*
Edited by Neville Burt, Reg Parker and Jacqueline Watts
©1997 John Wiley & Sons Ltd.

**Table 1**    Properties of the artificial muds used

| Sediment | | China Clay | Westwald Clay |
|---|---|---|---|
| particle size distribution | > 10 μm | 14 % | 5 % |
| | < 2 μm | 41 % | 84 % |
| C.E.C. [meq/100 g] | | 5.0 | 8.3 |
| density [kg·m³] | | 2593 | 2644 |
| specific surface area [m²·g¹] | | 29.9 | 1.22 |
| mineralogical composition | | mainly kaolinite | mainly kaolinite and muscovite |

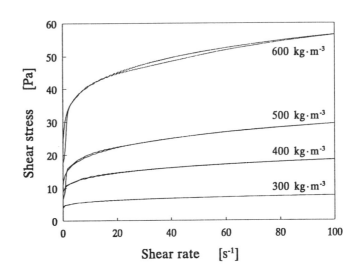

**Figure 1**    Flow curves of suspensions of Westwald Clay in saline tap water (salinity 5 ppt)

Various properties characterising these clays were determined and some of them are listed in table 1. For further specifications on these clays see De Wit (1994a,b).

The artificial muds were mixed with tap water in which sodium chloride was dissolved (salinity 5 ppt) to increase the flocculation and to eliminate the possible influence of small quantities of other chemicals on the properties of these muds, e.g. the settling velocity.

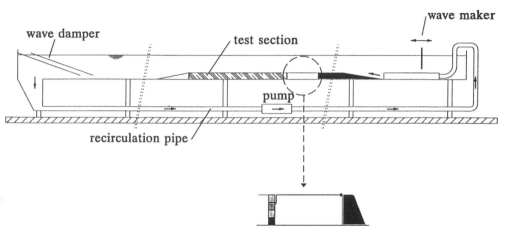

**Figure 2**    Sketch of the experimental set-up

The rheological behaviours of China Clay and Westwald Clay suspensions were determined using a Haake viscometer; model Rotovisco RV100 and measuring system CV100. Several suspensions of China and Westwald Clay were tested with suspended sediment concentrations ranging from approximately 100 to 600 kg·m⁻³. In these tests the shear rate increased from zero to a maximum value of 125 s⁻¹ in 3 minutes and then decreased in the same period to zero again. The temperature of the samples was 20 ± 0.5°C.

Some results of the rheological measurements on Westwald Clay suspensions are presented in Figure 1. For more information on the rheological properties of China Clay suspensions see De Wit (1994a). The dynamic viscosity and the yield strength of a suspension can be estimated using these results. These tests and the mineralogical composition showed that Westwald Clay was a more cohesive clay than China Clay.

## EXPERIMENTAL SET-UP

In the Hydromechanics Laboratory an existing flume was modified for the purpose. The flume was approximately 40 m long, 0.80 m wide and 0.8 m high. A sketch of the flume is shown in figure 2. At one end of the flume a mechanical wave maker was mounted in order to generate regular waves. At the other end of the flume a stainless steel wave-damper was installed to reduce wave reflection. The wave damper was constructed in such a way that the waves were dampened and a possible current would not be hindered. A recirculation pipe was installed below the flume to be able to generate a steady current. At the downstream end the fluid was withdrawn from the flume and subsequently the fluid passed an electromagnetic flowmeter and a centrifugal pump before it re-entered the flume.

The test section which held the sediment was 8.0 m long. The vertical endwalls of the test section were 0.20 m high and were formed by stacking four beams which were removable during an experiment. The downstream and upstream endwalls were connected to the bottom and the upstream cement false bottom, respectively, by 2.0 m long plates which were made adjustable by means of hinges. In this way the height of the test section could be adjusted during an experiment by removing a beam and lowering the free side of the plate.

Prior to the experiments, tests were carried out to measure the wave decay, the wave reflection and the velocity distribution in the flume. For this purpose a temporary false bottom, made of cement, was placed over the test section. Measurements were also made in this configuration when both waves and current were present. The results showed that there was no significant wave reflection and no significant wave decay above the closed test section. Furthermore, the velocity distribution for a steady current was almost uniform. For further information on these tests and the experimental set-up see De Wit (1994a).

## EXPERIMENTAL PROCEDURE

The artificial clays were thoroughly mixed with saline tap-water (salinity 5 ppt) in a mixing tank until their consolidation characteristics did not change any more. (China Clay had to mixed for at least 24 days and Westwald Clay for at least 21 days). Subsequently, the suspension was pumped to the test section, which was separated from the rest of the flume, and mixed again and then was allowed to consolidate. The experiments were initiated as soon as the consolidating mud layer was approximately 0.2 m thick.

Four miniature pore-pressure transducers (Druck PDCR 81) were fixed in a cross-section in the middle of the test section as soon as the mixing in the flume had stopped. The measured accuracy of the devices was $\pm$ 5 Pa. Furthermore, three electromagnetic current meters and six wave height meters were installed in the test section. Suspended sediment concentrations were measured by an optical method and by taking samples. The concentration in the bed was measured using a conductivity probe. The data were recorded on a personal computer and video recordings were made.

In all, five major experiments were made in the wave/current flume. Three of these experiments were made on China Clay and two on Westwald Clay. In all experiments the behaviour of and the processes in the mud were studied, and measurements were made for different settings of the wave height ($\leq$ 0.1 m) and water velocity ($\leq$ 0.2 m·s$^{-1}$). The wave period was 1.5 s and the initial water depth was 0.30 m. For a detailed overview of the experimental programmes of these experiments see De Wit (1994a, b).

## RESULTS

Prior to each experiment concentration profiles of the initial mud bed were measured. Some typical examples of concentration profiles in the bed for China and Westwald Clay are shown in figure 3. The average bed concentrations in the China Clay experiments were approximately 600 kg·m$^{-3}$. However, due to the unfavourable consolidation characteristics of Westwald Clay the average bed concentration was only 300 kg·m$^{-3}$.

When waves were generated it was observed that a layer of fluid mud was formed in each experiment. However, in the experiments on China Clay it was found that fluid mud was only formed when the wave height exceeded a threshold value. This value increased with the conso-

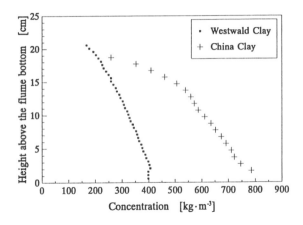

**Figure 3**    Typical concentration profiles of the initial China and Westwald Clay beds

lidation period.

An important role in the liquefaction process of cohesive sediments seems to be the shear stresses in the bed caused by the streamwise variation in wave-induced pressures on the bed (Suhayda, 1984). These shear stresses are sometimes overlooked although these stresses may be significant in several cases, which will be shown in the following.

A rough estimation of these pressure-induced shear stresses in a slab of mud can be made by assuming that the non-liquefied mud behaves as an ideal poro-elastic solid. Various researchers, e.g. Yamamoto (1978) and Mei (1982), have studied the response of a poro-elastic bed to water waves. Spierenburg (1987) studied in particular the response of a poro-elastic finite layer of sediment over a perfectly smooth or rough base. Using Spierenburg's model it can be calculated that the deviatoric shear stress at the surface of a China Clay bed, for instance, is approximately 20 Pa for a wave height of 2 cm, a wave period of 1.5 s, a water depth of 0.30 m and assuming the following characteristic values for the bed: thickness: 0.20 m, Poisson's ratio: 0.45 and shear modulus: 100 Pa. The initial concentration of the upper part of the bed was about 400 kg·m⁻³, see figure 3. The yield stress of a China Clay suspension with a concentration of 400 kg·m⁻³ is approximately 3.5 Pa (De Wit and Kranenburg, 1993), which is much lower than the stress generated by the pressure oscillations. Consequently, the bed will liquefy immediately after the generation of waves with a wave height of 2 cm, which agrees with the observations made. A similar result was found for the experiments on Westwald Clay.

As soon as a layer of fluid mud had been generated, the waves were significantly damped and the damping increased with the thickness of the fluid-mud layer. The damping was only little influenced by a current. A typical example of the wave decay measured in an experiment on China Clay is presented in figure 4. The analytical result also shown is discussed at the end of this section.

Three of the four miniature pore-pressure transducers installed were fixed at several levels in the bed, and the fourth was fixed just above the bed as a reference. Wave-averaged water pressure changes measured just above and in the Westwald Clay after the onset of liquefaction

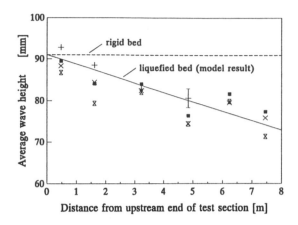

**Figure 4**   Wave heights measured in the test section and model result (China Clay, wave period 1.5 s, fluid-mud layer present, symbols refer to different instants)

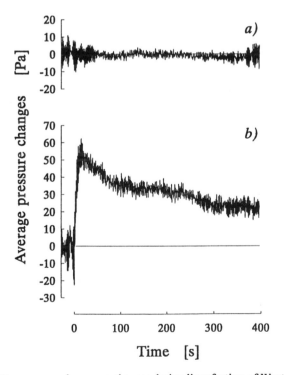

**Figure 5**   Wave-averaged pressure changes during liquefaction of Westwald Clay, a) pressure at 5 mm above the bed, b) pore-pressure at 55 mm below the bed surface

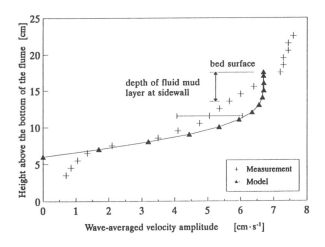

**Figure 6**    Average velocity amplitudes in and above a partly liquefied bed of China Clay.

are shown in figure 5. The pressure above the bed hardly changed during liquefaction. The pore pressures, however, showed a transient decrease which was probably caused by the break-up of the aggregate structure. Then a gradual build-up of an excess pore pressure was observed to compensate for the decreasing effective stress. The excess pore-pressure dissipated gradually with time probably caused by the high permeability of the low-concentrated Westwald mud. The pressure changes measured in China Clay showed a similar behaviour: a transient decrease followed by a build-up of an excess pore-pressure. However, this excess pore-pressure did not dissipate, but rose to a final value.

In the first two experiments on China Clay injections with dye near the bed surface seemed to indicate that the turbulence intensities in the overlaying water decreased when fluid mud was present. In order to validate these observations, electromagnetic current meters were used to measure turbulence intensities over non-liquefied and liquefied mud beds during the third experiment on China Clay. In preliminary experiments over a cement false bottom it was found that the root-mean-square (rms) velocities measured agreed with the results found by Nezu and Nakagawa (1993) for open channel flows. The average longitudinal velocities and rms velocities were measured in the middle of the test section for three settings of the flow rate. After these measurements a layer of fluid mud was formed by generating waves for approximately half an hour. Subsequently, the generation of waves was stopped and similar measurements were made. These measurements showed that although the velocity profiles measured prior to and after liquefaction were almost identical the rms velocities decreased approximately 20 per cent when fluid mud was present.

In the experiments electromagnetic current meters were also used to measure velocities in a partly liquefied bed. As soon as a significant layer of fluid mud had been generated due to wave action, an electromagnetic velocity meter was lowered into the mud layer. The calibration of the electromagnetic velocity meter was checked in towing tests on China Clay suspensions and saline tap-water prior to the experiments.

Examples of velocity measurements in and over the bed are shown in figures 6 and 7. In

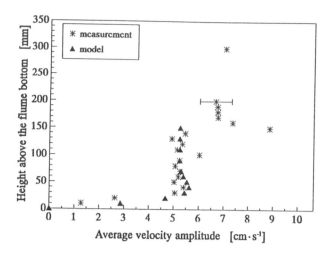

**Figure 7**     Average velocity amplitudes in and above a bed of Westwald Clay

figure 6 the velocity amplitudes measured in the third experiment on China Clay are presented. According to visual observations the bed was approximately 17.5 cm thick at that moment. The measured average velocity amplitudes indicated that in the centre of the flume the whole mud layer had been liquefied, whereas the thickness of the fluid-mud layer when judged at the glass sidewall was only 4 cm. In figure 7 a velocity profile measured in the second experiment on Westwald Clay is presented. Westwald Clay showed an adhesive affinity for the glass sidewalls of the flume. As a result a thin layer of approximately a few millimetres firmly sticked to the sidewall and no movement could be observed at all. However, the results of the velocity measurements showed that in the centre-line of the flume the entire bed liquefied.

Wave-induced velocities in the liquefied mud layer and the wave decay were also calculated using a modified version of a model due to Gade (1958). Gade's two-layer model was extended to account for arbitrary wavelengths. The upper layer is treated as a frictionless fluid and the lower layer, i.e. the liquefied mud, is considered to be a homogeneous fluid with a constant viscosity. The boundary layer approximation was adopted for the fluid-mud layer. The parameters used in the calculations were estimated from the results of the rheological measurements and the flow conditions during that particular phase of the experiment. The calculated velocity amplitudes are also shown in figures 6 and 7, and these results agree very well with the measured wave-induced velocity amplitudes. Dalrymple & Liu's (1978) model, which also includes the viscosity of the water layer, would give nearly the same results. The calculated wave decay is also shown in figure 4. For China Clay the calculated and measured wave decays agreed well. However, the calculated decrease in wave height for Westwald Clay underestimated the measured decrease by about 15 per cent.

## DISCUSSION AND CONCLUSIONS

The average bed concentrations in the experiments on Westwald Clay were rather low as compared with the experiments on China Clay because of the unfavourable consolidation characteristics of this clay.

A layer of fluid mud was generated due to wave action in each of the experiments on China and Westwald Clay. However, from the experiments on China Clay it was concluded that the threshold value of the wave height at which liquefaction occurred depended on the consolidation period of the mud: the threshold value increased with the consolidation period.

Although assuming the clays used being poro-elastic materials may be not quite appropriate, the calculated pressure-induced shear stress in the bed using Spierenburg's (1987) model are large so that these stresses play a major role in the liquefaction process of mud.

The wave height was damped significantly as soon as a layer of fluid mud was generated and the damping was only little influenced by a current. The wave decay calculated using the modified model of Gade (1958) agreed well with the decrease in wave height measured during the experiments on China Clay. In the experiments on Westwald Clay the calculated wave decay underestimated the measured decay.

Measurements made using electromagnetic current meters showed that the turbulence intensities decreased when fluid mud was present, which agrees with the observations made when dye was injected. However, more advanced measurements using laser doppler anemometry, for instance, are desirable to validate these measurements.

The wave-induced velocities in the bed calculated using the modified Gade model agreed very well with the measured velocity amplitudes. Consequently, it may be concluded that the fluid mud in the experiments behaved like a viscous fluid.

Measurements made at the onset of liquefaction showed a transient decrease in pore-pressure, possibly caused by the break-up of the aggregate structure, succeeded by gradual build-up of an excess pore pressure so as to compensate for the reduced effective stress.


**FURTHER DEVELOPMENTS**

This work is continued by examining the liquefaction process in small quantities of natural muds. An oscillating-water tunnel was built for this purpose. This set-up is a closed circuit in which velocity amplitudes and periods can be generated which are also encountered in the field. The mud compartment is 0.50 m long and 0.25 m wide. The thickness of the initial mud-layer is 0.20 m. Deposited and placed beds of natural and artificial muds are being tested in the experiments. The onset of liquefaction is determined for different settings of the period. For further details and results see De Wit (1995).


**ACKNOWLEDGEMENTS**

The authors wish to thank Delft Hydraulics for making available their Haake viscometer. This project was partly funded by the Commission of the European Communities, Directorate General for Science, Research and Development under MAST2 (G8 Morphodynamics research programme) and supported financially by Rijkswaterstaat. The project was conducted in the frame work of the Netherlands Centre of Coastal Research.

# REFERENCES

Dalrymple, R.A. and Liu, P.L.-F., (1978), *Waves over soft muds: a two-layer fluid model*, Journal of Physical Oceanography, Vol. 8, November 1978, pp. 1121-1131.

De Wit, P.J. and Kranenburg, C., (1993), *Liquefaction and erosion of China Clay due to waves and current*, Proceedings of the 23th International Conference on Coastal Engineering, Vol. 3, pp 2937-2948.

De Wit, P.J., (1994a), *Liquefaction and erosion of mud due to waves and current - Experiments on China Clay*, Report no. 10-92, Delft University of Technology.

De Wit, P.J., (1994b), *Liquefaction and erosion of mud due to waves and current - Experiments on Westwald Clay*, Report no. 2-93, Delft University of Technology.

De Wit, P.J., (1995), *Liquefaction of cohesive sediments caused by waves*, Ph.D. dissertation, Delft University of Technology.

Feng, J., (1992), *Laboratory experiments on cohesive soil bed fluidization by water waves*, Thesis, Report UFL/COEL-92/005, University of Florida, Gainesville, USA.

Gade, H.G., (1958), *Effect of a nonrigid, impermeable bottom on plane surface waves in shallow water*, Journal of Marine Research, Vol. 16, No. 2, pp. 61-82.

Lindenberg, J., van Rijn, L.C. and Winterwerp, J.C., (1989), *Some experiments on wave-induced liquefaction of soft cohesive soils*, Journal of Coastal Research, Special issue no. 5, pp. 127 - 137.

Maa, J. P.-Y. and Mehta, A.J., (1990), *Soft mud response to water waves*, Journal of Waterway, Port, Coastal, and Ocean Engineering, Vol. 116, no. 5, pp. 634 - 650.

Mei, C.C., (1982), *Analytical theories for the interaction of offshore structures with a poro-elastic sea-bed*, Proceedings BOSS' 82 Conference, pp. 358-370.

Nezu, I. and Nakagawa, H., (1993), *Turbulence in open-channel flows*, IAHR Monograph Series, A.A. Balkema Publishers, ISBN 90 5410 118 0.

Ross, M.A., (1988), *Vertical structure of estuarine fine sediment suspensions*, Ph.D. dissertation, University of Florida, Gainesville, U.S.A.

Sakakiyama, T. and Bijker, E.W., (1989), *Mass transport velocity in mud layer due to progressive waves*, Journal of Waterway, Port, Coastal, and Ocean Engineering, Vol. 115, No. 5, September 1989, pp. 614-633.

Spierenburg, S.E.J., (1987), *Seabed response to water waves*, dissertation, Delft University of Technology, The Netherlands.

Suhayda, J.N., (1984), *Interactions between surface waves and muddy bottom sediments*, Lecture notes on Coastal and Estuarine Studies, Vol. 14, Estuarine Cohesive Sediment Dynamics edited by A.J. Mehta, pp. 401-428.

Yamamoto, T., (1978), *On the response of a poro-elastic bed to water waves*, Journal of Fluid Mechanics, Vol. 87, part 1, pp. 193-206.

# 24　Mud fluidization by water waves

Y. LI[1] and A. J. MEHTA[2]
[1]*Moffatt & Nichol Engineers, Long Beach, California, USA*
[2]*Coast and Oceanographic Engineering Department, University of Florida, Gainesville, Florida, USA*

## INTRODUCTION

The transport of nutrients and contaminants between water and mud beds is strongly influenced by the wave-induced fluid mud layer on the bed. Various concepts have been proposed for wave-soft bed interaction and fluidization, considered here to mean liquefaction, of the bed. Foda et al. (1991) suggested a "soil piping" concept according to which resonance is initiated within vulnerable cavities in the fine-grained bed, followed by progressive failure of the soil matrix around such cavities. Feng (1992) studied cohesive bed fluidization under wave motion through an experimental investigation in which the growth of the fluid mud layer with time following the inception of wave motion was measured. Fluid mud was defined by the top layer thickness within which the effective normal stress was practically nil. In other words, fluidization was shown to occur by virtue of the loss of effective stress in the bed. Foda et al. (1993) applied a non-linear shear model to calculate the wave-induced mud fluidization depth at equilibrium.

Dalrymple and Liu (1978) compared results on water wave attenuation from their wave-soft bed interaction model to results from Gade's (1957) inviscid, shallow water model and found that despite Gade's neglect of viscosity in the upper fluid, and therefore absence of shear stress at the interface, both solutions agreed reasonably well with the same laboratory data. A conclusion can therefore be drawn that the principal mode of energy transfer to the mud is the normal stress, rather than the shear stress, due to wave working on the lower medium. Isobe et al. (1992) also found that of the two external wave-induced forces, namely that due to the gradient of dynamic pressure and that due to bed shear stress, the former is dominant for mud motion, which in turn is responsible for energy dissipation and associated surface wave damping. Selly (1988) observed that fluidization of a sandy bed by vibration occurs when the upward drag exerted by the pore water moving relative to particles exceeds the submerged weight of the grain. A simple mathematical model was provided by Das (1983), which treats the vibrating bed as a spring-mass system and yields an expression for the resonant frequency of vibration in terms of the bed thickness.

Two aspects of wave action on mud bed can be distinguished: "shaking" due to cyclic shear stress, and "pumping" due to cyclic normal stress. Inasmuch as pumping has been shown to be the dominant mechanism for mud motion, energy dissipation and wave damping, we will consider the main mechanism for fluidization of the mud bed to be the cyclic normal stress

*Cohesive Sediments.*
Edited by Neville Burt, Reg Parker and Jacqueline Watts
© 1997 John Wiley & Sons Ltd.

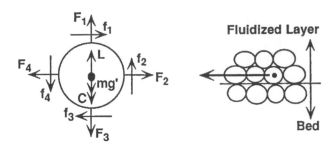

**Figure 1.** Forces on a particle. $F_1, F_2, F_3, F_4$=normal forces from adjacent particles; $f_1, f_2, f_3, f_4$= shear forces from adjacent particles; L=inertia force; $m_p$=particle mass; $g'$=reduced gravity; and C=cohesion.

acting on the bed which causes the bed to vibrate, while the shear stress is considered to be responsible for rupturing the bonds between the particles when the bed is critically stretched during vibration.

## MECHANISM OF FLUIDIZATION

While in a sandy bed pore fluid motion is essential for fluidization, in a typical mud bed the permeability tends to be so low that the bed behaves as a practically impermeable continuum with little motion of the pore fluid relative to the particulate matrix over the wave time-scale. Yet, under wave motion mud beds are known to fluidize, and a fluid mud layer of equilibrium thickness is observed to occur after a certain time depending on the initial conditions with respect to the mud-water system and the forcing wave characteristics. Consequently, a mechanistic framework for explaining mud bed fluidization must be based on a consideration of forces required to cause particle-particle separation, hence a loss of effective stress. Considering a particle (or a cohesive floc treated as an integral particle) in the mud under wave motion as shown in Fig.1, four types of forces must be recognized: (I) effective gravity or buoyancy in water, (ii) inter-particle cohesive forces, (iii) normal and shear forces due to contacts between adjacent particles, and (iv) inertia force due to vibration, L, under wave loading. Following others, e.g. Partheniades (1977), the influence of cohesion, C, can be approximately incorporated with the buoyant weight by defining an effective submerged weight, which depends on mud composition. Furthermore, the normal and shear forces (respectively F and f with subscripts in Fig.1) due to contacts between adjacent particles, which are almost randomly directed, are considered to have no tangible effect on bed fluidization as a first order approximation.

While the bed can sustain great pressure, it cannot withstand significant stretching forces. Thus, in the vertical direction, when the sum of the upward forces equals (or just exceeds) the sum of the downward forces, the bed is critically stretched, the connections between particles are broken and incipient bed fluidization occurs. We therefore select the criterion for fluidization at any point to be that instant when the effective submerged weight of the particle in water, $m_p g' + C$, where $g' = g[\rho(z) - \rho_1]/\rho_1$, is exceeded by the maximum upward force, $L_{max}$, due to wave-induced vibration. Here $m_p$=particle mass, g=gravity acceleration, $\rho_1$=water density and $\rho(z)$=the bed density at any depth z. This criterion, based on forces on a single particle, can be considered to reflect a corresponding condition causing the separation of the fluidized mud layer from the bed layer (Fig.1). We will further consider the bed to be a viscoelastic medium in examining its response to wave forcing.

## VOIGT AND EXTENSIONAL-VOIGT MODELS

For the commonly used Voigt model representing bed viscoelasticity the shear stress-strain relationship is:

$$T_{ij}' = 2GE_{ij}' + 2\mu\dot{E}_{ij}' \tag{1}$$

where: $T_{ij}'$=stress deviator, $E_{ij}'$=strain deviator, subscripts i and j indicate directions, ij indicates a second order tensor, G=elastic modulus, $\mu$=viscous coefficient, and the dot denotes derivative with respect to time. For bed fluidization as defined, we are interested in the corresponding extensional stress-strain relationship. Assuming the bed's shear viscoelasticity to be represented by Eq.1, its extensional viscoelasticity is characterized as follows:

The volumetric relationship between the mean normal stress and the mean normal strain of the bed material is assumed to be elastic (Malvern, 1969):

$$\text{tr}[T_{ij}] = 3K\text{tr}[E_{ij}] \tag{2}$$

where: $T_{ij}$=stress, $E_{ij}$=strain, K=the bulk modulus and tr=trace. Assuming the dynamic external pressure to be important only in the vertical (z or I=j=3) direction, Eq.2 reduces to

$$T_{33} = 3K\text{tr}[E_{ij}] \tag{3}$$

Next, the deviators of stress and strain are expressed as:

$$E_{ij}' = E_{ij} - \frac{1}{3}\text{tr}[E_{ij}]\delta_{ij} \quad ; \quad T_{ij}' = T_{ij} - \frac{1}{3}\text{tr}[T_{ij}]\delta_{ij} \tag{4}$$

where $\delta_{ij}$=kronecker delta. Then inserting Eq.4 into Eq.1 and setting ij=33 we obtain the isothermal uniaxial stress, $T_{33}(t)$, versus tensile strain, $E_{33}(t)$, relation for a material whose deviatoric response is described by the following equation:

$$T_{33} - \frac{1}{3}T_{33} = 2G\left(E_{33} - \frac{1}{3}\text{tr}[E_{ij}]\right) + 2\mu\left(\dot{E}_{33} - \frac{1}{3}\text{tr}[\dot{E}_{ij}]\right) \tag{5}$$

Inserting Eq.3 into Eq.5 then yields

$$\frac{2\mu}{9K}\dot{T}_{33} + \left(\frac{2G}{9K} + \frac{2}{3}\right)T_{33} = 2GE_{33} + 2\mu\dot{E}_{33} \tag{6}$$

Since mud in the saturated condition is incompressible, $K\rightarrow\infty$ (Das, 1983), and Eq.6 becomes

$$T_{33} = 3GE_{33} + 3\mu\dot{E}_{33} = G_nE_{33} + \mu_n\dot{E}_{33} \tag{7}$$

where $G_n$=extensional elastic modulus and $\mu_n$=extensional viscous coefficient. Hence, the bed viscoelasticity in the normal direction is described under the given assumptions by the extensional-Voigt model (Eq.7) in which the parameters are, $G_n=3G$ and $\mu_n=3\mu$, provided the shear viscoelasticity is characterized by Eq.1. When G=0, Eq.7 satisfies the uniaxial extensional relationship for a Newtonian fluid (Petrie, 1978).

## BED AS A SPRING-DASHPOT-MASS SYSTEM

**Mechanical Analog**

Given the bed to be an impermeable continuum, we can choose to represent it as a mechanical, spring-dashpot-mass (S-D-M) analog. The key aspect of this approach is to find the equivalent elastic and viscous coefficients and the equivalent mass of the bed. To obtain these quantities the application of a constant stress, F, to the bed surface at time t=0 is considered first. We assume that the vertical distributions of the bed density, $\rho(z)$, the elastic modulus, $G_n(z)$, and the viscous coefficient, $\mu_n(z)$, are known, as shown schematically in Fig.2. Unfortunately, an arbitrary distribution of these parameters is not amenable to an analytical solution of Eq.7. Two specific cases are therefore considered below.

### Case 1: $G_n$, $\mu_n$ and $\rho$ Independent of Depth (Uniform Bed)

Given h=bed thickness, k=equivalent elastic coefficient, c=equivalent viscous coefficient and m=equivalent mass (Fig.2), from Eq.7 we have

$$F = G_n \frac{d\zeta}{dz} + \mu_n \frac{d\dot{\zeta}}{dz} \tag{8}$$

where $F=T_{33}$ is the normal stress, $d\zeta(z,t)/dz=E_{33}$ is the normal strain, $\zeta(z,t)$=vertical displacement within the bed and $\zeta(0,t)$=vertical displacement at the bed surface. Since F is independent of time, the solution of Eq.8 is

$$\frac{d\zeta}{dz} = \frac{F}{G_n}(1-e^{-\frac{G_n}{\mu_n}t}) \tag{9}$$

Further, integrating Eq.9 from z to h we obtain:

$$\zeta(z,t) = \int_z^h \frac{F}{G_n}(1-e^{-\frac{G_n}{\mu_n}t})dz = \frac{F}{k}(1-e^{-\frac{k}{c}t})\left(1-\frac{z}{h}\right) \tag{10}$$

where $k=G_n/h$, $c=\mu_n/h$, and at the bed surface, z=0:

$$\zeta(0,t) = \frac{F}{k}(1-e^{-\frac{k}{c}t}) \tag{11}$$

Eq.11 in fact represents the response of a Voigt solid to loading by F, with a retardation time constant c/k. Thus the extensional stress-strain relationship for the bed (Eq.7) satisfies the Voigt model relationship with $k=G_n/h$ and $c=\mu_n/h$. Therefore, the bed can be simply represented as a S-D-M system with equivalent coefficients k and c. The equivalent mass is in general obtained from:

$$m\ddot{\zeta}(0) = \int_0^h \rho(z)\ddot{\zeta}(z)dz \tag{12}$$

where $\ddot{\zeta}(z)$= vertical acceleration profile. Then, from Eqs.10 and 12,

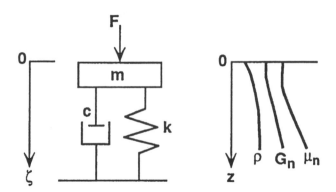

**Figure 2.** Bed simulation as a one-degree of freedom system with known vertical distributions of the bed density, $\rho$, elastic modulus, $G_n$, and viscous coefficient, $\mu_n$. Here m=equivalent mass, k= equivalent elasticity, and c=equivalent viscosity.

$$m = \int_0^h \rho(z)\left(1 - \frac{z}{h}\right)dz \tag{13}$$

and from Eqs.10 and 11 we obtain (for $\rho$ independent of z):

$$\zeta(z,t) = \zeta(0,t)\left(1 - \frac{z}{h}\right) \tag{14}$$

where $\Psi_a = (1-z/h)$ is a shape function, which relates the displacement at any depth, z, to the displacement at the bed surface (z=0).

**Case 2: $G_n$, $\mu_n$ and $\rho$ Increase with Depth (Stratified Bed)**
Here we present a case in which $G_n$ and $\mu_n$ increase with depth according to:

$$G_n = \alpha_1 e^{\gamma z} \quad ; \quad \mu_n = \alpha_2 e^{\gamma z} \tag{15}$$

where $\alpha_1$, $\alpha_2$ and $\gamma$ are mud-specific coefficients. Since the bed is assumed to be a Voigt solid, the solution steps are the same as in Case 1. Thus, we obtain

$$\zeta(z,t) = \int_z^{+\infty} \frac{F}{\alpha_1 e^{\gamma z}}(1 - e^{-\frac{\alpha_1}{\alpha_2}t})dz = \frac{F}{k}(1 - e^{-\frac{k}{c}t})e^{-\gamma z} \tag{16}$$

and $\zeta(0,t)$ is given by Eq.11 as well, where $k=\alpha_1\gamma$ and $c=\alpha_2\gamma$ are the equivalent S-D-M coefficients for the bed. The relationship between the displacement of the bed surface and that at any depth in the bed is obtained as

$$\zeta(z,t) = \zeta(0,t)e^{-\gamma z} \tag{17}$$

where $\Psi_b = \exp(-\gamma z)$ is a shape function, and the equivalent mass is obtained in the same general way as in Case 1 (Eq.12).

Thus far, we have considered two representations of the bed by an equivalent S-D-M system. Once $\zeta(0,t)$ is obtained, the response at any depth within the bed can be determined through the appropriate shape function. For Case 1 representing a uniform bed the bed depth, h, must be specified, and $G_n$ and $\mu_n$ must be determined through rheometry from which G and $\mu$ are obtained. Then $G_n = 3G$ and $\mu_n = 3\mu$. In the more common situation of a stratified bed (Case 2), it is necessary to determine $G_n(\rho)$ and $\mu_n(\rho)$. We will next examine the response of the bed to wave motion. In what follows, subscript 1 is used to denote the water layer, 2 indicates fluid mud layer, and 3 is for the bed.

## DYNAMIC RESPONSE OF BED

### Wave over Bed: One-Degree of Freedom

In Fig.2, instead of a constant external stress, F, acting on the bed, consider the wave-induced cyclic normal pressure, p (with angular wave frequency, $\omega$), expressed as

$$p = p_0 \sin \omega t = \frac{\rho_1 g \zeta_{01}}{\cosh k_{w1} h_1} \sin \omega t \tag{18}$$

where $p_0$=pressure amplitude, $\zeta_{01}$=wave amplitude, $k_{w1}$=wave number and $h_1$=water depth.

The dynamic equation for the one-degree of freedom system shown in Fig.2 is

$$m \ddot{\zeta}_3(0,t) + c \dot{\zeta}_3(0,t) + k \zeta_3(0,t) = p_0 \sin \omega t \tag{19}$$

where c, k and m are defined in Fig.2. Denoting the characteristic resonance frequency, $\omega_o = (k/m)^{1/2}$, and also $\beta = \omega/\omega_o$, $\xi = c/2m\omega$, and $\omega_D = \omega\sqrt{1-\xi^2}$ we obtain the solution:

$$\zeta_3(0,t) = e^{-\xi t}[A \sin \omega_D t + B \cos \omega_D t] + \frac{\frac{p_0}{k}[(1-\beta^2)\sin \omega t - 2\xi\beta\cos \omega t]}{[(1-\beta^2)^2 + (2\xi\beta)^2]} \tag{20}$$

where A, B are coefficients dependent on the initial conditions. Since the first term decays exponentially with time and the second term is the harmonic response sought, the solution is

$$\zeta_3(0,t) = \frac{p_0}{k} \frac{1}{[(1-\beta^2)^2 + (2\xi\beta)^2]^{1/2}} \sin(\omega t - \phi) \tag{21}$$

where $\phi = \tan^{-1}[c\omega/(k-m\omega^2)] = \tan^{-1}[2\xi\beta\omega/(1-\beta^2)]$.

### Wave over Fluid Mud above Bed: Two-Degrees of Freedom

Oftentimes a layer of fluid mud is initially present on the bed. Inasmuch as the rheological properties of fluid mud and the bed from which it is derived are different, the two layers of continua must be represented by two equivalent S-D-M systems in series, constituting the two-degrees of freedom system shown in Fig 3. In this case, whether or not the bed fluidizes will depend on the wave conditions, fluid mud properties and layer thickness, and the bed properties. The governing equations of motion for this are

$$m_2 \ddot{\zeta}_2 + c_2(\dot{\zeta}_2 - \dot{\zeta}_3) + k_2(\zeta_2 - \zeta_3) = p_0 e^{i\omega t} \tag{22}$$

$$c_2(\dot{\zeta}_2 - \dot{\zeta}_3) + k_2(\zeta_2 - \zeta_3) = m_3 \ddot{\zeta}_3 + c_3 \dot{\zeta}_3 + k_3 \zeta_3 \tag{23}$$

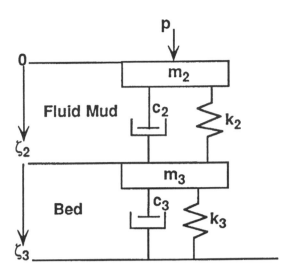

**Figure 3.** Fluid mud/bed layers represented by a two-degrees of freedom system. $m_2$, $m_3$= equivalent masses; $k_2$, $k_3$=equivalent elasticities and $c_2$, $c_3$=equivalent viscosities.

where $i=\sqrt{-1}$, $\zeta_2$ and $\zeta_3$ are displacements of the fluid mud surface and the bed surface, respectively, and the coefficients $m_2$, $m_3$, $k_2$, $k_3$, $c_2$ and $c_3$ are given in Fig.3.

As in Eq.21, the dynamic response sought is represented by the steady (harmonic) term. Thus, we express $\zeta_2=\zeta_{02}\exp[i(\omega t-\phi_2)]$ and $\zeta_3=\zeta_{03}\exp[i(\omega t-\phi_3)]$, where $\phi_2$ and $\phi_3$ are the respective phase shifts of the fluid mud and the bed displacement relative to $\zeta_1$, and $\zeta_{02}$, $\zeta_{03}$ are the corresponding displacement amplitudes. Then, solving Eqs.23 and 24 we obtain

$$\zeta_{02} = P_0\left[\frac{(k_3+k_2-m_3\omega^2)^2+(c_3\omega+c_2\omega)^2}{Z_3^2+Z_2^2}\right]^{1/2} \tag{24}$$

$$\zeta_{03} = P_0\left[\frac{(k_2)^2+(c_2\omega)^2}{Z_3^2+Z_2^2}\right]^{1/2} \tag{25}$$

where $Z_2=c_3\omega(k_2-m_2\omega^2)+c_2\omega[k_3-(m_3+m_2)]\omega^2$ and $Z_3=(k_3-m_3\omega^2)(k_2-m_2\omega^2)-(k_2m_2+c_3c_2)\omega^2$. If $c_2$ and $k_2$ approach zero, and $m_2$ is set equal to zero, we recover the solution for the one-degree of freedom system (Eq.22).

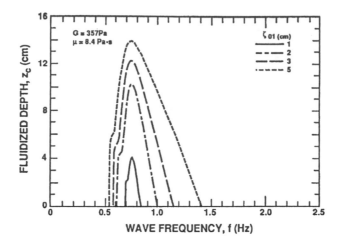

**Figure 4**. Fluidization depth variation with wave frequency and amplitude

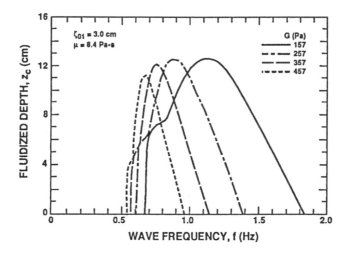

**Figure 5**. Fluidization depth variation with wave frequency and bed elastic modulus

## FLUIDIZATION DEPTH

### Model Results

From Eqs.14 and 18, we recognize that the vibratory displacement and acceleration decrease from the bed surface downwards. Therefore, the inertia force decreases downwards while the effective submerged weight of the particle either remains constant (uniform bottom) or increases (stratified bottom). Thus, there must exist an equilibrium depth down to which the bed is fluidized.

For the water-bed (one-degree of freedom) or the water-fluid mud-bed (two-degrees of freedom) system, the equilibrium depth to which the bed is fluidized, $z_c$, can be obtained via

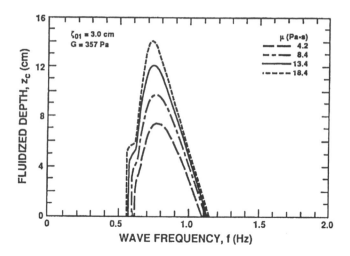

**Figure 6.** Fluidization depth variation with wave frequency and bed viscous coefficient

Eq.21 or Eqs.22 and 23 based on the criterion

$$g' = \ddot{\zeta}_3(z_c,t)_{max} \tag{26}$$

in which mud viscoelasticity is characteristically included (via $\ddot{\zeta}_3$), but the effect of cohesion in modulating $g'$ is ignored for simplicity. It is noteworthy that once the bed is fluidized, the broken inter-particle bonds apparently do not reform as long as wave motion continues (Feng, 1992). Therefore the calculated fluidization depth represents the maximum fluid mud layer thickness for the one- or the two-degrees of freedom system.

Figures 4, 5 and 6 are results for $z_c$ based on the criterion of Eq.26 for a bed/water system initially without the presence of fluid mud, with $h_1=19$ cm; $h_3=16$ cm; $\rho_1=1,000$ kg/m³; $\rho_3=1,170$ kg/m³; $\zeta_{01}=1, 2, 3,$ and 5 cm; wave frequency, f ($=\omega/2\pi$) range 0.4 to 4.0 Hz; G=157, 257, 357 and 457 Pa and $\mu=4.2, 8.4, 13.4$ and 18.4 Pa.s. These parameters are representative of the laboratory experiments of Feng (1992) described later.

The procedure involves calculation of $z_c$ at first for the one-degree of freedom system (Case 1). Then with this value of $z_c$, a second $z_c$ is calculated for the two-degree of freedom system (Case 1). Finally, the value of $z_c$ represented in the figures is the larger of the two values. In Fig.4, $z_c$ is observed to increase with the wave amplitude. For a given amplitude, $z_c$ attains a maximum, $z_{cmax}$, at the characteristic resonance frequency, $\omega_0$. In Fig.5, $z_{cmax}$ is observed to occur at increasing $\omega_0$ as G increases. Also, the greater the value of $\mu$, the smaller the $z_c$ (Fig.6). It is noteworthy that for high frequency waves, the wave-induced cyclic normal pressure tends to zero (Eq.18), while for low frequency waves, the inertia force, L, due to bed vibration becomes very small. Thus for both high and low frequency waves, the bed is not fluidized.

**Table 1.** Wave and bed conditions and calculated and measured fluidization depths.

| Test no. | Wave amplitude (cm) | Water depth (cm) | Initial bed thickness (cm) | Fluid mud thickness, $z_c$ | | |
| --- | --- | --- | --- | --- | --- | --- |
| | | | | Measured (cm) | Case 1 (cm) | Error (%) |
| 8 | 2.0 | 18.4 | 16.6 | 3.5 | 5.1 | 46 |
| 9 | 3.0 | 19.3 | 15.7 | 8.4 | 5.6 | -33 |
| 10 | 4.0 | 20.3 | 14.7 | 7.0 | 8.0 | 14 |
| 11 | 1.5 | 20.2 | 14.8 | 2.0 | 1.0 | -50 |

**Comparison with Data**

The experiments of Feng (1992) were conducted in a flume in which the water level was maintained at 35 cm above the bottom and the wave frequency was 1 Hz. The nominally 16 cm thick bed was composed of an aqueous mixture of attapulgite and kaolinite clays in equal proportion by weight and with $\rho_3 = 1,170$ kg/m$^3$. For each setting of wave amplitude the pore and total pressures were measured at different depths in the bed. The maximum depth below mud surface at which the effective stress (total pressure minus pore pressure) was equal to zero was considered to be $z_c$. In each test, $z_c$ increased with time from nil at the beginning as wave motion continued and seemingly approached an equilibrium value.

From the work of Mehta and Jiang (1993) involving the use of a Carri-Med CSL applied stress rheometer to determine G and μ as functions of mud density, we deduce G=357 Pa and μ=8.4 Pa.s for the chosen mud.

Results are compared with measurements in Table 1. In general, the computed values show an order of magnitude agreement with the data. As an overall measure of comparison we note that the mean error is -23%, which is attributed to model assumptions and also to experimental errors arising from the limited sensitivities of the pressure transducers, and the fact that test durations were insufficient to attain truly equilibrium values of the fluidized mud thickness (Feng, 1992).

**CONCLUSIONS**

The main observations from this work are as follows:

(1) The continuum bed, assumed to be characterized as a uniaxial extensional-Voigt solid, can be simply represented by an one-degree of freedom S-D-M system, whereby the displacement at any point in the bed is related to the bed surface displacement by an appropriate shape function. When fluid mud is present over the bed, a two-degree of freedom representation is appropriate.

(2) The thickness of the fluidized mud layer at equilibrium can be estimated by considering the balance of vertical forces on a particle (or floc) within the bed.

(3) Comparison between theoretically calculated and measured fluidization depths indicates an order of magnitude agreement. This agreement implies that applied shear rheometry can be used to determine mud-specific viscoelastic coefficients required to estimate the depth of fluidization.

## ACKNOWLEDGMENT

Funds for this study were provided by U.S. Army Engineer Waterways Experiment Station, contract DACW39-93-K-0008.

## REFERENCES

Dalrymple, R. A. and Liu, P. L-F., 1978. Waves over soft muds: A two-layer fluid model. *Journal of Physical Oceanography*, 8, 1121-1131.

Das, B. M., 1983. *Fundamentals of Soil Dynamics*. Elsevier, New York.

Feng, J., 1992. Laboratory experiments on cohesive soil bed fluidization by water waves. *M.S. Thesis, Coastal and Oceanographic Engineering Department, University of Florida, Gainesville.

Foda, M. A., Tzang S. Y. and Maeno, Y., 1991. Resonant soil liquefaction by water waves. *Proceedings of Geo-Coast'91*, Yokohama, Japan, 549-583.

Foda, M. A., Hunt, J. R. and Chou, H. T., 1993. A non-linear model for the fluidization of marine mud by wave. *Journal of Geophysical Research*, 98, 7039-7047.

Gade, H. G., 1957. Effects of a non-rigid, impermeable bottom on plane surface waves in shallow water. *M.S. Thesis*, Texas A & M University, College Station.

Isobe, M., Huyuh, T. N. and Watanabe, A., 1992. A study on mud mass transport under waves based on an empirical rheology model. *Proceedings of the 23rd International Conference on Coastal Engineering*, 3, ASCE, New York, 3093-3106.

Malvern, L. E., 1969. *Introduction to Mechanics of a Continuum Medium*. Prentice-Hall, Englewood Cliffs, New Jersey.

Mehta, A. J. and Jiang, F., 1993. Some observations on water wave attenuation over nearshore underwater mudbank and mud berm. *Report UFL/COEL/MP-93/01*, Coastal and Oceanographic Engineering Department, University of Florida, Gainesville.

Partheniades, E., 1977. Unified view of wash load and bed material load. *Journal of the Hydraulics Division of the American Society of Civil Engineers*, 103(9), 1037-1057.

Petrie, C. J. S., 1978. *Elongational Flows*. Pitman, London.

Selly, R. C., 1988. *Applied Sedimentology*. Academic Press, London.

# 25 Erosion and liquefaction of natural mud under surface waves

**H. VERBEEK[1] and J. M. CORNELISSE[2]**
[1]*Institute for Inland Water Management and Waste Water Treatment, Lelystad; presently at National Institute for Coastal and Marine Management, The Hague, The Netherlands*
[2]*Delft Hydraulics, Delft, The Netherlands*

In shallow lakes erosion of a mud bed is caused by surface waves. The wave orbital motion generates a shear stress and a load fluctuation which, depending on the stress and resistance of the bed, can result in liquefaction. Liquefaction in these circumstances may occur either as a direct response to the waves or it may occur after some time. In the first case the grain structure collapses and immediately an excess pore water pressure is created, while in the second case an excess water pressure is built up due to a gradual compaction of the structure. Both situations are investigated based on the literature and by means of wave experiments. Characteristic figures and relevant mud properties are given. For the sake of verification, the wave experiments are compared on the basis of mud properties and density profiles with in situ measurements. In addition to describing what happens on the bed, a brief discussion on the consequences for the critical shear stress for erosion and the erosion parameter is made.

## INTRODUCTION

The transport of micro-pollutants in Dutch rivers and lakes is largely determined by the transport of cohesive sediments to which the pollutants are attached. In recent years, the Ministry of Transport, Public Works and Water Management (Rijkswaterstaat) has paid considerable attention to the development of numerical models for determining the transport of

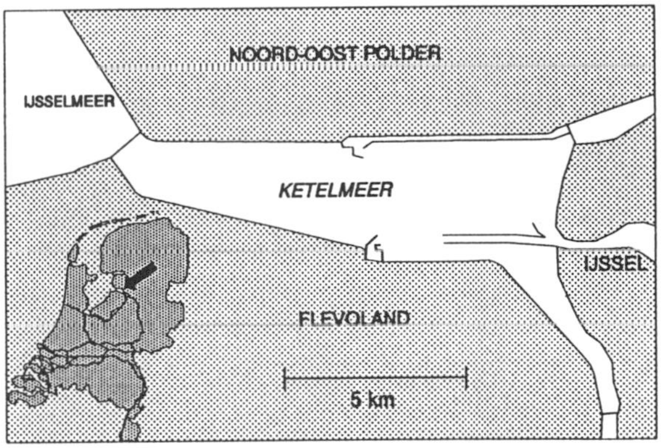

**Figure 1** The location of lake Ketelmeer in The Netherlands

pollutants attached to cohesive sediments to and from the bed, particularly in the Ketelmeer, a lake in the north east part of the Netherlands (fig. 1). This work has been supplemented with an investigation into the erosion and sedimentation characteristics under flow conditions of mud out of the Ketelmeer (Kuijper, *et al.* 1990).

*Cohesive Sediments.*
Edited by Neville Burt, Reg Parker and Jacqueline Watts
©1997 John Wiley & Sons Ltd.

In shallow lakes, such as the Ketelmeer, the wave climate is in general dominant for the erosion of the bed. In such circumstances it is difficult to describe the erosion of mud in terms of formulations derived for erosion under flow conditions. Waves can cause liquefaction (as well as erosion), which causes flux and entrainment. Liquefaction can be caused by waves as a direct effect or as an effect progressive in time (dynamic load). The present approach in numerical models is that shear stresses resulting from waves and flows are aggregated and that the erosion of mud is determined on the basis of parameters determined under flow conditions. Liquefaction is not included in this approach.

A combined experimental and theoretical investigation is started in order to find an answer to the question as to how erosion of mud under waves takes place. In this investigation, attention is paid to the influence of the bed structure and the physico-chemical parameters of the mud on erosion behaviour. In a preceding survey, the liquefaction of mud beds under waves is determined, as a function of the wave load (Lindenberg *et al.* 1988).

In this article an attempt is made to determine the circumstances under which liquefaction occurs, based on a number of parameters and characteristic figures, which are easy to determine. The aim is to construct an easy operational model with which liquefaction and erosion behaviour of a mud bed can be predicted for situations in which waves are dominant.

The method used is as follows. The circumstances in nature are observed and these circumstances are simulated with experiments in the laboratory. The various phenomena are then coupled to the properties of the bed. The bed properties are analyzed in more detail in the context of this investigation. An attempt is made to describe the influence of waves on a liquefied bed according to a viscoelastic model. The movements in the bed are descripted on the basis of the theory of Yamamoto (1983) for a non-liquefied bed. These two models are linked together.

## NATURAL CIRCUMSTANCES

Lake Ketelmeer lies at the mouth of the River IJssel, one of the delta branches of the River Rhine (fig. 1). It is 3800 hectares in size and has an average depth of 3 m. The lake is created in 1957 when a part of the IJsselmeer is cut off due to the formation of the Flevoland polder. The IJsselmeer itself is created in 1929 when the Zuiderzee was closed. The bed of the Ketelmeer is therefore made up of old layers of marine origin, covered with layers of sediment of fluvial origin. The uppermost layers originate in the IJssel and contain the most micro-pollutants. 70% of the bed of the Ketelmeer is composed of silt particularly in the western and central sectors. The average thickness of the deposit is 0.5 m, but in certain places (at the mouth of the IJssel and in sand dredging pits) it is much thicker. The sediment from the IJssel is recognisable by the high percentage of lutum (approx. 20%) and organic content (approx. 10%). There is a net sediment transport from the Ketelmeer to the IJsselmeer partly containing resuspended material. Estimates indicate that there is a total resuspension of 5 million tonnes per year, of which 15-20% is caused by shipping.

The flow speed in the Ketelmeer as a result of the discharge of the IJssel is very low. Wind driven circulation is the main transport factor in the lake. The average flow velocity is less than 0.1 m/s, except in the IJssel mouth during a high discharge. Fig. 2a indicates the frequency distribution of the wave height resulting from the wind speeds which occur. The maximum value lies between 0.15 and 0.25 m at an average water depth of 3 m.

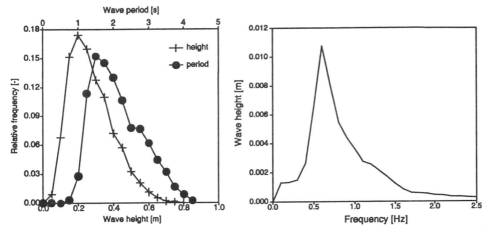

**Figure 2** (a) The wave height and wave period occurance in lake Ketelmeer and (b) the applied wave height spectrum in the laboratory

The relative wave height H/D=0.2, corresponding to a bed shear of 0.75 Pa, is exceeded during approx. 3% of the time. The average wave period lies between 1.5 and 2.0 s. This is in line with a wave length of 3.0 to 3.5 m. From this it follows that the waves which occur are between shallow and deep water waves. The JONSWAP spectrum, which is used in wave experiments with irregular waves, includes both types of waves (Fig. 2b).

## LABORATORY EXPERIMENTS

In 1991 and 1992, experiments are conducted in a wave flume at Delft Hydraulics on a natural mud bed, taken from the Ketelmeer (Cornelisse et al. 1992). The experiments are conducted in a 14 m long wave flume (with a test area of 8 m long) with a water depth of 0.25 m. The mud bed is approx. 0.05 m thick. The waves are generated on one side by means of a wave paddle. On the other side of the flume the waves are absorbed by a wave damper.

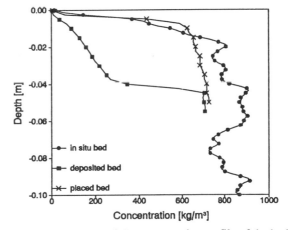

**Figure 3** The dry weight concentration profile of the bed in lake Ketelmeer and in the laboratory

The in situ density profile of the bed is determined by using an acoustic densitometer (Verbeek & Cornelisse, 1994). As indicated in Fig. 3, a bed formed from deposition of a suspension (called deposited bed) has a density profile that differs from a bed resulting under natural conditions. Furthermore, segregation occurs as the bed is formed. Therefore, it is also necessary to investigate a bed which has been placed directly in the flume. The mud of this placed bed is sieved roughly and spread out in the flume. In this way the aggregates which are present are maintained and no segregation occurs during the formation of the mud bed. The profile of the dry

355

**Table 1** The physico-chemical properties of lake Ketelmeer mud

| Mineralogical composition | | Chemical properties | | Physical properties | |
|---|---|---|---|---|---|
| organic content | 10 % | CEC | 11 meq/100g | $d_{50}$ | 18 μm |
| clay content | 50 % | Chlorinity | 310 mg/kg | Specific surface | 83.4 m²/g |
| quartz content | 30 % | pH | 7.63 [-] | dry content | 35 - 45 % |
| | | redox potential | -210 mV | viscosity μ | $10^{0.004\ C}$ mPa/s |
| | | SAR | 41 √(meq/l) | | (C in kg/m³) |

weight concentration and composition (see table 1) agree with the measurements in nature.

The wave experiments consist of single reproduction tests and multiple tests with a gradually increasing wave height. For the experiments regular waves with a wave period of 1.5 s and a wave height varying from 0.015 to 0.138 m are used. Irregular waves are also applied according to the JONSWAP spectrum with a peak wave period of 1.5 s (fig 2b). The significant wave height during these experiments is increased stepwise from 0.029 to 0.110 m. Irregular waves are used with the aim of analyzing whether a resonance frequency of the bed or aggregates in the bed can be found which will result in a weakening of the bed. During the experiments, the suspension concentration is determined with turbidity sensors. These sensors detect the erosion of the mud bed in situations of increasing wave load. In Fig. 4 the stabilised concentration profiles are presented as they occur after 45 minutes of waves.

In addition to wave height and suspension concentration, attention is paid, when conducting the wave experiments, to pressure fluctuations in the soil and to the horizontal movements on the water-bed interface. The pressure measurements are conducted with the help of pressure gauges in the wall and on the bed of the flume. In this way it is possible to determine the pore water pressure and the effective stress (Fig. 5). The variation in the effective stress due to the waves is almost zero (not shown).

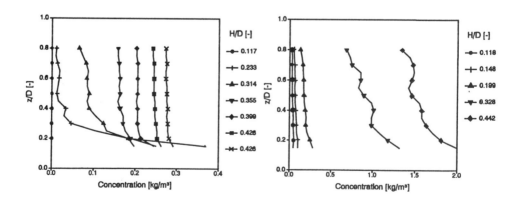

**Figure 4** Concentration profiles during the experiments with a gradually increasing wave height for (a) a placed bed and (b) a deposited bed

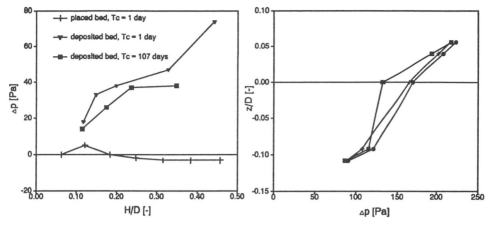

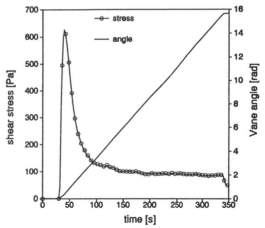

**Figure 5** (a) The mean excess pore water pressure for increasing wave loading and (b) the pressure amplitude in and above the bed for the case of a placed bed with H/D = 0.248

In the experiments with a placed bed no liquefaction behaviour occurred (excess pore water pressure $\Delta p \approx 0$ Pa). The increasing wave load did result in a gradual increase of the suspension concentration but this is mainly caused by floc erosion of the top layer of the mud bed. The erosion characteristics for a non-liquefied bed are calculated on the basis of these measurements with the use of the formulations of Partheniades for erosion. In contrary to this, the deposited bed did liquefy. After a brief consolidation period of 1 day, an high erosion flux occurred which is five times as great as that in the same experiments on a placed bed. When the consolidation period is extended to more than 100 days, the liquefaction continued to occur but no initial erosion occured. From this it can be concluded that the aggregates in the liquefied layer have still sufficient consistency to resist erosion.

It is found that the critical shear stress for erosion $\tau_{ce}$ decreases to almost zero as a result of liquefaction. The concentration profile above a placed bed shows a strong gradient close to the bed, caused by the erosion of large clusters. This concentration gradient is virtually absent above a liquefied bed. This is due to the disintegration of the structure in the bed due to liquefaction.

Since the phenomena are clearly linked to the type of bed, it is investigated whether rheological measurements will be able to indicate the difference. If so, it will be possible to describe bed types in situ. However, during a standard measurement of the flow curve, the aggregates are broken in the rheometer and only the stirred sample is tested. During the wave experiments on the placed bed, it is seen that it is precisely the structures and aggregates present which determine the strength and erosion

**Figure 6** The adapted test setup with the rheometer

357

**Table 2** The rheological parameters determined with the adapted rheometer test

| | | placed bed | | | deposited bed | | | |
|---|---|---|---|---|---|---|---|---|
| Tc [days] | $\sigma_y$ [Pa] | $\sigma_f/\phi$ [Pa] | $\sigma_f/\sigma_y$ [-] | $\sigma_v/\sigma_y$ [-] | $\sigma_y$ [Pa] | $\sigma_f/\phi$ [Pa] | $\sigma_f/\sigma_y$ [-] | $\sigma_v/\sigma_y$ [-] |
| 1 | 50 | 900 | 6.2 | 1.8 | 10 | 100 | 4.2 | 1.8 |
| 10 | 50 | 1250 | 9.0 | 1.8 | 10 | 400 | 5.8 | 1.8 |
| 40 | 50 | 1600 | 12.0 | 1.8 | 10 | 450 | 9.5 | 1.8 |
| 114 | 50 | 1600 | 12.0 | 1.8 | 10 | 450 | 15.0 | 1.8 |

rate. Therefore, a rheometer test is adapted in such a way that the circulation speed is sufficiently slow to obtain information on the non-stirred strength. This is shown in Fig. 6. The breakdown resistance and the resulting vane strength can be derived from this test.

In table 2 Tc is the consolidation period, $\sigma_y$ is the Yield strength, $\sigma_f$ is the failure strength, $\phi$ is the failure angle and $\sigma_v$ is the vane strength. For both types of bed the relation between yield and vane strength is the same and the yield value is independent of the consolidation period. The failure strength divided by the failure angle is constant after 40 days for a placed bed. For a deposited bed this value still increases. The same behaviour happens with the division of failure strength and yield strength.

By measuring the diameter of the area influenced by the vane and making the assumption that there is a linear relation between shear and deformation, the static value of the shear modulus can be approximated. This is presented in table 3.

## COMPARISON OF MODELS

The movements of the liquefied mud layer can be described with the help of the theory of McPherson (1980). This model is based on an elastic and a viscous response of the mud layer in the case of small disturbances. For this the Voigt model has been applied:

$$\frac{\partial^2 x}{\partial t^2} = -\frac{1}{\rho}\nabla p + v\frac{\partial}{\partial t}\nabla^2 x + \frac{G}{\rho}\nabla^2 x - g \tag{1}$$

In this x is the horizontal displacement, p the water pressure, $\rho$ the density, $v$ the viscosity, G the elasticity modulus, g the acceleration due to gravity and t the time. With application

**Table 3** Calculated shear modulus value for a deposited and a placed bed.

| bed type | consolidation period | shear modulus |
|---|---|---|
| deposited bed | 1 day | $0.09 \times 10^6$ Pa |
| | 114 days | $0.34 \times 10^6$ Pa |
| placed bed | 1 day | $0.73 \times 10^6$ Pa |
| | 114 days | $1.28 \times 10^6$ Pa |

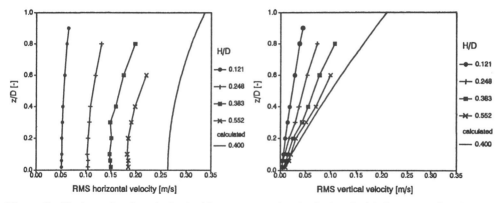

**Figure 7**  Horizontal and vertical velocities as measured and calculated with the model of McPherson

of a complex viscoelastic parameter $v_e=v-iG/\rho\omega$, where $\omega$ is the wave frequency and i the complex number $\sqrt{-1}$, the equation of motion (1) can be written as the linear Navier-Stokes equations for a viscous fluid. In Fig. 7 a comparison is made between the measurements in the case of a deposited bed after liquefaction and the theory of McPherson (1980). During the experiments it can be seen that the liquefied layer moved independently of the underlying fixed mud layer

Chou *et al.* (1993) conclude, based on laboratory measurements of the rheological properties of kaolin, that the properties of a bed depends on the wave induced strains. In the models of McPherson (1980) and others the sediment bed is assumed as a homogeneous viscous or viscoelastic medium. The laboratory results of Chou indicate that wave induced strains can significantly affect the properties of the bed, which can only be accounted for by a nonlinear model. First results of the model of McPherson indicate that there is an acceptable comparison between calculation and measurements except of a factor 2. The deviation is due to the shape of the waves, which is not sinosoidal.This leads to an exaggarated wave height. The use of a nonlinear model for the simulation of the movements in a liquefied bed is subject of study.

Yamamoto (1977) presented a model which calculates displacements and pressure in a bed under the influence of wave load. The solution found provides an interesting characteristic parameter, $c\lambda^2/\omega$ which compares the diffusion of the pore water with the wave diffusion. In this $\lambda$ is the wave number, $\omega$ the wave frequency and c the consolidation coefficient of the mud. In the natural material which is investigated it was found that $c\lambda^2/\omega$ $\approx 3\ 10^{-7}$. For the situation investigated, this means that there is an undrained situation. This situation is described by Yamamoto (1983) in which a characteristic is derived which gives a comparison between the energy dissipation due to percolation and Coulomb friction. With this it can be determined whether the bed can be approached using a two phase theory (based on effective stress) or on the basis of a single phase model (based on total stress). In the situation with a placed bed of Ketelmeer mud, with a permeability of 2.0 $10^{-10}$ m/s (constant head measurement, see Lambe and Whitham, 1969), the energy dissipation is mainly through Coulomb friction. It is therefore possible to work with a single phase model.

The comparison of displacement shown in Yamamoto (1983) and the measurements give a reasonable link-up (fig.8). Visual observation showed that the bed moves as if a leaf

spring is fixed at a certain depth in the
bed, with a length and a displacement
amplitude depending on the wave height.
By means of Particle Image Velocimetry
a more accured measurement will be
done and based on that the Yamamoto
model (1983) will be further tested.

## MODEL EXTENSION

In the above, the transition from a non-
liquefied bed to a liquefied bed is not
indicated. This information is necessary,
however, in order to describe what
happens in nature. An attempt to fill in
this gap is made, using the theory drawn
up by Spierenburg (1987).

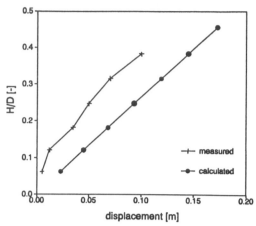

**Figure 8** Measured and calculated displacement at the surface of the bed as a function of the wave height

In addition to a direct influence of the stresses in a mud bed, waves will, over time, also
change the soil properties due to their dynamic load. An example of this is the compaction
of a sand bed due to a cyclic load. In the case of mud where a similar tendency for compaction occurs, permeability is considerably less (sand $10^{-3}$-$10^{-6}$ m/s and mud < $10^{-9}$ m/s). This
results in a build up of excess pressure and, linked to this, a decline of the effective stress.

If there is sufficient build up of pore water pressure, the effective stress will become
zero and the bed will liquefy. If after some time excess pore water pressure is achieved but
liquefaction did not occur, the tendency to compact continues to exist and consolidation will
result in the dissipation of the excess pore water pressure. A consequence is that a more
compact bed will be created which is less permeable and after some time will be stronger
than the original bed. The loading path which permits a material to permeate, determines
the ultimate strength, a strength which may be either higher or lower than the original
strength. In order to arrive at a comparison which describes the build up of compaction, the
premise is used that after each wave a deformation of the material occurs. In the case of a
certain number of waves $N_t$ an ultimate compaction will occur. In a situation in which N
< $N_t$ the subsequent period with waves will give only a part of the ultimate compaction. The
equation which embraces these effects is:

$$\epsilon_p = -P\frac{\hat{\tau}}{\sigma}N_t\left(1 - \exp\left(-\frac{N+N_0}{N_t}\right)\right) \qquad (2)$$

in which: $\epsilon_p$ is the plastic deformation, P is a material constant, $\tau$ the (deviator) shear stress,
$\sigma$ the isotropic grain pressure, N the number of waves, $N_t$ the number of waves for a
maximum compaction and $N_0$ the number of waves present in a preceding load (Fig. 9a)

The Péclet number (the ratio advective transport / diffusive transport: $vL/c_v$ with
respectively viscosity, characteristic length scale and consolidation coefficient) shows that
in the case of waves over a placed bed there is evidence of an undrained situation. This
corresponds to the results of Yamamoto (1983). In the case of a deposited bed the situation
will be between undrained and drained. In an undrained situation the sample volume will
not change and the plastic deformations induced by the cyclic load will need to be fully

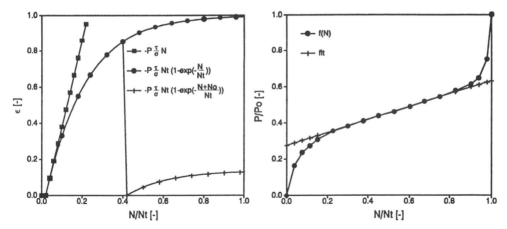

**Figure 9** (a) The influence of the wave loading on the excess pore water pressure (by Spierenburg) and (b) measurements on sand with $N_t = 3300$ cycles

compensated by elastic effects, resulting in excess pore water pressure. If the Hooke Law is applied to the elastic part, it is found that:

$$\epsilon_e + \epsilon_p = \frac{\Delta\sigma}{K} - P\frac{\hat{\tau}}{\sigma}N = 0 \qquad (3)$$

in which: K is the bulk elasticity module and $\epsilon_e$ the elastic deformation. By applying the compression equation of Terzaghi, it can be found that:

$$\Delta p = \Delta\sigma = PC_{10} \ln(10)\,\hat{\tau}\,N \qquad (4)$$

in which $\Delta p$ is the increase in pore water pressure and $C_{10}$ the coefficient of compression [-]. In Spierenburg (1987) the proposal is made to describe the increase in pore water pressure as:

$$\Delta p = \sigma_i\, N \tan(\beta) \qquad (5)$$

in which $\sigma_i$ is the initial grain pressure, $\beta$ the angle with which the linear part of the pore water pressure increases as f(N). In Figure 9b the result is schematically shown as an undrained cyclic triaxial test, in which $\Delta p$ is determined as a function of N.

For the experiments with Ketelmeer mud, N increases up to 5000, during which a very small increase of pore water pressure was observed. On the basis of extrapolation of this increase it may be expected that a placed bed has built up sufficient excess pore water pressure (to reduce the effectieve stress to zero) to become liquefied after several days.

During the experiments with a deposited bed it was established that the excess pore water pressure first decreases and subsequently the pore water pressure increases and the bed liquefies. A deposited bed liquefies almost immediately when waves are subjected. Over its entire height the bed is at its critical state. An additional pressure will cause a breakdown and liquefaction. In general in the case of a placed bed there will be a certain excess consolidation which can cause the elastic deformation in the bed to be transformed into excess pore water pressure. From this a compaction can result if the load period is not long enough. A subsequent consolidation can result in a stronger bed.

## CONCLUSION

During the investigation into the effect of surface waves on a mud bed it is established that, depending on the type of bed structure, liquefaction behaviour can occur. Measurements showed that an in situ bed is more comparable to a placed bed than a deposited bed. If an in situ bed is composed of strong aggregates or is already compacted by previous wave loading, than liquefaction does not occur directly. This is because the bed is not at its critical state line as is the case for a deposited bed. Only by accumulation of excess pore pressure build up liquefaction can occur in time.

A mud bed, created from suspension and which has not been subjected to a top load, will liquefy almost immediately after a wave load is placed on it. After the creation of a liquefied layer, the movements are simulated with the use of a viscoelastic two-layer model.

Due to liquefaction, the critical shear stress for erosion will decline and under comparable circumstances an erosion flux five times greater occurs. If there is sufficient flow this will result in a much larger transport than in the case of a non liquefied bed.

Since the theory for an elastic response does not link with the response, as observed in the case of a placed bed, a one phase soil mechanics model is used. This produces a good description for the movements in the mud bed. The sensitivity for liquefaction of a mud bed can be described by using the theory as described by Spierenburg.

It is shown that for erosion of a thin layer of a placed bed under waves the formulations derived under flow conditions can be used. When however the eroded layer will be thick or the duration of the wave loading yields to weakening the characteristics and mud properties will change. Without an extension to the erosion formulation numerical models do not incorporate liquefaction although it is needed. Before implementation of the proposed model a verification using field data of wave height and suspension concentration over a longer period with several wave loadings is necessary.

## ACKNOWLEDGEMENT

We like to acknowledge the Directorate-General for Public Works and Water Management (Rijkswaterstaat) and the European Community (MAST-1 programme), for providing the financial support to enable us to carry out this investigation. Further, we wish to thank Erik van Velzen, Frans de Vreede and Kees Koree for their help with many of the experiments.

## REFERENCES

H.T. Chou, M.A. Foda & J.R. Hunt (1993), *Rheological response of cohesive sediments to oscillatory forcing*, Nearshore and Estuarine Cohesive Sediment Transport, ed. A.J. Mehta, AGU.

J.M. Cornelisse, H. Verbeek, C. Kuijper & J.C. Winterwerp (1992), *Experiments on the influence of surface waves on natural mud*, Delft Hydraulics & Rijkswaterstaat, Cohesive Sediments Report 40, december 1992.

C. Kuijper, J.M. Cornelisse & J.C. Winterwerp (1990), *Erosion and deposition characteristics of natural muds, sediments from the Ketelmeer*, Delft Hydraulics & Rijkswaterstaat, Cohesive Sediments Report 30, november 1990.

T.W. Lambe & R.V. *Whitham, Soil Mechanics*, Wiley & sons, 1969.

J. Lindenberg, L.C. van Rijn & J.C. Winterwerp (1988), *Some experiments on wave-induced liquefaction of soft cohesive soils*, Journal of Coastal Research, SI (5), 127-137.

H. MacPherson (1980), *The attenuation of water waves over a non-rigid bed*, Journal of Fluid Mechanics, 97 (4), 721-742.

S.E.J. Spierenburg (1987), *Seabed response to water waves*, PhD thesis, Delft University of Technology, 112 pp.

H. Verbeek & J.M. Cornelisse (1995), *Consolidation of Dredged Sludge, measured by an acoustic densitometer*, Marine and Freshwater Research, 46 (1), 179-188.

T. Yamamoto (1977), *Wave induced instability in seabeds*, Coastal Sediments '77, 898-913.

T. Yamamoto (1983), *Numerical integration method for seabed response to water waves*, Soil Dynamics and Earthquake Engineering, 2 (2), 92-100.

# MODELLING

# 26 A review of cohesive sediment transport models

C. TEISSON

*Electricité de France, Laboratoire National d'Hydraulique, Chatou, France*

## Introduction

The prediction of transport of very fine sediments in coastal and estuarine areas is receiving a growing concern and numerical models are more and more asked to give reliable results, or are used as a decision tool, on three related issues :
- estimation of accumulations of sediment in navigational fairways and sheltered areas, to infer a cost of dredging,
- evaluation of disposal sites and fate of dumped materials at sea,
- water quality problems, where cohesive sediment play a major role by their faculty of adsoption of heavy metals and chemical substances.

Since the pioneer works in the 1970's, numerous cohesive sediment transport models have been developped, the most popular for engineering applications being depth integrated models, based on the conceptual framework of fig. 1. The last ten years have seen the emergence of efficient 3D models for estuarine applications or more complex situations (see e.g. Sheng, 1983; Markofsky et al., 1986; Hayter and Pakala, 1989; Le Normant et al., 1993).

The present paper proposes to review current practice of engineering cohesive sediment models (§ 1). As the well known bottleneck of these practical models lies in the description and implementation of cohesive sediment processes (§ 2), academic models have also been developped, most often dedicated to one process. These academic models, amongst them the two-phase flow model (§ 4), might show the way of what could be the modelling of cohesive sediment transport in a near future. Some identified gaps in our practice of modelling mud transport, such as validation of the softwares, long term techniques, biological and seasonal influences are addressed in § 3.

## 1. Modelling problem

The general equation of transport of suspended sediment is classicaly written :

$$\frac{\partial \overline{C}}{\partial t} + \overline{U}_i \frac{\partial \overline{C}}{\partial x_i} + \frac{\partial w_c \overline{C}}{\partial z} = -\frac{\partial}{\partial x_i}\left(\overline{u'_i c'}\right) + S\ (x, y, z) \tag{1}$$

$x_i = x, y, z$ and $U_i = U, V, W$

where     $\overline{C}$ = mean concentration of sediment in suspension,

           $\overline{U}_i$ = mean flow velocities,

*Cohesive Sediments.*
Edited by Neville Burt, Reg Parker and Jacqueline Watts
©1997 John Wiley & Sons Ltd.

CONCENTRATION, C or VELOCITY, u

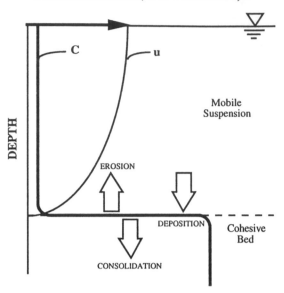

Fig. 1: "Classical" definition of sediment bed and suspension - related processes
(after Mehta, 1989a)

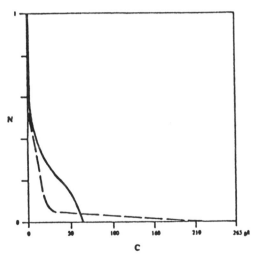

Fig. 2: Vertical profiles of concentration obtained with the two-phase flow model for
two kinds of sediment having the same settling velocity of 1 mm/s:
_____ : flocs of density 1070 kg/m$^3$
_ _ _ _ _ : non cohesive sediment of density 2650kg/m$^3$
(after Teisson, 1992; abcissa converted in massic concentration for sake of
clarity)

with instantaneous values :     $U_i = \overline{U_i} + u'_i$          $C = \overline{C} + c'$

$w_c$ = settling velocity of particles (mean relative velocity),

$S(x, y, z)$ = possible source of sediment within the domain (not to be confused with the source term at the bed boundary representing deposition or erosion).

The boundary conditions express that no sediment is passing through the free surface :

$$w_c \overline{C} + \overline{w'c'} = 0$$

whereas at the bottom the net flux of sediment corresponds to the flux due to deposition $Q_d$ or erosion $Q_e$:

$$w_c \overline{C} + \overline{w'c'} = Q_d + Q_e \qquad (2)$$

The turbulent flux of sediment  is generally determined through the eddy diffusivity concept :

$$\overline{u'_i c'} = -\Gamma_{si} \frac{\partial \overline{C}}{\partial x_i} \qquad (3)$$

Assuming that the velocity components $\overline{U_i}$ are provided by a decoupled hydrodynamic model, the closure of (1) requires information on the settling velocity $w_c$ , and expressions for the sediment flux across the bottom (2) and for the eddy diffusivity (3). To solve (1) it is also needed to prescribe initial conditions - which might not be an easy task in practical applications for the initial thickness and vertical density distribution of deposit, seldom known-, as well as flux of sediment entering the domain by lateral boundaries (incoming waters).

## 2. Implementation of cohesive sediment processes

So far, (1) is merely an equation for non cohesive suspended sediment transport. To become a cohesive sediment transport model, (1) should incorporate physical, and possibly biological processes which influence the transport of mud.

### 2.1 Choice of the variable C

Engineering models devoted to study mass fluxes are naturally based on massic concentration C, expressed in g/l. However academic models, amongst them the two-phase flow model, have shown that the volumetric concentration is a more relevant parameter controlling many processes such as hindered settling, damping by turbulence, aggregation.

This is illustrated on fig. 2, where the two profiles displayed are obtained for two sediments having the same initial bulk massic concentration in suspension, the same settling velocity (1 mm/s), under the same hydrodynamic forcing ($\overline{U}$ =0.5 m/s). The first sediment is a non cohesive sediment of mean diameter 30 m (settling velocity 1 mm/s after Stokes law). The second one represents macroflocs of diamer 160 m and density 1070 kg/m3 : these are realistic values for aggregates observed in the field, leading also to a settling velocity of 1 mm/s. With the same settling velocity and massic concentration, non cohesive sediment settles rapidly to the bed whereas the macroflocs form a lutocline with a highly concentrated layer. The difference between the two profiles obtained with a two-phase flow model is explained by the much larger volumetric concentration for the macroflocs (0.10 to 0.50 against 0.01 to 0.05 for sand), due to their low density value; this large volumetric concentration induces hindered settling of flocs due to drag reduction at such concentration and upward flux of water which tries to escape between the settling flocs, due to volume continuity of the two-phase medium. Fig. 2 thus outlines the role played by the size and density of flocs in the generation of fluid mud layer, and the central role played by volume concentration of aggregates, already mentioned by Krone (1986).

## 2.2 Aggregation

Aggregation is indirectly considered in engineering models through the dependency of the settling velocity on concentration :

$$w_c = K C^n \qquad (4)$$

expressing the fact that increasing concentration of particles will increase the settling velocity through the increased interparticle collisions. However this expression, useful at a macroscopic level, does not consider the nature of the floc thus created, neither the mecanisms responsible for collision and aggregation - brownian motion, differential settling and current shear - nor the response time needed for particles to flocculate. The knowledge about flocs has been considerably enhanced these past years (see Van Leussen, 1994) but is yet poorly reflected in modeling.

In order to improve (4) in models, Le Hir et al. (1993) proposed an empirical relationship between settling velocity, concentration and shear stress, in line with the conceptual diagram proposed by Dyer (1989). Malcherek et al. (1994) used satisfactorily the formulation proposed by Van Leussen (1994) :

$$w_c = KC^n \frac{1 + aG}{1 + bG^2} \qquad (5)$$

where a and b are empirical constants. The numerator takes into account the influence of low turbulence on the collision frequency which promotes flocculation while the denominator represents the influence of turbulence on the break up of flocs. G is the absolute velocity gradient defined as :

$$G = \sqrt{\frac{\varepsilon}{v}}$$

where $\varepsilon$ is the turbulent energy dissipation and $v$ the molecular viscosity.
However, in most of cohesive sediment transport models, the flocculation process is not explicitly represented: the particles are assumed to behave as individual ones rather than as aggregate ones i.e. flocs, whereas the transport characteristics of sediment flocs differ considerably as shown on Fig. 2.
Krishnappan (1991) proposed to use a coagulation equation to represent the flocculation process which expresses the rate of change of number of sediment particles per unit fluid volume in a particular size class, i, as :

$$\frac{\partial N(i, t)}{\partial t} = -N(i, t) \sum_{j=1}^{\infty} K(i, j) N(j, t) + \frac{1}{2} \sum_{j=1}^{\infty} K(i-j, j) N(i-j, t) N(j, t) \qquad (6)$$

where N (i, t) and N (j, t) are number concentrations of size classes i and j respectively at time t and K(i, j) is the collision frequency function which is a measure of the probability that a particle of size class i collides with a particle of size j in unit time. The collision frequency function K accounts for the various collision mechanisms.

This equation was applied, together with a multiclass equation to a sediment divided into 16 size classes, on a river stretch (Krishnappan, 1991) : flocculated material deposited very quickly whereas in a classical model sediment remained transported in suspension. However, Krishnappan did not consider break-up of flocs by turbulence, which could mitigate his results obtained only by treating the flocculation process. In any case, approach such as eq. (6) seems conceptually the most appropriate for cohesive sediments, but still require information about the physics of the processes, especially the break-up process. Eq. (5) may also be a simpler alternative, which need to be further explored.

## 2.3 Bed boundary definition

To define the transition from suspension to a bed has always been a delicate problem, in the field as well as in numerical models. The definition of this interface is of utmost importance in models, as fluxes of deposition and erosion will be specified at the level of the interface. Most often the bed is assumed to start at a certain density, for instance 1.2 for navigational purposes, or at a certain concentration $C_S$ (in the range 80 - 220 g/l), or at the gel point, the critical volume fraction ($\approx 0.01 - 0.1$) above which effective stress exists. Once defined, it is assumed that sediment which settles from the water column deposits to the bed with the bed concentration $C_S$. More generally deposition or erosion fluxes will induce a consequent variation of the bed elevation $z_b$ :

$$\frac{\delta(C_s z_b)}{\delta t} = Q_d + Q_e$$

where $C_S$ may depend on $z_b$.
The ambiguity of bed definition is one of the motivations to design an approach where this definition is circumvented (§ 4).

## 2.4 Deposition and erosion

Even if advection or dispersion of sediment (1) is the ultimate goal of the model, in fact these are the sink and source terms (2) and the induced distribution of sediment through the water column (3) which contributes to the budget of sediment, and consequently to the final output of the model.

The most popular law for deposition has been proposed by Krone :
$$Q_d = P_d \, w_c \, C \tag{7}$$
in which $P_d$ is the probability for particles to be sticked to the bed and not to be re-entrained by the flow, given by :

$$P_d = 1 - \frac{\tau_b}{\tau_{cd}} \qquad \text{for } \tau_b < \tau_{cd}$$

where $\tau_b$ is the bottom shear stress and $\tau_{cd}$ the critical shear stress under which deposition occurs.
Many erosion formula have been proposed, depending on the type of the bed, but generally taking the form :

$$Q_e = M \left[ \frac{\tau_b}{\tau_{ce}} - 1 \right] \qquad \text{for } \tau_b > \tau_{ce} \tag{8}$$

Generally bed properties exhibit changes with depth and time, and both M and $\tau_{ce}$ will change accordingly (which requires the use of a consolidation model).

More complicated laws for deposition have been proposed (Mehta and Parthenadies, 1975; Mehta and Lott, 1987), however their use in tidal conditions did not show a significant improvement compared to the easy to use Krone formula (Verbeek et al., 1993 ; Ockenden, 1992).

Empirical fitting of parameters of the deposition laws during model calibrations on field observations has generally led modellers to adopt values of settling velocities $w_c$ in the range of 1 mm/s or above (Le Hir et al., 1993 ; Sanford and Halka, 1993; Diserens et al., 1993), sometimes an order of magnitude larger than settling velocities observed in laboratory experiments; these large values are in agreement with recent in situ measurements of large estuarine flocs (Van Leussen, 1994).

Can erosion and deposition occur simultaneously? This question refer to the adopted values for critical shear stress for deposition and erosion. Accepting simultaneously erosion and deposition is a practical approach from some modellers, but Lau and Krishnappan (1994) clearly state, from their laboratory experiment, that reentrainment does not occur during cohesive sediment settling. On the other hand, Sanford and Halka (1993) refute "the paradigm of mutually exclusive erosion and deposition of mud" from field data and show that models excluding mutual erosion and deposition fail to reproduce field data, whereas the best fit is for model assuming continuous deposition (no critical shear stress for deposition).

Sanford and Halka (1993) put forwards three explanations for the apparent discrepancy between established laboratory research and practical approach based on field data: the variability of the bottom shear stress around its mean value $\tau_b$ could be larger in the field than in the laboratory experiment. This could allow simultaneous erosion and deposition at different locations or in succession at one location in the field, and not in the laboratory. Also, aggregates formed in the field might be much larger due to turbulence intensities, and much stronger, due to biogenic binding than those formed in the laboratoy, allowing rapid settling even at high shear.

Deposition laws and erosion laws issued from laboratory experiments might not be suitable in high turbid environments (Costa and Mehta, 1990). This could be due to stratifications effects, which may not be scaled appropriately in laboratory experiments (Galland et al., 1994). As a consequence of stratification effects in the field, the bottom shear stress could be reduced, enhancing the deposition of sediment, and slowing down in the same way erosion. This could be one explanation, amongst others, why engineering models which do not consider stratification effects on bottom shear stress sometimes underestimate deposition rates in the field or have to consider large values of settling velocity, to compensate this effect and to match the data.


*2.5 Multiclass equations*

Thanks to the increasing efficiency of computers, eq. (1) and (2) may be applied to multiple sediment classes, with corresponding multiple erosion and deposition behaviour (Mac Lean, 1985; Le Hir et al., 1993; Lavelle, 1993) with the sum all over the classes representing the total behaviour. Usually three size fractions are considered : sand, silt and clay. This is particulary relevant for water quality modelling, where contaminants behave quite differently depending on sediment size.

Modelling cohesive sediment as a sequence of particle classes present the difficulties of determining size specific critical erosion and deposition stresses. Although one could reasonably expect, from a conceptual point of view, that the results with a multiclass aproach are significantly improved compared to a one size model (Lavelle 1993), in practice, single particle class models sometimes agree better with in situ data (Ockenden, 1992; Sanford and Halka, 1993). The important contribution from multiclass equation should be to tackle with interactions between classes, such as aggregation, break-up, differential settling, which too few models consider (Krishnappan, 1991). Here more information about the physics of these phenomena are required, which should reduce the empiricism vehicled with the one size class model approach.

## 2.6 Modelling mixture of mud and sand

The composition of bottom sediment in estuaries is often a mud and sand mixture, which is entrained, deposited and consolidated under tidal conditions. The modelling of mud/ sand mixture is a particular case of the multiclass equations (see above), but is usually carried out on mud and sand treated as separated sediments without interactions. Promising knowledge has been gained very recently through field and laboratory experiments about the behaviour of mud-sand mixtures. For a low percentage of sand Williamson and Ockenden (1993) found that erosion formula (8) could hold for erosion of mud/ sand mixture: the presence of sand reduces significantly the erosion rate M; the critical shear stress for erosion increases with density, which is a function of consolidation or sand content. However, other investigations conducted with a low percentage of fines in the mixtures did not show a similar trend (Torfs, 1993). In a fully interactive way of modelling erosion of mud sand mixture, Chesher and Ockenden (1994) attempted to relate M and $\tau_{ce}$ in (8) to the percentage of sand and obtained a strong reduction of the mud transport. Toorman (1992) found in practice that it suffices to consider only two fractions to obtain good numerical reproduction of density profiles for settling and consolidation of mixtures.

This example of modelling mixture of mud and sand shows the gap which separates a straightforward treatment of two distinct sediment classes from a complex approach of two interactive sediment classes.

## 2.7 Consolidation

Consolidation is a very important parameter as it affects indirectly erosion through increased shear strength. Engineering cohesive sediment transport models generally represent the bed by multiple layers, whose parameters (density, thickness) are empirically fitted to consolidation curves. Consolidation and sedimentation theories have been proposed several decades ago (Gibson, 1967; Kynch, 1952), but it is only since the 1980's that these separate approaches have been brought closer, thanks to meritorious efforts to unify theories and notations between hydraulic and soil mechanics (Schiffman et al., 1985 ; Alexis et al, 1992 ; Toorman, 1994). However, consolidation models have been found very sensitive to the constitutive equation between effective stress, permeability and void ratio which are the closure equations for such models. Much effort is presently devoted to derive suitable laws for these parameters. The computational effort to solve this 1DV advection-diffusion type equation is negligible when applied for validation to a settling column, but becomes problematic when one wants to treat consolidation at each node of a mesh of a 2DH model.

The transition from suspension to consolidation is not clear. Consolidation theories deal with void ratio (volume of fluid divided by volume of solid), where grains have a density of 2650 kg/m$^3$. How flocs of density close to unity in the water column could possibly be transformed in particles of such a density in the bed, which is the way consolidation models consider the solid phase ? This point deserves further attention.
It should be noted that all these theories do not deal with bed shear strength and require an additional empirical relationship between bed density and erosion threshold for parameterizing the sediment erodibility :

$$\tau_{ce} = \alpha \, \rho_b^\beta$$

As bed density appear as the linking parameter between consolidation and erosion process, consolidation models should aim at reproducing vertical profiles of density within the bed and should not be only validated on the temporal decrease of the total bed thickness.

### 2.8 Eddy diffusivity and turbulence closures

Classically, the eddy diffusivity for suspended sediment is related to the eddy viscosity through an empirical relationship

$$\Gamma_s = \phi \, \beta \, v_t = \phi \, \Gamma_{sn}$$

where $v_t$ is the eddy viscosity in neutral conditions, $\beta$ is the inverse of the Turbulent Schmidt number. $\phi$ accounts for stratification effects and is related to the Gradient Richardson number : $\phi = (1 + a \, R_i)^b$ with a and b empirical constants. $\Gamma_{sn}$ is the eddy diffusivity in neutral conditions.

Empirical values of $\phi$ and $\beta$ can be found in Van Rijn (1989) for generally non cohesive suspended sediment. Satisfactory results in reproducing generation and evolution of lutoclines have been obtained by Wolanski et al. (1989), Ross et Mehta (1989), Smith and Kirby (1989). However, these simple models, requiring a large amount of data to tune the $\phi$ and $\beta$ parameters of the eddy diffusivity, do not offer the appropriate theoretical framework to investigate the dynamics of sediment laden flows, especially the strong interactions near the bed. There are some evidences that in high turbid environment in the field, the presence of the sediment affects the bottom shear stress and the bottom processes (Costa and Mehta, 1990). This influence is now tentatively reproduced by academic models such as Reynolds stress models (Sheng and Villaret, 1989; Teisson et al., 1992) and the turbulence damping near the bed may be sufficiently large to cause the formation of lutocline (Teisson et al., 1992). However, this effect is seldom encountered in laboratory experiments, even for concentration in the range, or above those observed in the field. Galland et al. (1994) put forwards by a scale analysis that laboratory experiments may not fulfill the similitude on stratification effects, due to the limited water depth in flume and carousels compared to the field.

### 2.9 Turbidity currents, fluid mud and wave effects

Historically, laboratory research has concentrated on the effect of tidal currents on erosion and deposition, as well as on settling and consolidation, and cohesive sediment transport models have attempted to reflect this knowledge. The growing environmental concern about contaminant cycling is causing a shift in emphasis from deeper channels to the shallow waters and tidal flats, where water waves rather than currents dominates (Foda et al, 1993). Since the 1980's, the complex interaction between waves and mud bed -bed weakening, turbidity generation and waves attenuation- has encouraged important theoretical and laboratory studies. A review of the various academic models can be found in Chou et al., 1993, which conclusions are reproduced hereafter. Most of the developed models belong to the category of linear interaction models where the properties of the bed material are given as input data. Recently Foda et al. (1993 ) presented a non linear model for the fluidization of marine mud by waves, based on laboratory experiments. The model assumes that the viscoelastic properties of the bed depend on the wave height: elastic at low forcing, viscous at high forcing, viscoelastic in between. This implies that the material properties should not be set a priori but modelled as a fonction of the wave height.

Engineering models generally apply eq. (8) for erosion of a mud bed by waves , following Maa and Mehta (1987) experiments, where $\tau_b$ is replaced by the peak bed shear stress under wave only. In combined wave and current, wave-current interaction term is considered in the total stress $\tau_b$. The expression for the vertical eddy diffusivity is modified due to the presence of waves; stratification effects have to be included to account for highly concentrated layers.

The other way of generation of fluid mud layer is through rapid settling of suspended sediment. Too few engineering models represent the generation and displacement of fluid mud, a very important feature in estuaries. A formulation of the various exchange between a fluid mud layer, the consolidated bottom and the flowing waters has been proposed by Odd and Cooper (1989).

Thus, despite recent numerous laboratory experiments and academic models, these topics do not receive yet a sufficient attention in usual engineering cohesive sediment transport models.


## 3. Some identified gaps in our practices of modelling cohesive sediment transport

### 3.1 Validation and calibration of cohesive sediment models

Validation of a model is viewed as the formulation and substantation of explicit claims about applicability and accuracy of the computational results (Standard Validation Document, 1994).

Generally, academic cohesive sediment models include a well documented validation step on the sole process they aim at representing. The validation of most of cohesive sediment transport softwares, as for them, is readily carried out on suspended sediment data case : the Rouse equation for instance, or source terms in a current, where analytical solutions exist. Therefore, these are merely the advection-dispersion terms, or the suspended transport, which are validated, and not the full cohesive sediment model. In that sense, cohesive sediment transport models do not fulfill the above specifications for the validation of softwares: their predictive capabilities rely on the formulation of cohesive sediment transport processes, but these processes, due to their intrinsic variabilities, and site specificities, cannot be adequately treated in a validation step, which should have a general scope. It is therefore difficult to set up a database against which cohesive sediment models could be validated, as it exists for non cohesive sediment (Zyserman, 1992). The intercomparison exercise of various engineering mud models on the same data sets (Hamm et al., 1994) is a first valuable attempt to quantify the accuracy of such models.

The weakness of the validation step of cohesive sediment softwares could be compensated by the calibration step, which should be part of any application of the software. However, the usual scarcity of field data, the inherent variability of the field render this calibration step also a non trivial task. Mac Anally (1989) calls the attention to "the myth" of the two-step calibration-verification procedure for sediment models and claims that there does not exist an optimum number for the data sets against which a model could be said calibrated. Using two data sets, one for calibration, one for verification, is by no means a proof of the future predictability of the model.

Inversely, very simple models can reveal themselves as valuable tools when relying on extensive field data sets.

Modellers have sometimes to use values of parameters, such as settling velocities, which deviate from data obtained in laboratory experiments, or even to set a value to some parameters (critical shear stress for deposition, for instance) without any information on its value for the site under study. Especially in these cases, a sensitivity analysis should be carried out to give error bounds on model results.


## 3.2 Long term approach

Due to the low value of settling velocities of cohesive sediment, bed evolution observed in the field are generally slow, except in areas where fluid mud is present, and long term calculations, over one year or more, are often required. The same temporal scales concern the fate of dumped materials or water quality problems. Compared to long term techniques for non cohesive sediment (Latteux, 1994), long term approach for cohesive sediment is even more complex. Bed consolidation introduces a new time scale between the hydraulic time scale and the morphological time scale. Processes are numerous (flocculation, turbidity maximum and fluid mud) and more or less well quantified. Chronology of events play an important role (see the hysteresis effect during the neap-spring cycle). The variability of input affects cohesive sediment transport in a larger way than non cohesive sediment : for instance the tide condition will have a larger effect on cohesive sediment than on non cohesive sediment. Thus, it is dubious that a single tide could represent the whole tide climates as it might be achieved for non cohesive sediment (Latteux, 1994).

If situation is only driven by deposition processes (siltation in a harbour or settling basin), the absence of erosion problem authorizes the use of extrapolation of deposition rates, and input filtering. But if both deposition and erosion processes are seen as essential (typical situations of estuaries, where deposition occurs during neap tide and erosion in spring tide) the consolidation and resulting erosion processes lead to a real coupling of the various time scales of hydrodynamics, consolidation and morphodynamics. Villaret and Latteux (1992) using a very simple zero model for flow and suspended sediment, but with bed consolidation and sinusoïdal modulation of currents over a tide and neap spring cycle found, that after a transition of some neap spring cycles, the bed turned out to be characterized by a thin active layer with a periodic behaviour (neap spring cycle period) over a more and more consolidated bed underneath.

A lagrangian approach for suspended sediment transport, as it reduces numerical dispersion during long term modelling, could be a way to be deeper investigated (Futawari and Kusada,1993).

### 3.3 Parameterization of biological effects and seasonal influences

In addition to the physical processes tentatively modelled in § 2, it is recognized that biotic processes influence as well the behaviour of cohesive sediment. The biogenic binding of aggregates, or the biogenic stabilization of the resistance bed to erosion (Patterson ,1994) are suspected to alter drastically cohesive sediment properties, but the variety and complexity of the mechanisms hinder the development of suitable biological indicators to be used in modelling. Seasonal influences even complicate a little more the problem of long term techniques already mentioned: the temporal variability of temperature, salinity also affect the sedimentation, and is usually not considered in modelling. The likely importance of biological effects and seasonal influences on the predictability of the models remain to be assessed.

### 4. The two-phase flow approach for cohesive sediment transport processes.

Numerous models and studies on cohesive sediment transport have relied on the schematic description of the environment displayed on fig.1. Such an environment- a low homogeneous concentration over a well defined consolidating bed- is the predilection domain for application of depth integrated models. The usual methodology is to propose a decoupled approach between hydrodynamic modelling, sediment transport modelling and consolidation - most often empirical- modelling. This fairly attractive approach, widely used, can no more be proposed in high turbid environment (fig. 3): here, the vertical structure of the sediment concentration profile is strongly affected by flow sediment interactions. It displays a continuous evolution from the water surface to the bed, through a variety of processes (free, flocculated, hindered settling) and properties (newtonian or non newtonian).

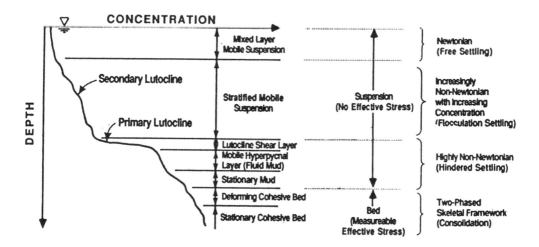

**Fig. 3. Concentration profile from the surface down to the settled bed (after Mehta, 1989b)**

The two-phase flow approach has the ambition, or at least offer the theoretical framework to tackle all the processes described on fig. 3, in a unified approach. The rigorous formalism of the two-phase equation is widely used in industrial domains and is penetrating the environmental water-sediment flow problem. It has been used as a theoretical starting point to derive simplified equations for turbidity currents (Brors, 1991), fluid mud (Le Hir, 1994), sedimentation and consolidation (Toorman, 1994). A standard two-phase flow model has been applied to sedimentation, turbulence damping (Teisson et al., 1992) and consolidation (Teisson et al., 1993) and has given a thorough insight into the various processes. Processes of aggregation and break-up, function of the level of turbulence in the water column, are currently tentatively reproduced.

Addressing the proper parameters (floc size and density, rate of aggregation and break-up, effective stress) to describe these processes in a unified view, the two-phase flow approach should help in reducing empiricism in formula commonly used and encourage further experimental work on these parameters.

## Conclusion

Cohesive sediment transport models should rely first of all on the most accurate hydrodynamic models:
- due to their low settling velocity, cohesive sediment are advected back and forth over large horizontal distances during each tidal cycle, and their residual displacement is highly dependant on the hydrodynamic field,
- important aspects of cohesive sediment transport processes occur on tidal flats: this pleas for an accurate treatment of the hydrodynamic of these areas during covering and uncovering,
- in harbours and enclosed basins, gyres control deposition but are strongly dependant of the empirical values of the dispersion coefficients: higher order turbulence closures are requested for such applications,
- the description of incipient deposition or erosion is based on a threshold formula on the shear stress, i.e. flow velocity, with a quadratic dependance on the flow velocities: small errors in the flow field can induce large changes in the patterns of deposition or erosion.

Validation of cohesive sediment transport models is problematic due to the variability of the processes. Well documented reference data cases do not exist, although numerous data sets exists for some specific processes.

The intrinsic nature of cohesive sediment - the floc , and the basic related processes of aggregation and break-up - is not enough considered in cohesive sediment transport. Multiclass equations should improve models results provided interactions between classes are considered.

Processes description and modelling lack for adequate knowledge at high concentration in suspension and flow-sediment interactions.

Because of, or thanks to these limitations, cohesive sediment transport models will never be ready-to-use models and call on the expertise of the modeller gained in this field.

## References

Alexis, A., Bassoullet, P., Le Hir, P. and Teisson, C., 1992. Consolidation of soft marine soils: unifying theories, numerical modelling and in situ experiments. Proc. of 23rd Int. Conf. on Coastal Engineering Venice (I), pp2949 - 2961.

Brors, B., 1991. Turbidity current modelling. Ph. D. Thesis, Univ. of Trondheim, 165 p.

Chesher T.J. and Ockenden M.C., 1994. Numerical modelling of mud and sand mixtures.INTERCOH, 94.

Chou, H.T., Foda M.A. and Hunt J.R., 1993. Rheological response of cohesive sediments to oscillatory forcing. Nearshore and Estuarine Cohesive Sediment , Ed. Mehta A.J., Transport Coastal and Estuarine Studies n° 42, AGU, pp 126- 147.

Costa, R.G. and Mehta, A.J., 1990. "Flow-fine sediment hysteresis in sediment-stratified coastal waters". Proc. 22nd Int. Conf. on Coastal Eng., Delft, 2047-2060.

Diserens, A.P., Ockenden, M.C. and Delo, E.A., 1993. Application of a mathematical model to investigate sedimentation at Eastham Dock, Mersey Estuary. Nearshore and Estuarine Cohesive Sediment , Ed. Mehta A.J., Transport Coastal and Estuarine Studies n° 42, American Geophys. Union, pp 486-503.

Dyer, K.R., 1989. Sediment processes in estuaries: future research requirements. JGR 94 (C10), pp 14327 - 14339.

Foda M.A., Hunt J.R. and Chou H.T., 1993. A non linear model for the fluidization of marine mud by waves. Journ. of Geoph. Res., vol.98, C4, pp 7039 - 7047.

Futawari T. and Kusuda T., 1993. Modeling of suspended sediment transport in a tidal river. Nearshore and Estuarine Cohesive Sediment Transport . Ed. Mehta A.J., Coastal and Estuarine Studies n° 42, A.G.U., pp 504 - 519.

Galland J.C., Laurence D. and Teisson, C., 1994. Simulating turbulent vertical exchanges of mud with a Reynolds stress model. INTERCOH, 94.

Gibson, R.E., Englund, G.L. and Hussey, M.J.L., 1967. The theory of one-dimensional consolidation of saturated clays. I. Geotechnique, 17 : p. 261-273.

Hamm, L., Chesher T., Fettweis M., Pathinara K. and Peltier E. (1994). An intercomparison exercise of cohesive sediment transport models. INTERCOH'94.

Hayter, E.J. and Pakala C.V., 1989. Transport of inorganic contaminants in estuarial waters. J. Coastal Res., Special Issue n° 5, pp 217 - 230 .

Krone, R.B., 1986. The significance of aggregate properties to transport processes. Lecture notes on Coastal and Estuarine Studies, vol. 14, ed. A.J. Mehta, pp 66 -84.

Krishnappan, B.G., 1991 Modelling of cohesive sediment transport. Int. Symp. on the transport of suspended sediments and its mathematical modelling, Florence, September 1991, pp 433 - 466.

Kynch, G.J., 1952. A theory of sedimentation . Transactions Faraday Society, 48 : p. 166- 176.

Latteux, B. 1994. Techniques for long term morphological simulation under tidal current action. Submitted to Marine Geology.

Lau Y.L. and Krishnappan, B.G.,1994. Does reentrainment occur during cohesive sediment settling? Journal of Hydraulic engineering, vol. 120, N°2, pp 236 - 243.

Lavelle, J.W., 1993. A model for estuarine sedimentation involving marine snow. Nearshore and Estuarine Cohesive Sediment Transport . Ed. Mehta A.J., Coastal and Estuarine Studies n° 42, A.G.U., pp 148 - 166.

Le Hir, P., Bassoullet P. and L'Yavanc, J., 1993. Application of a multivariate transport model for understanding cohesive sediment dynamics.Nearshore and Estuarine Cohesive Sediment Transport . Ed. Mehta A.J., Coastal and Estuarine Studies n° 42, A.G.U., pp 467- 485.

Le Hir, P., 1994. Fluid and sediment integrated modelling: basis of a 2DV code and applcation to fluid mudflows in a macrotidl estuary. INTERCOH'94.

Lenormant, C. , Lepeintre, F., Teisson, C., Malcherek, A., Markofsky, M. and Zielke W., 1993. Three dimensional modelling of estuarine processes. MAST Days and Euromar market. Office for official publications of the European Communities, Luxembourg, pp 223 - 233.

Maa, P.Y. and Mehta, A.J., 1987. Mud erosion by waves: a laboratory study. Cont. Shelf Res., 7(11/12), 169-1284.

Mac Anally, W.H., 1989. Lessons from ten years experience in 2D sediment modeling. Proc. of Int. Symp. on Sediment Transport Modeling. New Orleans 14-18 Aug. 1989, pp. 350-355.

Mac Lean S.R., 1985. Theoretical modelling of deep ocean sediment transport. Marine Geology, 66, 243 - 265.

Malcherek, A., Markofsky, M. and Zielke W., 1994. Numerical modeling of settling velociy variations in estuaries. To be published in Archiv für Hydrobiologie. Beihefte Ergebnisse der Limnologie (in english).

Markofsky, M. , Lang, G. and Schubert R., 1986. Suspended sediment transport in rivers and estuaries. Lecture notes on coastal engineering studies, Physics of shallow estuaries and bays, Ed. van de Kreeke, J., PP 210-227.

Mehta, A.J. and Partheniades, E., 1975 An investigation of the depositional properties of flocculated fine sediments Journal of Hydraulic Research, Vol 13, n° 4 pp 361/381.

Mehta, A.J. and Lott, J.W., 1987. Sorting of fine sediment during deposition Coastal Sediments, Vol 1, proc. of conf. on advances in understanding of coastal sediment processes, New Orleans.

Mehta, A.J., 1989a. On estuarine cohesive sediment suspension behavior. Journ. of Geophysical Research, vol. 94, n°C10, pp 14 303-14314, October 15,1989.

Mehta, A.J., 1989b. Fine sediment stratification in coastal waters. Third National Conference on Dock& Harbour Engineering, 6-9 december 1989, Surathkal, 487-491.

Ockenden, M.C., 1992. Application of a distributed sediment deposition model to laboratory and field data. HR Wallingford report. MAST EC Contract n°35.

Odd, N.V.M. and Cooper, A.J. 1989. A two-dimensional model of the movement of fluid mud in a high energy turbid estuary. High Concentration Cohesive Sediment Transport, Journ. of Coastal Research, Special Issue n°5, pp 185-193.

Parchure, T.M. and Mehta, A.J., 1985. "Erosion of solft cohesive sediments deposits". J. of Hydraulic Engineering, Vol. 111, N° 10, 1308-1326.

Patterson D.M., 1994. A review of biological effects on substrate response. INTERCOH, 94.

Ross, M.A. and Metha, A.J., 1989. On the mechanics of lutoclines and fluid mud. J. Coastal Res., Special Issue n° 5, pp. 51-61.

Sanford L.P. and Halka J.P., 1993. Assessing the paradigm of mutually exclusive erosion and deposition of mud, with examples from upper Chesapeake Bay. Marine Geology, 114, pp 37 - 57.

Schiffmann, R.L., Pane, V. and Sunara V., 1985. Sedimentation and consolidation. In "Flocculation, sedimentation and consolidation", Proc. of the Eng. Foundation Conf. (Ed. Moudgil & Somasundaran). Sea Island, Georgia.

Sheng, Y.P., 1983. Mathematical modeling of three-dimensional coastal currents and sediment dispersion : model development and application. Technical report CERC -83-2. U.S. Army WES, Vicksburg, Mississipi.

Sheng, P. and Villaret, C., 1989. "Modelling the effect of suspended sediment stratification on bottom exchange processes". J. of Geophysical Res., Vol. 94, N°C10, 14,429-14,444.

Smith, T.J. and Kirby R., 1989. Generation, stabilization and dissipation of layered fine sediment suspensions. J. of Coastal Research, Special Issue n°5, pp 63-73.

Standard Validation Document for computational models. Definition and Guidelines, 1994. To be published by IAHR.

Teisson, C., Simonin, O., Galland, J.C. and Laurence, D., 1992. Turbulence and mud sedimentation. Proc. of 23rd Int. Conf. on Coastal Engineering Venice (I), pp2853-2866.

Teisson C. and Simonin O., 1993. Simulating turbulent vertical exchanges of mud concentration and bed consolidation with a two-phase flow model. EUROMECH 310, Le Havre, 13-17 September 1993

Toorman, E., 1992. Modelling of fluid mud flow and consolidation. Ph. D. Thesis : Katholieke Universiteit Leuven.

Toorman, E., 1994. A unified theory for sedimentation and consolidation. Subm. to ASCE J. Geotechn. Eng.

Torfs, 1993. Erosion of sand/mud mixtures. MAST G8M report. Kath. Univ. Leuven (B).

Van Leussen, W, 1994. Estuarine macroflocs and their role in fine-grained sediment transport. Ph. D. thesis, Utrecht University (NL), 488 p.

Van Rijn, L.C., 1989. Handbook. Sediment transport by currents and waves. Delft Hydraulics. Report H 461.

Verbeek, H., Kuijper C. and Cornelisse J.M. and Winterwerp J.C., 1993. Deposition of graded natural muds in the Netherlands. Nearshore and Estuarine Cohesive Sediment Transport . Ed. Mehta A.J., Coastal and Estuarine Studies n° 42, A.G.U., pp 185-204.

Villaret C. and Latteux B., 1992. Long term simulation of cohesive sediment bed erosion and deposition by tidal currents. Proc. Int. Conf. Comp. Modelling of Seas and Coastal regions, Southapton, UK.

Williamson H.J. and Ockenden M.C., 1993. Laboratory and field investigations of mud and sand mixtures. Advances in Hydro-Science and Engineering, Sam S. Y. Wang ed., vol.1, pp 622- 629.

Wolanski, E., Chappell, J., Ridd, P. and Vertessy R., 1988. Fluidization of mud in estuaries. J. of Geoph. Res., vol. 93 (C3) pp 2351-2361

Zyserman, J. A., 1992. A critical review of available data for calibration and/or verification of sediment transport models. Proc. of 23rd Int. Conf. on Coastal Engineering Venice (I), pp2567 - 2580.

# 27 Modelling multiphase sediment transport in estuaries

D. H. WILLIS[1], and N. L. CROOKSHANK[2]
[1]*National Research Council Canada, Ottawa, Canada*
[2]*Canadian Hydraulics Centre, Ottawa, Canada*

## NUMERICAL BED SEDIMENT MODULE, CUMBSED

Modelling only cohesive sediment transport is not enough in many estuaries. These estuaries also have significant deposits of sand which are selectively eroded and deposited during the tidal, or seasonal cycle. Three examples are —

*Cumberland Basin, Bay of Fundy* — Exhibits a single, but significant, bar of fine sand at Low Water on the inside of a 180° bend

*Liverpool Bay* — Predominantly fine sand, but with enough mud in the upper reaches to make it interesting

*Miramichi Inner Bay* — A wide shallow, muddy bay cut off from the Gulf of St Lawrence by a chain of sandy barrier islands

All three have been the subjects of recent bed sediment numerical model studies at the National Research Council (Crookshank et al 1993/09, de Margerie et al 1993, Willis 1991/03, and Willis et al 1993/03). A fourth such study cannot be named or described at the client's request; and a fifth is reported in Willis et al (1995). The sixth is presently (1996) underway

For Cumberland Basin, the National Research Council expanded its numerical mud model (Willis 1985) to include both sand and mud phases (Willis 1991/03). At the same time, the model was rewritten as a 'bed sediment module' to be coupled to a numerical hydrodynamic model with advection and dispersion of two components. Three of the six studies have been with MIKE21AD, the 2-dimensional, finite-difference model of the Danish Hydraulics Institute. For the Miramichi Inner Bay study, an inerodible 'bedrock' layer was added

The resulting bed sediment module features:

- Erosion and deposition of (mostly) mud and (mostly) sand

- Recalculation of suspended concentrations

- Mixing of the two phases on the bed and in the water

- Consolidation, including strengthening, with both time and overburden

- Layering, the last 5 layers at each model node

*Cohesive Sediments.*
Edited by Neville Burt, Reg Parker and Jacqueline Watts
©1997 John Wiley & Sons Ltd.

- An inerodible underlayer

- Wave activity as an additional stirring function.

The module is based on 'excess shear' relationships —

— for mud erosion (Partheniades 1962)                    (1)

$$\frac{dm}{dt} = M_p \frac{(\tau_c - \tau)}{\tau_c}$$

— and mud deposition (Krone 1962)                    (2)

$$\frac{dm}{dt} = Cw \frac{(\tau_s - \tau)}{\tau_s}$$

where   $m$        = mass of sediment on the bed

$t$        = time

$\tau$        = bed shear

$\tau_c$        = critical shear for erosion

$\tau_s$        = critical shear for deposition

$M_p$        = erosion rate at twice $\tau_c$

$C$        = mean sediment concentration in suspension over the depth of the layer in contact with the bed, and

$w$        = fall velocity of sediment

Both Equations (1) and (2) are for cohesive mud. Equations of similar form are used for the non-cohesive sand phase, using a critical shear for deposition approximately equal to that for erosion determined from the Shields' diagram. Sand erosion is only used on beds which are clearly non-cohesive sand — where the proportion of sand in the bed layer is greater than some critical value (87%, Torfs 1994) — and when the sand transporting capacity of the flow (Ackers and White 1973, modified for waves by Willis 1979) is not exceeded. Sand deposition occurs whenever there is sand in suspension, assumed independent of mud deposition, but mixed with the mud into a new surface layer

— the Migniot relation for critical shear for erosion of mud — (3)

$$\tau_c = N \varrho_s{}^M$$

where $\varrho_s$ = bulk density of sediment on the surface of the bed, and

M, N = constants to be determined

— and the Terzaghi consolidation relations — (4)

$$u = 0.964 T_v{}^{0.415} \qquad T_v = \frac{t C_v}{P^2}$$

where u = degree of consolidation

$C_v$ = a consolidation coefficient to be determined

P = length of the drainage path

These form a simple numerical framework for interpolating and extrapolating the observed hydrodynamic behaviour of the estuary sediment, specified by some 20 sediment parameters given in Table 1. Ideally, these parameters would be determined in the field or the laboratory, using the natural water and sediment, in annular flumes, settling columns, and the like

## 0-DIMENSIONAL MODEL

For modelling sedimentation in three of the estuaries, the National Research Council has used CUMBSED in conjunction with the 2-dimensional, finite-difference numerical hydrodynamic model MIKE21AD, of the Danish Hydraulic Institute. For Cumberland Basin, CUMBSED was developed using a quasi 3-dimensional hydrodynamic model belonging to ASA Consulting Limited. For the two projects not reported here, the 2-dimensional, finite-element model, TELEMAC-2D was used. Although all three hydrodynamic engines offer excellent representation of flows and sediment transport, none could be described as fast enough for the trial-and-error required in calibrating a bed sediment model. We therefore developed a 0-dimensional tidal model as part of developing CUMBSED

The 0-dimensional model provides an accurate representation of flow and waterlevel variations at a single point — from which CUMBSED can calculate bed shears, erosion and deposition, and sediment transport — over 5 complete tidal cycles, 62 hours. One such run on a PC takes about 20 s. More than $1.5 \times 10^6$ tidal cycles were run in checking CUMBSED for bugs during its development. Because of its speed, the 0-dimensional model has become the (one cylinder) engine of choice in calibrating specific estuaries, and has evolved into quite a powerful tool

**Table 1.** Hydrodynamic Sediment Properties, Miramichi Inner Bay

| PROPERTY | CALIBRATED VALUE |
|---|---|
| Fully Consolidated Mud Bed Concentration | 1 kg/kg |
| Concentration of Freshly Deposited Mud | 0.35 kg/kg |
| Mud Consolidation Coefficient | $10^{-5}$ m²/s |
| Mud Particle Size | 20 μ |
| Sand Particle Size | 250 μ |
| Increase in Mud Bed Concentration with Depth of Overburden | 0.125 kg/kg/m |
| Kinematic Viscosity of Water | $1.2 \times 10^{-6}$ m²/s |
| Exponent in Migniot Equation (3) | 0.7 |
| Mud Erosion Rate at Twice Critical Shear | $2 \times 10^{-4}$ kg/m²/s |
| Sand Erosion Rate at Twice Critical Shear | $2 \times 10^{-2}$ kg/m²/s |
| Coefficient in Migniot Equation (3) | 0.0017 N $m^{0.1}$ /$kg^{0.7}$(Units depend on value of exponent) |
| Critical Sand Fraction, delimiting mud and sand | 0.8 kg/kg |
| Specific Gravity of Water | 1025 kg/m³ |
| Specific Gravity of Mud | 2100 kg/m³ |
| Specific Gravity of Sand | 2650 kg/m³ |
| Critical Shear for Erosion and Deposition of Sand | 0.175 Pa |
| Critical Shear for Deposition of Mud | 0.075 Pa |
| Fall Velocity of Mud | $4 \times 10^{-5}$ m/s |
| Fall Velocity of Sand | 0.05 m/s |

## MODEL CALIBRATION

The traditional calibration procedure was simple. In the absence of measurements of the 19 properties of Table 1 — plus extras, such as the initial bed consolidation, background wave conditions and suspended sediment concentrations — we adjusted each by trial-and-error until CUMBSED reproduced a measured sedimentation event in the estuary. In the absence of a measured sedimentation event, we assumed no net bed change under 'normal' conditions: deposition balanced by erosion and consolidation. Such was the case for the first three estuaries, where sediment sizes and fall velocities could be measured in the laboratory, but quantitative behaviour on the seabed was unknown. The fourth estuary provided several measured sedimentation events. It should be noted that in all four studies reported here, the problem — both in the calibration and the future — was deposition. The fifth estuary (Willis et al 1995) was a mud erosion problem, and did not change the conclusions

At the time of writing the original of this paper, the NRCC Canadian Hydraulics Centre was part of an Institute for Mechanical Engineering. One of the benefits of this organization was that one of us received a course in Design of Experiments (Taguchi 1984) from the Canadian Supplier Institute. Design of Experiments is a statistical technique most often used by industrial and mechanical engineers to solve problems of variability on production lines. The Taguchi Method is said to be a favourite of the Ford Motor

Company; the Canadian Supplier Institute began as a subsidiary of Ford

Essentially, the engineer identifies the parameters affecting product quality, and estimates high and low values — and sometimes a best estimate — of each parameter. Experiments are then conducted on the production line using combinations of the parameter values to identify the combination resulting in the minimum variability, or error or rejection rate. If there are 15 parameters, with high and low values of each, it would require $15^2 = 225$ experiments to test all possible combinations. Design of Experiments identifies the 16 key experiments needed to determine the effect of each value of the 15. It is usually necessary to repeat the 16 experiments at least once, using only the most important parameters and fine-tuning the high and low values towards a best estimate

We successfully adapted this, at a low level by Taguchi's standards, to the calibration of the fourth model, and retrospectively to the other three including the Miramichi Inner Bay. All that was required was identification of 14 of the unknown hydrodynamic sediment properties of Table 1 and estimating high and low values. The 15th parameter was kept as a dummy, to identify any 'interaction' between the Migniot exponent, M, and coefficient, N. Instead of a production line, we used the 0-dimensional model to attempt to reproduce measured rates of erosion, deposition or stability of the bed. The results are given in Table 2

## RESULTS

For each estuary, the 16 experiments were run 4 times. Water depth, maximum velocity and target deposition were different each of the 4 times, covering the range to be found in the estuary. The results in Table 2 summarize the 64 experiments for each estuary

Before looking at the results, something needs to be said about the drawbacks of Design of Experiments: Because we are substituting 16 experiments for the full-factorial 196 needed to investigate all possible high and low combinations of our 14 factors, we must make some allowance for the fact that all factors are not independent. For example, the Migniot Coefficient, Factor 9, might be related to the Migniot Exponent, Factor 6. An increase in one could lead to a decrease in the other, to describe a given sediment. For the $L_{16}$ experiment design which we used, an interaction between Factors 6 and 9 would interfere with the results for Factor 15. Hence Factor 15 was left empty. This Migniot Interaction turned out not important, with an overall rank of only 12 in importance to the erosion/deposition process

One can imagine other, perhaps more important interactions —

2-4     Full Consolidation with Consolidation Coefficient. The effects of this interaction would appear with those of Factor 6, the Migniot Exponent

3-11    Density of freshly deposited mud with the threshold shear at which that deposition takes place. These effects will appear with those of Factor 8, the erosion rate of sand at twice the critical shear

Table 2. Summary of Calibration Results

| FACTOR | UNITS | CUMBERLAND | | LIVERPOOL | | MIRAMICHI | | ALL ESTUARIES | | |
|---|---|---|---|---|---|---|---|---|---|---|
| | | RANK | BEST | RANK | BEST | RANK | BEST | RANK | BEST | STD |
| 1 Significant Wave Height | m | 12 | 0.1 | 15 | 0.1 | 12 | 0.1 | 13 | 0.09 | 0.02 |
| 2 Full Consolidation | kg/kg | 10 | 0.5 | 9 | 1.2 | 14 | 1 | 9 | 0.8 | 0.3 |
| 3 Deposition Density | kg/kg | 14 | 0.15 | 5 | 0.15 | 9 | 0.35 | 10 | 0.24 | 0.09 |
| 4 Consolidation Coefficient | - | 4 | 0.000001 | 12 | 0.001 | 10 | 0.000001 | 7 | 0.0002 | 0.0004 |
| 5 Consolidation with Overburden | kg/kg/m | 6 | 0.25 | 4 | 0.0625 | 4 | 0.25 | 2 | 0.22 | 0.09 |
| 6 Migniot Exponent | - | 9 | 1 | 2 | 3.5 | 2 | 1.5 | 2 | 1.85 | 0.97 |
| 7 Mud Erosion at twice Critical | kg/m$^2$/s | 2 | 0.0001 | 13 | 0.001 | 6 | 0.0001 | 4 | 0.0003 | 0.0004 |
| 8 Sand Erosion at twice Critical | kg/m$^2$/s | 4 | 0.001 | 10 | 0.04 | 10 | 0.01 | 7 | 0.01 | 0.02 |
| 9 Migniot Coefficient | ? | 6 | 0.017 | 6 | 0.000004 | 4 | 0.002 | 6 | 0.006 | 0.007 |
| 10 Percent Sand for sand behaviour | kg/kg | 8 | 90 | 1 | 40 | 2 | 90 | 4 | 75 | 0.21 |
| 11 Mud Deposition Threshold | Pa | 2 | 0.03 | 11 | 0.2 | 7 | 0.04 | 6 | 0.07 | 0.08 |
| 12 Top Layer Consolidation | m | 1 | 9990 | 3 | 2 | 1 | 9990 | 1 | 7490 | 4320 |
| 13 Background Mud Concentration | kg/kg | 13 | 0.001 | 8 | 0.0001 | 13 | 0.0001 | 13 | 0.0006 | 0.0004 |
| 14 Background Sand Concentration | kg/kg | 11 | 0.000001 | 14 | 0.000003 | 15 | 0.000003 | 15 | 0.000002 | 0.000001 |
| 15 Interaction between 6 and 9 | - | 15 | | 7 | | 8 | | 11 | | |

388

In practice, the Design of Experiments technique is used in successive iterations to eliminate the effects of these interactions, to narrow down the number of important factors to be adjusted, and to optimize their values for a given estuary

Let's look at some of the rather surprising *less important factors*, those with ranking greater than 10. These include —

*Background Sand Concentration* — This was not a factor, even in Liverpool Bay, where sand predominates on the bed

*Background Mud Concentration* — Mud predominates in the other estuaries

*Significant Wave Height* — We didn't expect this to be a major factor, since we were looking at 'normal' wave heights which were less than 0.3 m in all estuaries. On Liverpool Bay we had found that some ambient wave activity was necessary to get *any* sand movement. Clearly the amount of ambient wave activity is not critical, as long as there is some and it does not dominate

Now for the *more important factors* —

*Consolidation of the Top Layer* — There is no doubt that the initially specified consolidation of the surface layer of the bed is the most important factor. And in all the estuaries except Liverpool, it needed to be fully consolidated, lending support to those modellers who deal only with resuspension of fresh cohesive deposits

*Consolidation with Overburden* — This is consolidation with depth below the bed, rather than with time. All estuaries except Liverpool favoured high values. Liverpool Bay is the least active, with the coarsest sediment, and needed encouragement to transport

*Migniot Exponent* — This was an important factor for both Liverpool and the Miramichi, where current velocities, and hence bed shears, are smaller. The Migniot Coefficient was not so important. Nevertheless, the variability in the Migniot Exponent was rather small for all the estuaries, as indicated by the small standard deviation, so that at present (1996) it has been assumed constant at 0.7

*Percent Sand for Sand Behaviour* — This is another factor affecting bed erosion — the percentage of sand required in the bed before it behaves as if it were all sand. Torfs (1994) found about 87% sand was required, since a small amount of cohesive mud can act as a binder for a large quantity of sand. Once again, all the estuaries except Liverpool demanded the higher value over 80%. Torfs' 87% has been assumed for further studies

*Mud Erosion at Twice Critical* — As expected, the coefficient in Equation (1)

Finally, the standard deviations — the last column of Table 2 — offer some insight into the stability of the processes, or the universality of the numbers. 8 factors showed standard deviations less than the mean: significant wave height, full consolidation, deposition density, consolidation with overburden, Migniot exponent, percent sand for sand behaviour, top layer consolidation, and background sand and mud concentrations

# THE MIRAMICHI INNER BAY

There is a difference in 'calibrated' values for the Miramichi in Tables 1 and 2. Table 1 presents the results of the traditional trial-and-error calibration reported in Crookshank et al (1993/09); Table 2, the Design of Experiments calibration

The Miramichi is probably the third most studied estuary in Canada, behind the Fraser and the St Lawrence, making it the ideal candidate for extra attention here. In fact we first attempted to model Miramichi mud almost 20 years ago (Willis et al 1977). At that time, we carried out laboratory measurements of critical shears for erosion which stood us in good stead in 1993, and formed the basis for the traditional calibration of Table 1

The Miramichi is a wide shallow estuary, cut off from the Gulf of St Lawrence by a chain of barrier islands [Figure 1]. It is not only one of the prime salmon fishing rivers on the east coast of North America, but it is also an industrial region with pulp mills and ocean vessel navigation. Our most recent modelling exercise was aimed at assessing the stability of a dredged spoil dump located within the Inner Bay [Figure 1]

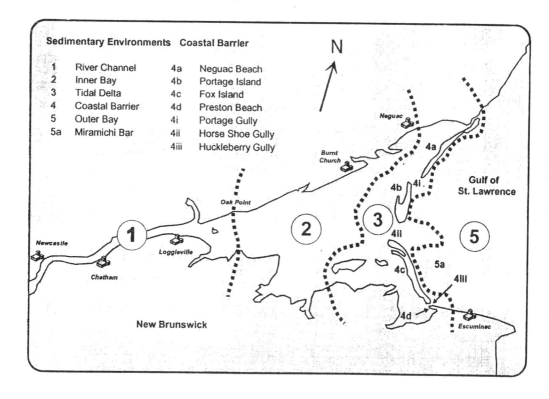

**Figure 1.** Miramichi Inner Bay

The mud is interesting, although field researchers seem unable to agree on whether it is diatomaceous or whether it forms 'fluid mud' or a 'fluff layer'. The most recent work (Brylinsky et al 1992) suggests that it isn't and doesn't. The dredged spoil is even more interesting, in that it was the least stable sediment in the estuary during, and immediately after dumping, but quickly consolidated to become the most stable. The reviewer has asked us to speculate on the reasons for this rapid consolidation; the following is therefore pure speculation —

*Method of Deposition* — The natural deposition is by flocculation with gentle settling velocities, resulting in a bed structure made up of disconnected interstitial water and narrow drainage paths. The spoil dump is by turbidity current, in which the flocs are broken up, and connected water surrounds each particle. Drainage paths would therefore be shorter and wider in the dump, with faster drainage and consolidation times

*Segregation* — The different bed materials — sand and mud in the case of the model — settle at different fall velocities if not trapped in a turbidity current. In the extreme, the coarse sand will settle on the spoil dump and the mud will be carried away with the water to deposit elsewhere. It was this fear — in fact the fear that the fines would never deposit — that fuelled the present studies

The model consisted of MIKE21AD hydrodynamics and CUMBSED, on a 200 m square grid. Figure 2 shows a typical erosion/deposition pattern after 24 hours — almost a complete almost diurnal spring tidal cycle, with measured winds and minimum waves — and Figure 3, the corresponding suspended sediments. Figure 4 shows output from the MIKE21 advection/dispersion module of a point source tracer on the dredged spoil dump

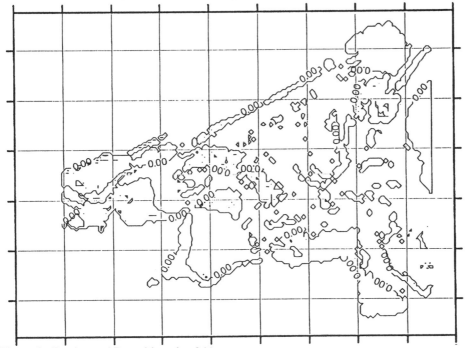

**Figure 2.** Erosion and Deposition after 24 hours

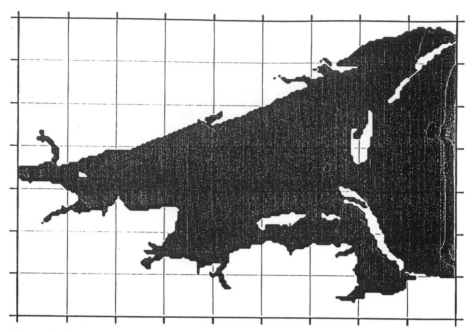

**Figure 3.** Suspended Sediment after 24 hours

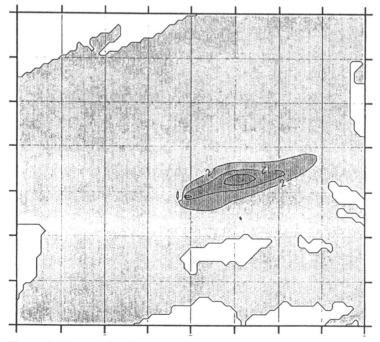

**Figure 4.** Tracer Advection and Dispersion

# CONCLUSIONS

1. With limited funds available for field and laboratory support of an estuarine sediment model, the higher ranked items of Table 2 should be addressed. Particular attention must be paid to the bed consolidation and erodibility, especially at the surface; especially if erosion, rather than deposition, is the problem to be studied

2. If the problem to be modelled is deposition, it is probably sufficient to assume a fully consolidated bed and model only resuspension of recently deposited sediment

3. Background suspended sediment concentration is not worth worrying about, except as it affects stability of the hydrodynamic engine at the boundaries or at start up

4. A 0-dimensional model, used with Design of Experiments, can determine the value of hydrodynamic sediment properties within an order of magnitude

5. We believe Conclusions 1 through 4 to be of general application to models based on Equations (1) to (4)

6. For the sixth model, currently in progress (1996), only the following 10 properties need be optimized: top layer consolidation, consolidation with overburden, full consolidation, consolidation coefficent, erosion rates at twice critical shear, mud fall velocity, Migniot coefficient, critical shear for deposition of mud, and density of freshly deposited mud

7. Although some wave activity is necessary to the successful operation of the bed sediment module, CUMBSED, sediment movement is insensitive to wave heights that are of a lesser order than the water depth

# REFERENCES

Ackers, P And W R White. 1973. Sediment Transport: New Approach and Analysis. Proceedings of the American Society of Civil Engineers, Hydraulics Division, No HY11, November, pp 2041-2060

Brylinsky, M, J Gibson, G R Daborn, C L Amos, And H A Christian. 1992. Miramichi Inner Bay Sediment Stability Study. Acadia Centre for Estuarine Research, Acadia University, Publication No 22, 200 pp

Crookshank, N L, D H Willis and J Zhang. 1993/09. Hydronumerical Modelling of Sedimentation in Miramichi Inner Bay. National Research Council Canada, Institute for Engineering in the Canadian Environment, Coastal Engineering Program Controlled Technical Report No IECE-CEP-CTR-003, 53 pp

de Margerie, S, M MacNeill, D L DeWolfe and D H Willis. 1993. Numerical Modelling of Tidal Basin Dynamics. Proceedings 11th Canadian Hydrotechnical Conference, Canadian Society for Civil Engineering, Vol I, pp 505-514

Krone, R B. 1962. Flume Studies of the Transport of Sediment in Estuarial Shoaling Processes. University of California, Hydraulic Engineering Laboratory and Sanitary Engineering Research Laboratory, June

Parthenaides, E. 1962. A Study of Erosion and Deposition of Cohesive Soils in Salt Water. University of California, PhD Thesis

Taguchi, G. 1984. System of Experimental Design. Referred to in course manuals of the Canadian Supplier Institute Inc.

Torfs, H. 1994. Erosion of Mixed Cohesive/Non-Cohesive Sediments in Uniform Flow. Proceedings INTERCOH '94. This volume

Willis, D H. 1979. Sediment Load under Waves and Currents. National Research Council Canada, *DME/NAE Quarterly Bulletin*, No 1979(3), October, pp 1-17

Willis, D H. 1985. A Mathematical Model of Mud Transport. Proceedings, Canadian Coastal Conference '85, National Research Council Associate Committee for Research on Shoreline Erosion and Sedimentation (ACROSES), pp 205-221

Willis, D H. 1991/03. A Bed Sediment Module for Numerical Hydraulic Models. National Research Council Canada, Institute for Mechanical Engineering, Hydraulics Laboratory Technical Report No TR-HY-039, NRC No 32116, 65 pp

Willis, D H (with N L Crookshank and R R Johnson). 1977. Miramichi Channel Study Hydraulic Investigation. National Research Council Canada, Division of Mechanical Engineering, Hydraulics Laboratory Technical Report No LTR-HY-56, 2 Vols

Willis, D H, V Barthel, N L Crookshank and J Zhang. 1993/03. Hydronumerical Modelling of Sedimentation in Liverpool Bay. National Research Council Canada, Institute for Mechanical Engineering, Coastal Zone Engineering Controlled Technical Report No IME-CZE-CTR-005, 272 pp

Willis, D H, T D Faure, and N L Crookshank. 1995. Calibration of Multiphase Sediment Transport in Estuaries. Proceedings of the 12th Canadian Hydrotechnical Conference, Canadian Society for Civil Engineering, Vol 1

# 28 Numerical modelling of mud and sand mixtures

**T. J. CHESHER and M. C. OCKENDEN**
*HR Wallingford Ltd, Wallingford, Oxon OX10 8BA, UK*

## ABSTRACT

A combined mud and sand mixtures 2DH numerical model has been developed from existing separate models. Interactions between the two sediment types determined from previous laboratory and fields measurements have been incorporated. The model was applied to a 1-D channel tests case and to a study of the Mersey Estuary, where a significant reduction in the entrainment and transport of both types of sediment was predicted.

## INTRODUCTION

Recent fields and laboratory studies undertaken by HR [Williamson and Ockenden (1992), Williamson (1993a,1993b)] have sought to identify the main effects of the presence of sand on the bed on the relative transport of mud. These investigations have concentrated on the in-situ measurement of the erosion and entrainment of mud from a mixed sand/mud bed. Interpretation of the effects of sand mixed into the mud layer has been carried out in terms of a modification to the classical theory of functions defining the threshold stress and the erosion rate for mud for erosive conditions.

This paper describes the development and application of an integrated mud and sand transport model from existing well-established separate models for each sediment type, incorporating the various interpreted interactions measured in the field. Using this numerical model the effects of the mud/sand interaction on the relative transport of each sediment type are investigated in detail by simulating a simple 1-D channel with constant discharge and a more a specific application to the Mersey Estuary.

Results from the two test case scenarios indicated a strong reduction in the entrainment and transport of both types of sediment when the full interactions were included. In particular, the application to the Mersey Estuary gave rise to a net tidal residual transport of mud in the upper part of the Estuary that *reversed direction* on inclusion of the mud/sand interactions. These results highlight the importance of the use of a fully interactive model, since an appreciation of, for example, dredging requirements would yield markedly different estimations depending on the type of model used.

*Cohesive Sediments.*
Edited by Neville Burt, Reg Parker and Jacqueline Watts
©1997 John Wiley & Sons Ltd.

## INTERPRETATION OF MUD/SAND INTERACTIONS FROM LABORATORY AND FIELD DATA

The following conclusions were drawn from the previous work undertaken in the research programme:

- sand content on the bed increases the mud consolidation rate
- sand content on the bed increases the shear strength of the mud bed
- sand content on the bed reduces the erosion rate of the mud
- once suspended in the water column mud and sand appear to act independently
- deposition of mud and sand appears to be independent.

The field and laboratory work undertaken in this study has concentrated on the effects of a relatively small proportion of sand on the relative transport of mud, and therefore these results have been interpreted in terms of a modification to the existing theory for mud transport.

Williamson (1993a) suggests modifications to the erosion rate for mud, $m_e$ and the erosion shear strength for mud, $\tau_e$ to account for the sand content according to:

$$\tau_e = K \, \rho_d^L \tag{1}$$

where

$\rho_d$ is the dry density (kg/m³)
K, L are constants
$\rho_d = f(\% \text{ sand, time}) \tag{2}$

The measured relation between the density of the bed material and $\tau_e$ is presented in Figure 1, indicating an increase in $\tau_e$ from the mud-only value with increasing sand content.

The erosion of mud could be described by:

$$S = m_e \, (\tau - \tau_e) \tag{3}$$

where

S    is the rate of supply of sediment eroded (kg/m²/s)
$\tau$    is the bed shear stress (N/m²)
$m_e = f \, (\% \text{ sand}) \tag{4}$

Using this definition Ockenden and Delo (1988) noted a decreasing erosion rate with increasing sand content for erosion tests in a mud carousel. Williamson and Ockenden (1992) also measured a lower erosion rate for mud beds with sand added and Williamson (1993a) observed a similar trend, noting that the addition of sand decreased the erosion rate by as much as an order of magnitude.

Williamson (1993a) also identifies the presence of distinct layering of mud and sand deposits and postulates differential depositional events as the main mechanism. However it is also concluded that more research is required to gain a better understanding of the hydrodynamic and sedimentological conditions controlling this process.

Collins (1989) provides additional information describing the effect of a relatively small proportion of mud on the entrainment of sand. This effect was interpreted in terms of the threshold flow velocity for sand erosion, where it was found that the threshold velocity increases according to the mud content. These results are shown in Figure 2.

From these analyses it was concluded that interactions between the mud and the sand fractions could be interpreted in terms of properties relating to the *entrainment* of each type of sediment from the bed.

## MODEL DEVELOPMENT

The objective of the development of the numerical model was to combine existing, well-established mud and sand transport models into one integrated model and to incorporate new design algorithms to describe the various interactions that will affect the relative transport of each sediment type.

The new model was based on MUDFLOW-2D and SANDFLOW-2D from the HR tidal modelling suite TIDEWAY. These models have been used extensively for a wide variety of engineering studies including coastal defence construction design and spoil disposal impact assessments. Such studies require careful model calibration and validation, and therefore it was considered prudent to use these models as a reliable basis for the new model in order that attention was concentrated on the particular aspects of the study relating to the mud/sand interactions. Further details of these models are described in Cole and Miles, 1988 and Miles, 1991.

## INCORPORATION OF ALGORITHMS FOR MODELLING MUD/SAND MIXTURES

The new model was developed from two separate models and hence is formulated according to two sediment concentration equations with mud/sand interactions rather than a single equation defining the entrainment and transport of a mixed concentrate.

For the mud module, where erosive conditions are described by equation 3, $\tau_e$ and $m_e$ were modified from constants to relationships dependent on the dry density of the bed, which is directly related to the relative proportions (by weight) of each sediment on the bed. The data analyses indicated an increase in $\tau_e$ from the purely mud bed value of $\tau_{emud}$ with increasing sand content, as well as an increase in $U_t$ for purely sand beds with increasing mud content. Best fit lines fitted through the data from Williamson (1993a) and Collins (1989) indicated a peak in the erosion strength, here defined as $\tau_{emax}$ of about 0.5N/m$^2$ for a mixed bed with a mud content of about 20% of the total bed by weight and in the absence of any other information this value was adopted for the purposes of this study. This value of $\tau_{emax}$ is also in broad agreement with the data presented in Soulsby (1994). Accordingly, $\tau_e$ was assumed to vary with the proportion of mud to sand on the bed as described in Figure 3.

The effective $m_e$ is assumed to vary linearly from its mud-only value, $m_{emud}$ to zero for a purely sand bed, as described in Figure 4. This effective erosion rate can be interpreted in terms of a constant $m_e$ valued which is rescaled in proportion to the amount of mud to total deposits on the bed.

For the sand module the threshold velocity, $U_t$ for mixed beds was assumed to vary according to the same $\tau_e$ relationship for mud over mixed beds after specification of a suitable Darcy Weisbach friction factor, $f_s$. This represents the cohesion of sand particles to the muddy bed. Hence the $\tau_e$ relationship given in Figure 3 also represents the variation in $U_t$, according to:

$$U_t = (8\tau_e/\rho f_s)^{1/2} \tag{5}$$

where $\tau_e$ ranges from $\tau_{esand}$ for purely sand beds up to $\tau_{emax}$, to $\tau_{emud}$ for mud-only beds as shown in Figure 3, and $\tau_{esand}$ was defined according to:

$$\tau_{esand} = \rho f_s U_{ts}^2/8 \tag{6}$$

where $U_{ts}$ is the threshold velocity over sand beds only. In this way consistency was achieved between simulations with and without the inclusion of the mud/sand interactions.

Clearly the model predictions will depend on the specification of these interaction parameters. As a result it should be stressed that the aim of this study was to assess the relative importance of the mud/sand interactions in order to improve the model accuracy. Further field and laboratory data would be of considerable benefit for validation of the existing model framework, as well as in improving the representation of the interactions.

Layering of the bed deposits was not represented for a number of reasons. Firstly, it was concluded in the previous studies that more field and laboratory work is required to gain a better understanding of the physical conditions associated with this process. Secondly, the formulation of a layered bed *to represent the discrete sediment strata* was outside the scope of this study, since this would require the capability of storage of immense amounts of data bearing in mind the typical layer thicknesses and total depth of deposits. Consequently, the deposits of mud and sand are assumed to be completely mixed.

Consolidation of the mud deposits is represented by prescription of a higher erosion strength than threshold stress for deposition. A more sophisticated representation of consolidation in terms of the bed density with respect to time would improve the model accuracy although for the design timescale for this model (typically a tidal cycle) this formulation was considered adequate.

## APPLICATION AND ASSESSMENT OF THE NUMERICAL MODEL

Initial tests to assess the behaviour of the new model were based on a 2km long channel case under steady flow conditions. Following on from this the model was used to investigate the sediment transport in the Mersey Estuary, set up for a spring tide simulation including spatial variations in the initial bed deposits.

## CHANNEL TEST CASE

Tests were run for steady flow conditions with a depth of 18.6m and a mean velocity of 0.7m/s. The initial conditions describing the sediment properties comprised equal proportions of mud and sand deposits of 100kg/m$^2$ and no sediment in suspension. No sediment was introduced at the input boundary, so that these conditions represent the familiar 'leading-edge' test case. The other sediment-related parameters were:

| sand module | | mud module | |
|---|---|---|---|
| grain size, $D_{50}$ | $= 86\mu m$ | deposition stress, $\tau_d$ | $= 0.08$ N/m$^2$ |
| settling velocity, $\omega_s = 4.4$mm/s | $\tau_{emud}$ | | $= 0.1$ N/m$^2$ |
| $U_{ts}$ | $= 0.17$m/s | settling velocity, $w_s$ | $= 1.0C$mm/s (C in g/l) |
| (f = 0.01) | | minimum $w_s$, $w_{smin}$ | $= 0.1$mm/s |
| | | $m_e$ | $= 0.007$kg/m$^2$/s per N/m$^2$ |

As described above $\tau_e$ varies from the value given above for model cells containing mud only, to $\tau_{esand}$ as defined in equation 6, with a peak value of 0.5 N/m$^2$ at a mud deposits content of 20%.

The model was run for a period of 40 hours, storing values of bed deposits and suspended concentrations every half hour. The flow conditions were assumed constant, with no feedback of bed level changes following erosion. The test was repeated with the interaction between the mud and sand fractions switched off, in order to demonstrate the results that would be obtained if using separate mud and sand transport models. Concentration distributions with downstream distance after one hour (to allow for the removal of transients) for both tests are presented in Figure 5. Inclusion of the interaction gives rise to a reduction in the mud concentration of a factor four at the downstream end, and a three-fold reduction in the sand concentrations.

The mud/sand interactions are dependent on the relative proportions of mud to sand on the bed, and therefore, differential erosion between the two sediment types will feed back to affect the subsequent bed erosion. This is demonstrated in Figure 6, which shows bed deposit time-histories at a point half-way along the channel. This highlights the non-linear response when the interactions are included, and also demonstrates the large overestimation in erosion of both mud and sand if the sediments are considered in isolation. In particular, without the interaction all the mud deposits are removed from this point (resulting in zero mud concentration) after about 8 hours, whereas the fully-interactive model predicts approximately one quarter of the original deposits remaining after 40 hours.

## APPLICATION TO THE MERSEY ESTUARY

Having assessed the behaviour of the new model to a relatively simple channel case a more complex application was undertaken to simulate the sediment transport in the Mersey Estuary, where there are large spatial variations in the mud and sand deposits. The aim of this study was to assess the sensitivity of the model predictions to the mud/sand interactions with special attention to the net tidal residual sediment fluxes.

As in the channel case the emphasis was concentrated on the sediment transport and hence the model geometry, bathymetry and validated flow fields were taken from a separate study of the Mersey, which is fully documented in HR Wallingford, 1991. The model bathymetry is presented in Figure 7. Tests were carried out for a spring tide, and time-histories of the currents at specified monitoring stations (also shown in Figure 7) are given in Figure 8. Representative values for the sediment-related parameters were also taken from this previous study, and are shown below.

| *sand module* | *mud module* | |
|---|---|---|
| $D_{50}$ = 250µm | $\tau_d$ | = 0.1 N/m$^2$ |
| $\omega_s$ = 25mm/s | $\tau_{emud}$ | = 0.2 N/m$^2$ |
| $U_{ts}$ = 0.4m/s | $w_s$ | = 1.0Cmm/s (C in g/l) |
| (f = 0.1) | $w_{smin}$ | = 0.1mm/s |

Initial conditions describing the sediment fields were generated by specifying initial deposits of mud and sand of 1000kg/m$^2$ over the entire model area and running the model (in non-interactive mode) for a period of 10 tides. By this means, and in contrast to the channel case, the deposits were spatially distributed over the model domain, so that deposits in cells ranged from all mud to all sand to no deposits at all (hard bed). Figures 9 and 10 show these initial distributions of mud and sand, and indicate broad agreement with the natural state of the estuary, including hard bed areas in The Narrows, mud deposits largely confined to the periphery and upper reaches, and a mixed distribution of sand deposits. During this period, and for all subsequent model tests, the sediment fluxes at the seaward boundary were specified on the flood tide according to saturation loads and a constant 400ppm for mud concentrations, in accordance with the previous study.

Clearly, running the model in interactive mode to define the initial distribution (or by using any other method) would have resulted in a different initial condition. However, this technique was considered appropriate as a means of describing these start conditions, in order to study the more relevant aspects of the mud/sand interaction.

As with the channel study, two tests were carried out (with and without the interactions), each one starting from these initial conditions, for a single tidal period.

The time-history station locations shown in Figure 7 were chosen according to the initial proportions of mud and sand on the bed, and these are presented in the table below.

| Station | A | B | C | D | E | F | G | H |
|---------|---|---|---|---|---|---|---|---|
| Initial sand deposits (kg/m$^2$) | 0 | 2053 | 1177 | 1961 | 224 | 796 | 719 | 853 |
| Initial mud deposits (kg/m$^2$) | 0 | 0 | 869 | 609 | 616 | 1145 | 965 | 1082 |

Time histories through the tide of mud concentrations and deposits, with and without the mud/sand interaction are given in Figures 11 and 12. These figures highlight the strong reduction in the mud   concentrations, and subsequent reduction in erosion when the interaction is included.

Looking more closely upstream the net tidal sediment flux vectors in the vicinity of Ince Bank were analysed, and the mud fluxes are shown in Figures 13 and 14. These, again, show a reduction in the net tidal sediment fluxes, but the most important aspect is related to the mud fluxes in this area that indicate, *in this case*, a reversal from upstream to downstream on inclusion of the interaction. This is  a consequence of the reduction in suspension of mud on the flood tide (cf Figure11).

Clearly this residual flux reversal is not a general result which can be applied to all model studies, although it highlights the effect of the interaction, and indicates that in mixed sediment environments the sediment transports should not be considered in isolation.

## CONCLUSIONS AND RECOMMENDATIONS

Results from the two test cases indicated a strong reduction in the entrainment and transport of both types of sediment when the full interactions were included.  In the case of the uniform channel study, it was found that for the non-interactive model all the mud was removed from the bed after some eight hours, whereas with the interactive model the erosion of mud from the bed was reduced to such an extent that there was still approximately one quarter of the initial deposits remaining after 40 hours of simulation.

The application to the Mersey Estuary gave rise to another significant result whereby it was found that the net tidal residual transport of mud in the upper part of the Estuary *reversed direction* on inclusion of the mud/sand interactions.  Clearly, in this case the results highlight the importance of the use of a fully interactive model, since an appreciation of, for example, dredging requirements would yield markedly different estimations depending on the type of model used.

Clearly the model predictions will depend on the specification of the various interaction parameters. As a result it should be stressed that the aim of this study was to assess the relative importance of the mud/sand interactions in order to improve the model accuracy. Further field and laboratory data would be of considerable benefit for validation of the existing model framework, as well as in improving the representation of the interactions.

## ACKNOWLEDGEMENTS

The authors gratefully acknowledge financial support for research underpinning this work from the UK Department of the Environment under Research Contract PECD 7/6/194. Allied work funded by the Commission of the European Communities Directorate General for Science, Research and Development under Contract Number MAS2-CT92-0027, forming part of the MAST G8M Coastal Morphodynamics community research programme, is reported here as well.

## REFERENCES

**Collins M B (1989)** The Behaviour of Cohesive and Non-Cohesive Sediments. Proceedings of the International Seminar on the Environmental Aspects of Dredging Activities. Nantes, France, Nov 27-Dec 1, pp15-32.

**HR Wallingford (1991)** Mersey Barrage. Feasibility Study - Stage III. 2D Mathematical Modelling of Tidal Flows and Sedimentation. HR Wallingford Report EX2303.

**Cole P and Miles G V (1983)** A two-dimensional model of mud transport. ASCE J, Hyd. Eng. Vol 109 No. 1

**Miles G V (1991)** Transport of sand mixtures. Euromech-262 Sand Transport in Rivers, Estuaries and the Sea. Soulsby and Bettess (eds)

**Ockenden M C and Delo E A (1988)** Consolidation and erosion of estuarine mud and sand mixtures - an experimental study. HR Wallingford Report SR 149.

**Soulsby R L (1994)** Manual of Marine Sands. HR Wallingford Report SR 351.

**Williamson H J (1991)** Tidal Transport of Mud and Sand Mixtures. Sediment Distributions - A Literature Review. HR Wallingford Report SR 286.

**Williamson H J and Ockenden M C (1992)** Tidal Transport of Mud and Sand Mixtures. Laboratory Tests. HR Wallingford Report SR 257.

**Williamson H J (1993a)** Tidal Transport of Mud/Sand Mixtures. Field Trials at Blue Anchor Bay. HR Wallingford Report SR 333.

**Williamson H J (1993b)** Tidal Transport of Mud/Sand Mixtures. Field Trials at Clevedon, Severn Estuary. HR Wallingford Report SR 372.

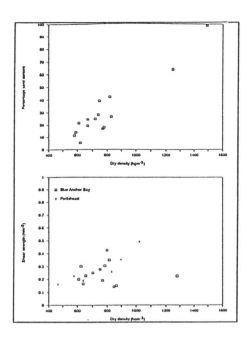

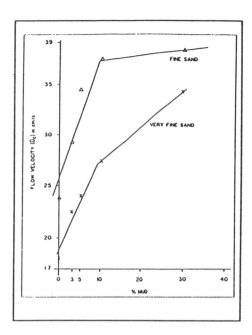

Figure 1    Sand content vs density and shear strength vs density Reproduced from Williamson (1993a)

Figure 2    Threshold of mud/sand mixtures under unidirectional flow Reproduced from Collins (1989)

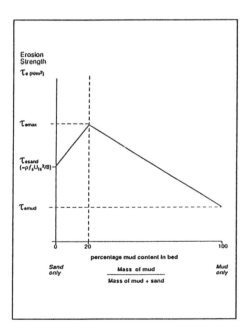

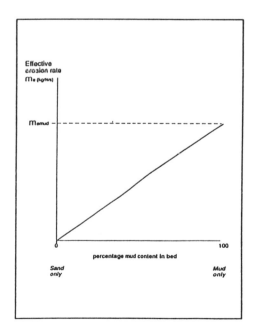

Figure 3    Representation of the effect of the proportion of mud to sand deposits on $\tau_e$

Figure 4    Representation of the effect of the proportion of mud to sand deposits on $m_e$

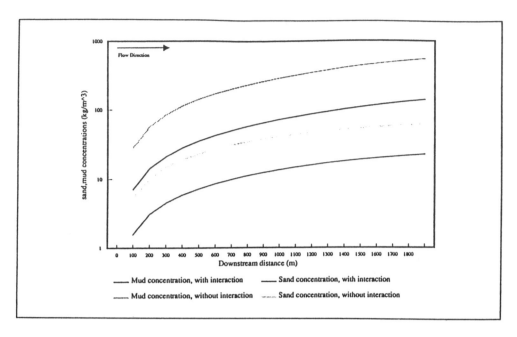

Figure 5    Channel test case.  Initial suspended sediment concentration distribution along channel
with and without mud/sand interaction

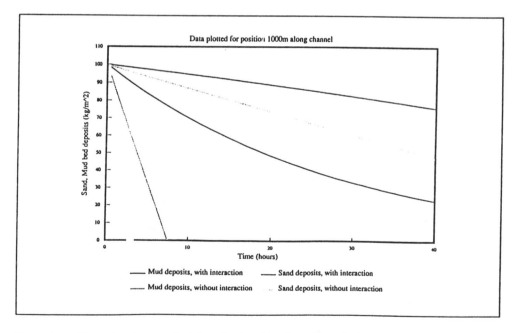

Figure 6    Channel test case.  Bed deposits time histories with and without mud/sand interaction

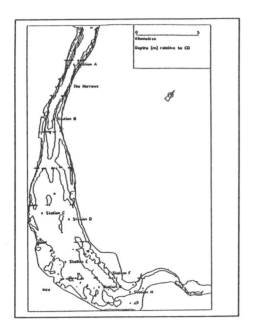

Figure 7    Mersey Estuary model bathymetry
            showing time-history station locations

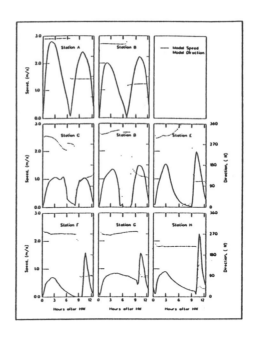

Figure 8    Mersey Estuary currents
            Spring tide

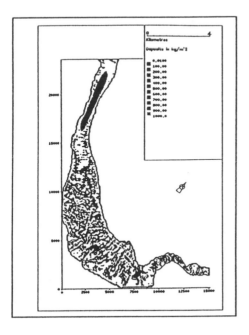

Figure 9    Mersey Estuary currents
            Initial distribution of sand on bed

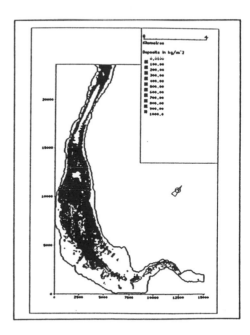

Figure 10   Mersey Estuary
            Initial distribution of mud on bed

405

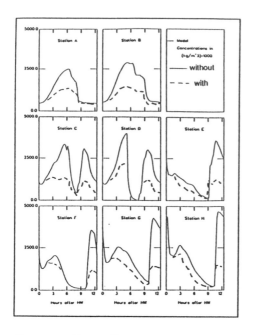

Figure 11   Mersey Estuary mud concentrations
with & without mud/sand interaction

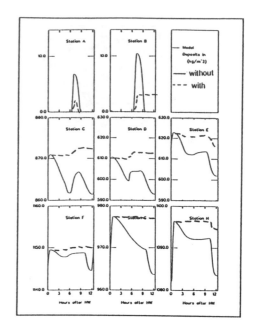

Figure 12   Mersey Estuary mud deposits
with & without mud/sand interaction

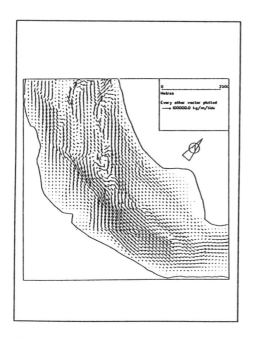

Figure 13   Mersey Estuary net tidal mud fluxes
Without mud/sand interaction

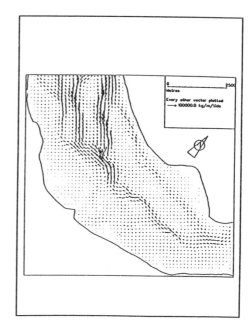

Figure 14   Mersey Estuary net tidal mud fluxes
With mud/sand interaction

# 29  Mass fluxes in fluid-mud layers on an inclined bed

T. KUSUDA, R. WATANABE and H. YAMANISHI

*Department of Civil Engineering, Kyushu University, Japan*

## INTRODUCTION

Fluid-mud formation and transport following settling of suspended solids in muddy tidal rivers and estuaries play an important role in narrowing of river cross section and in shoaling of navigation channels. Shoaling of navigation channels in estuaries has been studied intensively and is known to be mainly caused by horizontal transport of suspended solids as fluid-mud, not by direct settling of suspended solids to the channels (Odd and Cooper, 1989, for instance). Narrowing of the tidal rivers on a daily basis depends on the rate of the formation of stationary bed-mud. Although much work has been done on fluid-mud transport and formation processes (Wolanski *et al.*, 1992, for instance), fundamental elements involved in the processes such as concentration profiles and dispersion rates in fluid-mud layers and deposition rates to bed-mud have not been fully understood yet. The purpose of this study is to discuss concentrations at the upper and lower lutoclines in the growing phase of fluid-mud layers, a constitutive equation and characteristics of the mass flux, especially vertical one, in fluid-mud layers based on previous experimental results about fluid-mud transport on an inclined bed in a quiescent state (Kusuda et al., 1993) to make the simulation of fluid mud movement possible.

## EXPERIMENTAL MATERIAL, APPARATUS and METHODS

The experimental apparatus, methods and test material are as follows (see Kusuda *et al.*, (1993) for further reference): The material was obtained at the Kumamoto Port in Ariake Bay located in the western part of Japan. The grain size distribution of the test material is 48% in the clay range, 47% in the silt range, and 5% in the sand range. The mean diameter is 0.060 mm, its specific gravity and ignition loss are 2.65 and 12.2%, respectively. All experiments were conducted in salt water of specific gravity, 1.025. Settling velocity of the test material in a quiescent state was obtained by column tests using salt water of specific gravity 1.025. The column is 10 cm in diameter and 1 m in height. A flume employed in the experiments is 2 m high, 2 m wide, and 0.2 m thick. An inclined bed is installed in it, whose slope is adjustable from 0 to 1:473 (0.523 rad). Sampling tubes are installed in 4 rows, 0.30, 0.60, 0.90, and 1.20 m away from the upstream end, at points, 0.7 and 1.20 m above the bottom of the flume and from 0 to 5 cm from the inclined bed surface at a distance of 2.5 mm. The movement of fluid- mud on the inclined

*Cohesive Sediments.*
Edited by Neville Burt, Reg Parker and Jacqueline Watts

bed was recorded by a video camera with a close-up lens through the wall and analyzed by observation. Fluid-mud was sampled through small tubes and the concentrations of suspended solids were measured.

## MODELING THE FLUID MUD MOVEMENT

The momentum conservation of a suspension like fluid-mud moving along an inclined bed under a quiescent condition can be expressed in several ways. A basic method is to express each set of the   momentum equations for both solid and liquid phases. Another is to express the momentum equations for the suspension and for either one of the phases. The latter is adopted here so that it decreases correlation terms effectively and makes easy to obtain a constitutive equation. The following assumptions are introduced for simplicity of modeling the movement of fluid-mud:

1. The fluid-mud layer is so thin that the inner pressure is hydrostatic;
2. The Reynolds number of movement of the fluid-mud is less than unity, so that all acceleration terms in the Navier-Stokes equationsare negligible; and
3. The slope of an inclined bed is much less than unity.

Under these assumptions, two-dimensional continuity and momentum equations of the fluid-mud flow are as follows (refer to **Figure 1**):

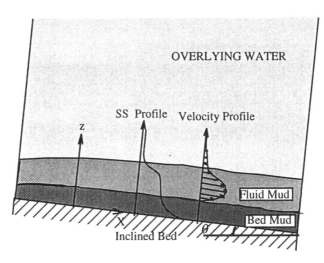

**Figure 1**    A fluid-mud layer on an inclined bed

Continuity equation:

$$\frac{\partial u}{\partial x} + \frac{\partial w}{\partial z} = 0 \tag{1}$$

Momentum equation

$$-\frac{\partial P}{\partial z} + Rg\left(C - C_a\right) = 0 \tag{2}$$

$$\frac{\partial \tau}{\partial z} = -\mu_a \frac{\partial u}{\partial z} = \frac{\partial}{\partial z}\left(u\, F_d\right) - Rg\left(C - C_a\right)\sin\theta + Rg\frac{\partial}{\partial x}\int_z^{H(x)} C\, dz \tag{3}$$

$$R = \frac{\rho_s - \rho_l}{\rho_s} \tag{4}$$

where $P$ is the hydrostatic pressure, $C$ is the concentration of fluid-mud, $C_a$ is the concentration of ambient fluid, $\tau$ is the shear stress along the inclined bed, $u$ and $w$ are the velocities of fluid-mud in the $x$ and $z$ directions, respectively, $\mu_a$ is apparent viscosity, $F_d$ is the flux of suspended solids, $g$ is the acceleration of gravity, $\theta$ is the slope of the inclined bed, $H$ is the influence height of the fluid-mud layer, $\rho_s$ is the density of suspended solids, $\rho_l$ is the density of liquid, $z$ is the axis perpendicular to the inclined bed, positive upward, and $x$ is the axis along the inclined bed.

A constitutive equation relates the shear stress, $\tau$, with the flow velocity, $u$ from velocity and concentration profiles in fluid-mud layers.

The boundary conditions to solve Eqs. 1, 2 and 3 are as follows:

> $u$ and $w$ are 0 for $z = $ infinite;
> $u$, $w$ and $P$ are continuous at the interface between overlying
> water and fluid-mud; and
> $u = 0$ and $w$ and $P$ are continuous at the interface between
> fluid-mud and bed-mud.

The mass conservation equation for the solid phase is as follows:

$$\frac{\partial C}{\partial t} + \frac{\partial(uC)}{\partial x} + \frac{\partial((w - w_s)\,C)}{\partial z} = \frac{\partial}{\partial z}\left(K\frac{\partial C}{\partial z}\right) \tag{5}$$

where $w_s$ is the settling velocity of suspended solids and $K$ is the dispersion coefficient, both of which are obtainable from experiments.

As the boundary conditions in Eq. 5, the settling velocity and the concentration of suspended solids are to be given at the interface between overlying water and fluid-mud layer and at the interface between fluid-mud and bed-mud layers. A method to calculate the former was given by Smith and Kirby (1989) by use of the local flux Richardson number. One for the latter is explained and discussed in the following section.

The dispersion coefficient is one of the control factors of the vertical mass flux in the fluid-mud layer. Under a steady-state condition, the dispersion coefficient is obtainable by Eq. 6. When the concentration profile above and in the fluid-mud changes little, that is, the settling flux of suspended solids is almost equal to the deposition flux of fluid- mud, the dispersion coefficient is easily obtained by Eq. 7.

$$K = \left[ w_{sa} C_a - w_s C + \frac{\partial}{\partial x} \int_z^\infty u\, C\, dz \right] / \frac{\partial C}{\partial z} \tag{6}$$

$$K = \left( w_s C - F_d \right) / \frac{\partial C}{\partial z} \tag{7}$$

where $C_a$ and $C$ are the concentration of suspended solids far above the fluid-mud and that at a point where $K$ is calculated, respectively, $W_{sa}$ and $W_s$ are the settling velocities of suspended solids as a function of $C_a$ and $C$, respectively, and $F_d$ is the flux to bed-mud.

## EXPERIMENTAL RESULTS and DISCUSSION

### Settling test
Settling velocities of suspended solids at the initial stage were obtained as Eq. 8,

$$w_s = 7.3 \times 10^{-4} \left( 1 - 5.0 \times 10^{-3} C \right)^{6.6} \tag{8}$$

where $ws$ is the settling velocity of suspended solids $(m \cdot s^{-1})$ and $C$ is the concentration of suspended solids $(kg \cdot m^{-3})$. The mass conservation on settling in a quiescent state is expressed as a wave equation of the first order (Kynch, 1952).

$$\frac{\partial C}{\partial t} + \frac{\partial (w_s C)}{\partial z} = \frac{\partial C}{\partial t} + \frac{d(w_s C)}{dz} \frac{\partial C}{\partial z} = 0 \tag{9}$$

The propagation speed of concentration is equal to $d(w_s C)/dC$, so that when $C$ is 26.3 $kg \cdot m^{-3}$, the speed becomes 0. This means that the concentration of fluid-mud at the upper lutocline in the growing phase is larger than this concentration. All experimental results agree well with this theoretical conclusion.

The minimum solid fraction at the surface of deposits was 0.012, that is, 31.8 $kg \cdot m^{-3}$. This seems to be the critical concentration for effective stress generation, so that the concentrations of the bed-muds are to be larger than this critical concentration.

### Fluid-mud transport experiments
Table 1 shows slope of the inclined bed and initial concentrations of suspension in the experiments. Some are added to the results of the previous experiments.

Table 1 Experimental conditions and results

| RUN | Slope of inclined bed (rad) | Initial concentration of suspension (kg·m⁻³) | Thickness of mobile fluid-mud layer(mm) | Max. velocity (cm·s⁻¹) | Highest conc. of mobile fluid-mud (kg·m⁻³) | Bed mud rising rate (mm·s⁻¹) |
|---|---|---|---|---|---|---|
| 1 | 0.043 | 17.1 | 5 | 0.128 | 52.0 | 0.024 |
| 2 | 0.049 | 11.1 | 4 | 0.0564 | 27.3 | 0.013 |
| 3 | 0.049 | 34.3 | --- | 0.0025 | --- | --- |
| 4 | 0.097 | 3.2 | 5 | 0.244 | 21.3 | 0.007 |
| 5 | 0.097 | 9.9 | 4 | 0.228 | 34.2 | 0.012 |
| 6 | 0.097 | 22.5 | 5 | 0.156 | 41.7 | 0.014 |
| 7 | 0.181 | 4.3 | 5 | 0.258 | 15.0 | 0.005 |
| 8 | 0.181 | 24.4 | 4 | 0.271 | 44.1 | 0.012 |
| 9 | 0.196 | 14.3 | 5 | 0.148 | 33.3 | 0.005 |
| 10 | 0.196 | 27.5 | 6 | 0.149 | 50.0 | 0.007 |
| 11 | 0.196 | 63.0 | 5 | 0.078 | 91.2 | 0.006 |
| 12 | 0.245 | 4.2 | 3 | 0.281 | 7.8 | 0.002 |
| 13 | 0.245 | 7.8 | 4 | 0.264 | 21.5 | 0.004 |
| 14 | 0.245 | 16.5 | 6 | 0.182 | 44.1 | 0.005 |
| 15 | 0.245 | 24.5 | 6 | 0.153 | 45.8 | 0.005 |
| 16 | 0.245 | 64.0 | 4 | 0.053 | 96.7 | 0.006 |
| 17 | 0.523 | 27.1 | 4 | 0.131 | 44.3 | 0.000 |

*Constitutive equation*

The constitutive equation, which combines stress with flow velocity, is obtainable based on experimental results. After determining velocity and concentration profiles in the experiments, deformation rates were calculated. Shear stresses in the fluid-mud layers were also calculated by setting shear stress to be 0 at the level of the maximum velocity.

The apparent viscosity, $\mu_a$ in the mobile fluid-mud layers is related to the deformation rate, D as shown in Fig. 2. The plastic viscosity, $\mu_p$, is defined as the viscosity for an infinitely large deformation rate. The normalized $\mu_p$ by the viscosity of water, $\mu_w$ is related to the solid fraction (concentration), $1 - \varepsilon$, in the fluid muds based on experiments as follows:

$$\frac{\mu_p}{\mu_w} - 1 = 3.8 \times 10^3 (1 - \varepsilon)^{1.7} \tag{10}$$

so that the apparent viscosity is reduced as shown in Fig. 3.

$$\mu_a = \mu_p \left(6.1\, D^{-0.66} + 1\right) \tag{11}$$

Then $\mu_a$ becomes

$$\mu_a = \mu_w \left(6.1\, D^{-0.66} + 1\right)\left\{3.8\times10^3(1 - \varepsilon)^{1.7} + 1\right\} \qquad (12)$$

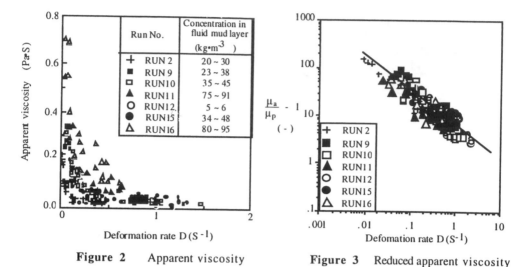

Figure 2    Apparent viscosity          Figure 3    Reduced apparent viscosity

*Deposition rate to bed mud*

The depositional flux of suspended solids from fluid-mud to bed-mud determines the rising rate of a bed-mud surface. When the flow of overlying water has high energy, the bed-mud might be eroded. On the other hand, when the flow is weak, the deposition of suspended solids takes place. The regime of the depositional flux is divided into two subregions around the critical deformation rate of deposition, $D_{cd}$ as shown in Figure 4a. One is the subregion in which the deformation rate of deposition is smaller than the critical deformation rate of deposition and the other is larger than that. The former corresponds to simple hindered settling and the latter to a slow collision process with bed-mud. The flux in the latter region has been considered as naught, the precise measurement in this study, however, indicates that the flux is small, not naught. The non-dimensionalized flux is shown in Figure 4b. In this figure, two sets of data, Runs 11 and 16, in which the initial concentrations are higher than the critical one for effective stress appearance, do not fall on the line of Run 2 data because of the underestimation of hindered settling velocity. Conventionally, the deposition flux or velocity is expressed in a non-dimensional form with the critical shear stress. The critical deformation rate is, however, employed in this expression instead because the shear stress is a function of concentration and the lift force might be proportional to the deformation rate of liquid. The force exerted on each particle is different from the shear stress. When flocs collide each other and transmit momentum, the effective stress may change these situations.

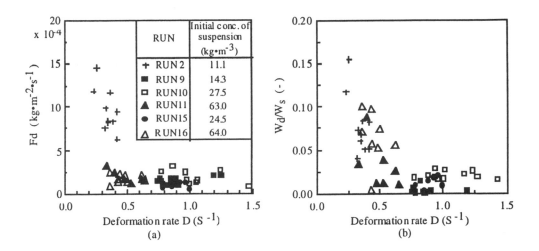

**Figure 4**    Dimensional and non-dimensional depositional fluxes to bed-mud

These date, except the two sets of the data, are formulated as a function of deformation rate as follows:

$$\frac{w_d}{w_s} = \left(1 - \frac{D}{D_{cd}}\right) \qquad D < (1 - \alpha)\, D_{cd} \qquad\qquad (13)$$

$$\frac{w_d}{w_s} = \alpha \left(\frac{D}{D_{cd}}\right)^{-1.2} \qquad D_{ce} > D \geq (1 - \alpha)\, D_{cd} \qquad\qquad (14)$$

$$\alpha = 0.1 \qquad\qquad (15)$$

where $D_{cd}$ is nearly equal to 0.4, $w_d$ is deposition velocity, and $w_s$ is settling velocity in a quiescent state.

A depositional flux is given by a product of the settling velocity and the concentration at the surface of bed mud. The concentration is obtainable through the method mentioned below.

*Dispersion coefficient*

When a fluid-mud layer is very thin, the dispersion coefficient in the layer tends to increase toward the bed-mud surface because eddies at the interface between fluid-mud layer and overlying water may mix the fluid-mud. Since this transient state is seen only within several minutes from the beginning of an experiment, this is out of scope for this study. The dispersion coefficient of "wall turbulence" is conventionally expressed as a function of $z^2D$ (Hinze, 1975), where $z$ is a distance from the bed surface to a plane. According to this expression and considering effects of the concentration of suspended solids in fluid-mud, $K$ is reduced as shown in Figure 5 and as follows:

413

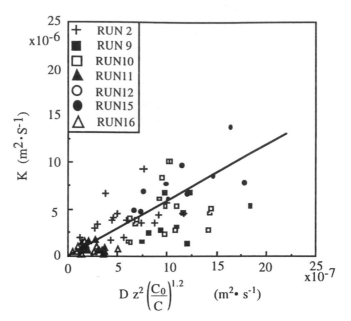

**Figure 5**  Dispersion coefficient in fluid-mud layer

$$K = 4.6 \, z^2 \, D \left(\frac{C_0}{C}\right)^{1.2} \tag{16}$$

where $C_o$ is a characteristic concentration which is taken here as the concentration of effective stress generation, $31.8 \, kg \cdot m^{-3}$. The dispersion coefficient for the upper region above the zero shear stress plane of a fluid-mud layer is expressed in a similar way as Eq.17. This means that the virtual origin is $1.2H$ away from the zero shear stress plane.

$$K = 4.6 \left(2.2H - z\right)^2 D \left(\frac{C_0}{C}\right)^{1.2} \tag{17}$$

Figure 6 indicates the Schmidt number, $\mu_a / \rho_a / K$ at $z = 0.001$ m. The Schmidt number is approximately one order of magnitude around $1 s^{-1}$ of deformation rate. The apparent viscosity, $\mu_a / \rho_a$ increases with an increase in concentration, but the dispersion coefficient, $K$ decreases with an increase in concentration, so that $K$ is not a constant.

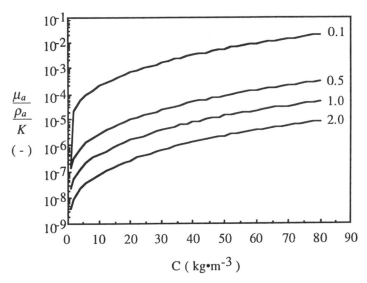

**Figure 6**    The Schmidt number in fluid-mud layers.

## CONCLUSIONS

The apparent viscosity, dispersion coefficient, mass flux to bed-mud, and two critical
concentrations on lutocline are presented based on experimental results on movement of fluid
mud on an inclined bed. Simulation on the movement of fluid-mud and on the formation of
bed-mud on an inclined bed are made possible.

## REFERENCES

Hinze, J. O.(1975) Turbulence, McGraw-Hill, p.617.
Kusuda,T., Watanabe, R., Futawatari, T., and Yamanishi, H.(1993) Fluid-mud movement on
an inclined bed, Coastal and Estuarine Studies No.42,AGU,Washington D.C.,pp.504-519.
Kynch, G. J. (1952) A theory of sedimentation, Transactions of Faraday Society, Vol.48,
pp.166-176.
Odd, N.V.M. and Cooper, A. J. (1989) A two-dimensional model of the movement of fluid mud
in a high energy turbid estuary, Journal of Coastal Research, SI(5),pp.185-193.
Smith, T.J. and Kirby, R.(1989) Generation, stabilization and dissipation of layered fine
sediment suspension. Journal of Coastal Research, SI 5(5), pp.63-73.
Wolanski, E., Gibbs, R.J., Mazda, Y., Mehta, A. and King, B.(1992) The role of turbulence in
the settling of mud flocs, Journal of Coastal Research, Vol.8, No. 1, pp.35-46.

# 30    Fluid and sediment "integrated" modelling application to fluid mud flows in estuaries

P. LE HIR

*IFREMER, BP 70, 29280, Plouzané, France*

## INTRODUCTION

Fluid mud patterns are common features in estuaries. In such areas the deposition rate is often high, the fresh deposits have not enough time to consolidate and a fluid mud can develops. According to  Sanchez and Grovel (1994), the occurrence of fluid mud can be correlated to the deposition rate : thus, when it is low, in bays for instance, a first quick settling occurs and fluid mud is seldom observed. The variations of the fluid mud are strongly related to the tidal regime as it grows up at slack water, especially during neap tides, when the bottom shear stresses are low.

For instance in the Loire estuary (France), the thickness of the fluid mud can reach several metres, in a concentration range of 10 to 200 kg.m$^{-3}$. A first 1D model of suspended sediment transport, cross-sectionally averaged, has reproduced satisfactorily these accumulations of fluid mud in the Loire estuary, as well as their resuspension as a turbidity maximum on spring tides (Le Hir and Karlikow, 1991). In the conditions of simulation (low fresh water input), salinity stratifications were supposed negligible and the turbidity maximum was mainly due to the asymmetry of the tide propagation, which was well simulated by the model. In this computation, the bottom was splitted into 2 layers of sediment: the consolidated mud and the fluid mud, characterized by two given shear strengthes. In a second model, the sediment has been finely discretized, in order to account for consolidation processes (Le Hir and Karlikow, 1992). But in either case the stability of fluid mud on neap tides was got by means of a slight shear strength, which is questionable considering the low density of the fluid mud. Actually, the Bingham behaviour of the fluid mud should probably be replaced by a viscous one , but in this case how the mass exchange between the sediment and the water column can be parameterized ? In addition, the turbulence damping due to large stratifications in the fluid mud was not accounted for and this process may explain the delay of mud resuspension as well as consolidation processes do. Lastly, the horizontal shift of the fluid mud had been neglected.

The fluid mud behaviour has also been investigated by using two-layered models that involve a specific friction at the interface and can generate "turbidity currents" within the fluid mud layer (*e.g.* Odd and Owen, 1972; Futawatari and Kusuda, 1993). However such models generally assume a (given) constant density of the fluid mud. In fact the vertical structure of the water+mud system presents a continuous increase of the sediment concentration from clear water until dense muds, with generally one or several large gradient layers, the so-called **lutoclines** (Kirby,1988). A new trend of modelling,

*Cohesive Sediments.*
Edited by Neville Burt, Reg Parker and Jacqueline Watts
©1997 John Wiley & Sons Ltd.

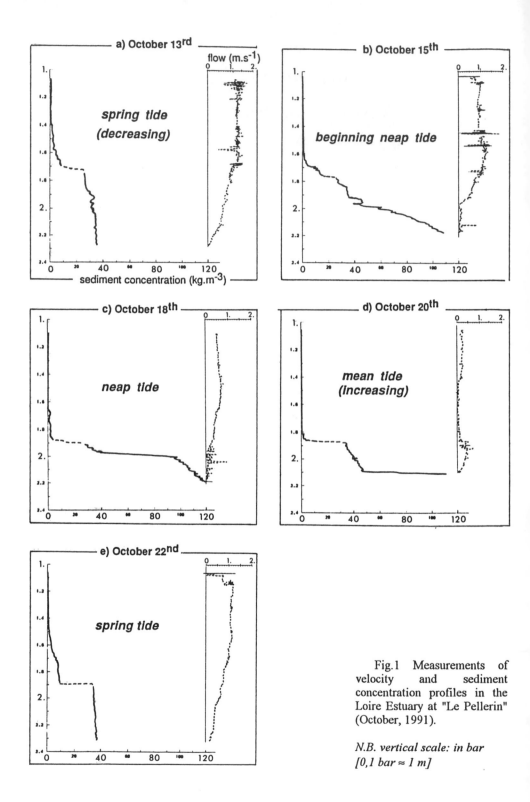

Fig.1 Measurements of velocity and sediment concentration profiles in the Loire Estuary at "Le Pellerin" (October, 1991).

*N.B. vertical scale: in bar [0,1 bar ≈ 1 m]*

introduced by Teisson *et al.* (1992), consists of simulating the water+mud complex as a whole, considering the deposition or erosion fluxes as internal vertical exchanges in the model: for instance, using a very elaborated two-phase flow model, the last quoted authors successfully reproduced the constitution of a fluid mud layer and its resuspension when increasing the flow rate.

The aim of the present modelling is to follow this *integrated* approach through a simple vertical model (1DV), to evaluate its abilities and limitations, especially by analysing the effects of different mixing processes on fluid mud behaviour, then to extend the technique in a 2DV frame, allowing the simulations of turbidity patterns on a bottom slope or in estuaries. For both models a refined vertical discretization enables to compute the flow as well as the sedimentation within the fluid mud. Results are to be compared to measurements of simultaneous vertical profiles of velocity and suspended sediment concentration conducted in the Loire estuary.

## OBSERVATIONS OF FLUID MUD IN THE LOIRE ESTUARY

An experiment on the fluid mud in the Loire estuary has been achieved in October 1991 (Karlikow *et al.*,1992): it consisted of monitoring the evolution of sediment features during a fortnightly tidal cycle at a fixed location. From a series of 75 flow and turbidity profiles (measured in the range 0-130 kg.m$^{-3}$ with an O.B.S. sensor), a selection of 5 typical profiles (quite representative for the concentration) is presented on figure 1. This figure shows that:

- before neap tide (1a), a thick but low-concentrated (30 kg.m$^{-3}$) layer of fluid mud can be observed, flowing at a rather high rate (around 0.5 m.s$^{-1}$); above it, the suspended sediment concentration remains large, meaning a strong mixing due to turbulence,
- on neap tide (1b and 1c), the consolidation begins and the previous lutocline is less marked, while a new layer appears just above,
- when the tidal amplitude increases (1d), the bottom layer seems to disappear, either because its concentration becomes out of range, or because it becomes "fluidized" within the previous intermediate layer,
- finally on spring tide (1e), the profiles look like the first ones and the cycle is completed.

We do not want to generalize these features, but are interested in local processes that these measurements point out. The successive curves show some coherence in the concentration profile evolution, justifying a global investigation of the water+sediment mixture.

## DESCRIPTION OF THE INTEGRATED MODELLING APPROACH

### General features

The classical method for computing fine (cohesive) sediments dynamics consists in solving the advection-dispersion of suspended sediment together with the resolution of hydrodynamics (depth-integrated or not), while the bottom sediment budget is managed separately, with possible integration of a consolidation algorithm. Naturally, both computations are coupled, but the exchanges between the two domains remain formulated empirically (*e.g.* in Mehta *et al.*, 1989). As for the integrated approach, it considers the whole water/sediment column as a continuous medium, more or less particle loaded. In this case, erosion or deposition terms are reduced to vertical fluxes of particles anywhere in the computational grid. These net fluxes depend on the settling velocity and the dispersive capacity of the mixture.

The former depends itself on the characteristics of particles, which can be flocculated, and on their distribution ; the so-called hindered settling process, which occurs for high concentrations, accounts for the slowing of water between particles when the void ratio is low. This upward pore water movement exactly compensates the settling of particles : thus the continuity induces a relationship between the hindered settling velocity of sedimentologists and the permeability used in soil mechanics (Schiffman *et al.*, 1985). In an integrated model dealing with fluid mud, the hindered settling domain has to be extended to the upper range of permeability.

As for the diffusion term it is mainly related to the turbulent mixing in the fluid, which is influenced by stratifications. In fluid mud the effects of density gradients on turbulence damping can become large, but on the other hand the "molecular" viscosity is likely to increase, thus reducing the mean flow and correlatively the mixing of sediment. More generally, a dense fluid (mud) can loose its Newtonian behaviour and the rheological property of the mixture has to be considered in the integrated approach. In any case the erosion process is converted as first an entrainment of a (fluid) mud layer and then its diffusion in the overlying fluid layers.

## Model formulation

When dealing with differential movements of water elements and sediment particles or aggregates that can interact, modelling should be undertaken in the frame of multiphase fluid dynamics. That means a system of one mass conservation equation and one momentum equation (with 3 components) for each phase to be solved (see Teisson *et al.*, 1992). However, the following simplifying assumptions apply to natural flows of suspended sediment and fluid mud in estuaries:

- particles and water have the same horizontal velocities, so that only one horizontal momentum can be written (2 components in a general case, only one for a 2DV model); consequently horizontal flows of pore water within a mud, for instance in tidal flats, cannot be considered.

- the acceleration terms (spatial and temporal) can be neglected in the vertical direction: this induces a quasi-hydrostatic pressure and simplifies the vertical momentum equation for each dispersed phase, which reduces to the expression of the relative vertical velocity of particles in the fluid (the "settling" velocity) resulting mainly from a balance between gravity, drag forces, and possibly turbulent and particulate stresses (see further).

- all phases are incompressible (this assumption could be removed in case of flocculation modelling).

Under these conditions, the multiphase and the continuous phase formulations are only slightly different: the former accounts for interactions between phases (mass exchanges, collisions, contacts, drag) more explicitly, but the main difference lies in the formulation.

With these assumptions, 2 codes have been developed : a 1DV multiphase model for process studies and a 2DV one phase model for applications in estuaries. As previously said, physics are nearly the same for both models, so that only the first one will be briefly described here.

The main additional hypothesis of the 1DV is that horizontal gradients are null or given for any quantity : practically, only horizontal pressure gradients are considered, as they constitute the main hydrodynamical forcing. Thus the model should fit the reality only in open seas where vertical profile of velocity and concentration are uniformly distributed in horizontal directions, or in close columns (*e.g.* laboratory settling columns).

In addition, for safe of mass conservation and coherence with assumptions on velocity gradients, the free surface variations are not considered: this important feature is justified when only processes are dealt with.

Applying the above hypotheses and after some algebra, the initial set of equations becomes (Cugier, 1993):

-for each particulate phase k, a continuity equation :

$$\frac{\partial}{\partial t}\alpha_k + \frac{\partial}{\partial z}(\alpha_k w_k) = \frac{\Gamma_k}{\rho_k} \tag{1}$$

where $\alpha_k$ is the volumetric concentration and $\Gamma_k$ the interfacial mass transfer rate between the phase k and others. This term, often zero as below, can represents a water adsorption by particles, or more generally exchanges between water and several classes of particles in flocculation processes.

- 2 equations for the horizontal components of the momentum balance :

$$\frac{\partial U}{\partial t} + \frac{1}{\rho}\sum_k \alpha_k \rho_k w_k \frac{\partial U}{\partial z} = -\frac{1}{\rho}\frac{\partial P}{\partial x} - \frac{1}{\rho}\frac{\partial}{\partial z}\left[\sum_k \left(\alpha_k \rho_k \overline{u'_k\, w'_k} + T_k^{xz}\right)\right] + fV \tag{2}$$

$$\frac{\partial V}{\partial t} + \frac{1}{\rho}\sum_k \alpha_k \rho_k w_k \frac{\partial V}{\partial z} = -\frac{1}{\rho}\frac{\partial P}{\partial y} - \frac{1}{\rho}\frac{\partial}{\partial z}\left[\sum_k \left(\alpha_k \rho_k \overline{v'_k\, w'_k} + T_k^{yz}\right)\right] - fU \tag{3}$$

"generalized" turbulent stresses     "generalized" viscous stress

→ turbulence closure     → rheology.

with $\quad \rho = \sum_k a_k \rho_k \quad$ and $\quad \sum_k a_k = 1$)

- a vertical dynamic balance equation for each particulate phase :

$$F_k = (\rho - \rho_k)\, g + \left(\sum_l \Gamma_l w_l - \frac{\Gamma_k w_k}{\alpha_k}\right) + \frac{\partial}{\partial z}\sum_l \left(\alpha_l \rho_l \overline{w_l'^2} + T_l^{zz}\right) - \frac{1}{\alpha_k}\frac{\partial}{\partial z}\left[\alpha_k \rho_k \overline{w_k'^2} + T_k^{zz}\right]$$

drag   gravity       effect of mass exchanges                 effective stress force   (4)<br>force

This last equation, together with the continuity equation, is also the basis of consolidation models, as pointed out by Teisson *et al.,* (1993) : the particulate stresses $T_k^{zz}$, the gradients of which reduce the effect of gravity and thus the consolidation itself, are nothing else than the effective stress, whereas the permeability, which represents the possibility of pore-water to flow through the particles, should be related to the drag force.

The system of equations (1) to (4) needs a closure for the vertical velocities and the viscous and turbulent stresses modelling. In fact the drag force involves a mean relative velocity between the dispersed phase and the fluid that can be written:

$$Wrel_k = W_k - W_f - Wd_k \tag{5}$$

where $W_f$ is the fluid velocity and $Wd_k$ a drifting velocity

$Wd_k$ results from the correlation between the turbulent movement of the fluid and the distribution of particles (Teisson *et al.*, 1992). $Wd_k$ can be seen as the mean value of the turbulent fluid velocities weighted by the presence of particles. Thus it can be evaluated from the concentration gradient through the classical concept of turbulent diffusivity $\gamma_t$ (see "turbulence closure", below).

$W_f$, the fluid velocity, is deduced from the continuity relationship:

$$W_f = - ( \Sigma_{k \neq f} \alpha_k . W_k ) / \alpha_f \tag{6}$$

### Turbulence closure

For modelling the turbulent Reynolds stress tensor, the concept of eddy viscosity ($v_t$) is applied for all the phases together. For saving computational costs, we choose a simple "mixing length" model, taking into account the damping of turbulence by stratifications (either salt or suspended sediment induced) by the use of a local gradient Richardson number Ri and empirical damping functions, leading to :

$$v_t = l^2 . \partial u / \partial z \tag{7}$$
$$l = l_o . e^{-a.Ri} \tag{8}$$
$$l_o = \kappa . z . (1 - z/h)^{1/2} \tag{9}$$
$$Ri = -g(\partial \rho / \partial z) / \rho_o (\partial u / \partial z)^2 \tag{10}$$
$$\text{and} \quad \gamma_t = v_t . e^{-b.Ri} \tag{11}$$

In the following applications of the 1DV model, coefficients a and b are equal to .5 and .25 respectively, while the Richardson number is constrained to a maximum value of 5.

### Sediment behaviour

The sediment characteristics are involved three times :
- The "settling velocity" is computed from the equation (4), and will depend on the particle characteristics, their density, their distribution (and thus the permeability)... according to the expression of the drag force. It will also depend on the vertical component of the particulate stress tensor (that can be formulated according to an hypothetic constitutive relationship).
- Flocculation processes can be considered, either as mass exchanges between fluid and several particulate phases as previously seen, or as size and density variations of a unique particulate phase and compensating fluid exchanges (Teisson *et al.*, 1992). In the followings, these processes are not accounted for.
- The rheological characteristics determine the shear stresses in equations (2) and (3) : if turbulent terms become negligible for high concentrated suspensions, on the opposite the "viscous" stresses in the mixture fluid/particles, $\Sigma_k T^{xz}_k$ and $\Sigma_k T^{yz}_k$, increase.
In the present state of the models, the soft mud is supposed to behave as a viscous fluid, the viscosity increasing with the concentration according to a power law (*e.g. in* Toorman, 1992) :

$$v = c (\Sigma_{k \neq f} \rho_k \alpha_k)^d$$

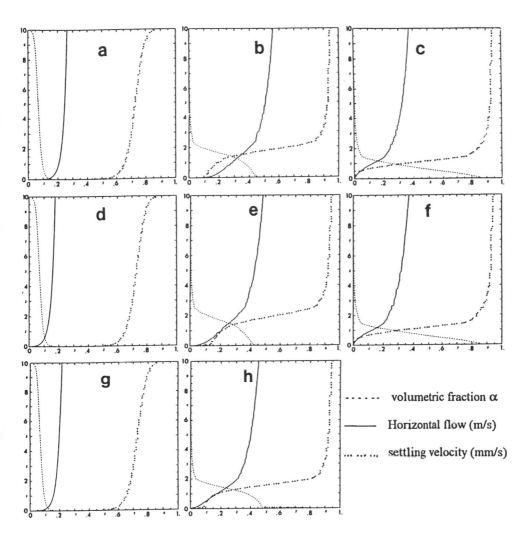

Fig.2: Equilibrium profile of concentration and flow velocity,
  computed with the 1DV model
a) no viscous increase in the mud, no turbulence damping by stratifications
b) no viscous increase in the mud, turbulence damping (a=.5, b=.25, $Ri_{max}$=5)
c) viscous behaviour of the mud (c=.003, d=3), same turbulence damping as in b)
d)*idem* case a) with a rugosity length of 1cm instead of 1mm
e)*idem* case b) with a rugosity length of 1cm instead of 1mm
f) *idem* case c) with a rugosity length of 1cm instead of 1mm
g) same viscous behaviour of the mud as in c), no turbulence damping
h) *idem* case c) with c=.001 instead of .003

In the applications below, the power is d=3 (Toorman, 1992) and c varies from 1 to 3.10$^{-3}$.

## Numerical considerations

Both models need a fine vertical resolution for simulating sharp gradients : in the order of 1mm near bottom for the 1DV. For saving computing time, the vertical resolution ot the 2DV model is only 10 cm ; criteria for selecting the best resolution have not yet been investigated. The equations are solved according to implicit finite differences methods in the vertical direction, while the horizontal convective terms in the 2DV model are computed with the positive definite advection scheme of Bott (*in* Le Hir and Karlikow, 1991).

## APPLICATIONS OF THE 1DV MODEL

### Equilibrium profile in a steady current

First the 1DV model is used for the study of a steady flow regime, and a sensitivity analysis to the various diffusion terms. The common input of the test cases are a constant forcing represented by a surface slope of 10$^{-5}$, a water height of 10m, a particle density of 1070 kg.m$^{-3}$, a (flocculated) particle size of 160 μm and an initial uniform vertical volumetric concentration α = .07. The figure 2 shows the equilibrium profiles of velocity and concentration for different sets of viscosity and diffusivity parameters (turbulent or not), and two given rugosity lengthes. These parameters have considerable effects on mean flow: the viscous behaviour of the fluid mud increases the bottom friction, while the turbulence damping by stratifications reduces its effect and induces a strong surface flow. Moreover, this latter process is necessary to enable the development of a fluid mud layer, the viscosity enhancing the bottom stratification. Only in this last case, the theoretical bottom friction is proved to have no effect on velocities and consequently on concentration profiles (Fig. 2c & 2f). Also it has been noticed that the smaller the internal friction, the longer the time to get an equilibrium ( 38 hours in case b instead of 22 hours in case a). The settling velocities are also plotted on the figure : the effect of hindered settling for high concentrations is obvious.

### Simulation of a fortnightly tidal cycle

The second test consists in applying a sinusoïdal forcing, modulated according to a spring/neap tidal cycle, to the same water column as in the steady regime, with the diffusivity parameterization of case 1c. After few tidal cycles, a quasi periodic regime is reached. Starting with a uniform volumetric fraction of .07 at slack time for a mean tidal amplitude, a fluid mud layer has clearly grown up on following neap tides (Fig.3a), showing a marked lutocline, and is easily eroded during the spring tides (Fig.3b). Within a tidal period the computed concentration profiles show larger variations than the observed ones in the Loire estuary : the difference may come from the turbulence modelling or from a non-viscous behaviour of the mud in the field, and calibration has to be carried on.

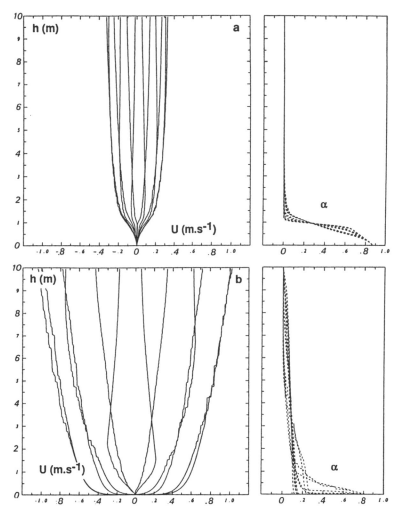

Fig. 3 Evolution of velocity and concentration profiles during a fortnightly tidal cycle, computed with the 1DV model (1curve/hour). **a) neap tide b)** following **spring tide**

## APPLICATIONS OF THE 2DV MODEL

### Sliding of a fluid mud pattern on a sloping bed

In order to explore the abilities of the integrated modelling, the becoming of dredged cohesive material suddenly released over a sloping bottom has been simulated with the 2DV model. The figure 4 shows the initial settling of materials, the constitution of a fluid mud layer and its sliding on the bottom. Also the structure of velocities during the sliding is presented, showing the turbidity current within the fluid mud, the entrainment of the overlying water and the continuity induced circulation above the fluid mud pattern. A sensitivity analysis concerning the parameterization should be necessary, as well as the validation by field or laboratory measurements as those from Kusuda *et al.* (1993).

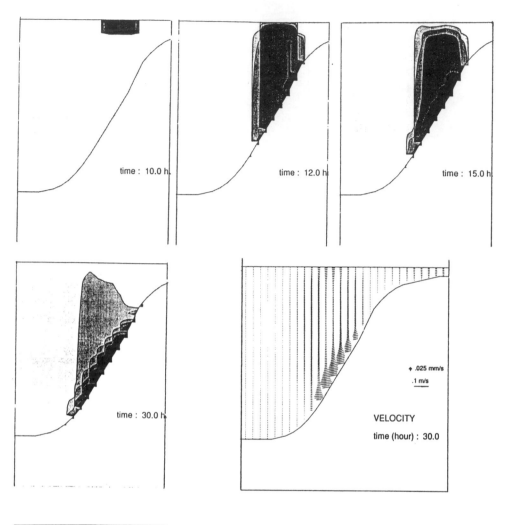

time : 10.0 h

time : 12.0 h

time : 15.0 h

time : 30.0 h

♦ .025 mm/s

.1 m/s

VELOCITY

time (hour) : 30.0

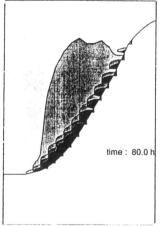

time : 80.0 h

Fig.4    Simulation of a dredged cohesive
material discharge over a sloping bed, in
quiescent water (around 15m depth).

a) initial conditions
b) and c) settling
d) and e) sliding
f) example of density current

*N.B. Attention has to be paid on the scales
distortion and on the poor horizontal resolution
(2km) regarding the reasonable vertical one
(0.1m) : this induces some diffusion and
artificial patchnesses in the drawings.*

**Simulation of turbidity patterns in a macrotidal estuary**

Finally the simulation of a turbidity maximum on spring tides and its partial deposition as fluid mud on neap tides has been successfully attempted, for a schematic estuary similar to the Loire estuary (with a length of 100 km and a tidal amplitude of 5 m). The sediment was initially uniformly distributed in suspension from the mouth over 60 km. The figure 5 gives some evidence of the vertical spreading of suspensions on spring tides and the increase of fluid mud thickness on neap tide (last isocontours are 30 and 60 kg.m$^{-3}$). The neap tide pattern is more upstream than the spring one, probably because of a preferential settling during high water slacks. Note that downstream, the turbidity patterns are related to trapped sediment generated by a sudden increase of depth, whereas the upstream structures represent the turbidity maximum.

## DISCUSSION AND CONCLUSION

The aforerepresented applications of the integrated modelling for fine sediment dynamics show that the approach is very promising. It seems appropriate to simulate the formation and evolution of fluid mud patterns. Either with a simplified multiphase formulation or only with a one phase model, the computational frame enables to consider the main physical processes, as settling, hindered or not, turbulence damping by stratifications, consolidation, viscosity increase... In particular, the computation of an equilibrium profile with a 1DV model showed the need to account for the role of fluid mud on bottom and internal friction when looking for the mean current for a given external forcing.

However, several problems remain, as the likelihood of numerical instabilities when the shear stresses are low, or the computational costs induced by the need of very fine vertical resolutions, which means that the approach is not usable for engineering applications at the moment. Actually, it constitutes a fine tool for investigating physical processes, especially if it is validated by measurements, what should be our future work.

Among the developments to be done, the turbulence closure could probably be improved, by means of the resolution of transport equations for the turbulent kinetic energy (TKE) and a turbulent scale (k-L model) or the TKE dissipation rate (k-ε model). But attention should be paid on the less importance of advection processes - especially in the 1DV model ! - regarding the role of turbulence destruction by stratifications, which is most often empirically related to a gradient Richardson number. That means that a real progress may require a second order turbulence closure modelling.

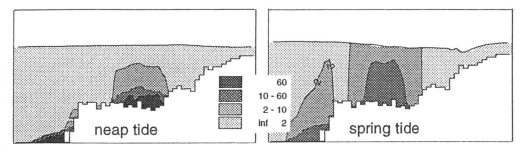

Fig.5 Simulation of a turbidity maximum and fluid mud patterns in a schematic estuary

Another domain of investigation for the integrated approach is the rheological property of a soft mud : a viscoelastic or a Bingham behaviour and even thixotropy could be tested, after some adaptation of the models (Toorman, 1992).

## ACKNOWLEDGEMENTS

This work was carried out as part of the G8 Coastal Morphodynamics programme. It was funded jointly by IFREMER (in the frame of the National Programme for Coastal Oceanography, PNOC) and the Commission of the European Communities, Directorate General for Science, Research and Development, under contract n°MAS2-CT-92-0027.

## REFERENCES

Cugier P., 1993. Elaboration d'un modèle numérique simplifié en vue de la simulation des processus verticaux dans la tranche d'eau et les sédiments superficiels. *Rapport de DEA, Univ. Marseille/IFREMER. (in french)*

Futawatari T. & Kusuda T., 1993. Modeling of suspended sediment transport in a tidal river. *In A. Mehta (ed.) Nearshore and Estuarine Cohesive Sediment Transport, Coastal and Estuarine Studies N°42, A.G.U.*

Karlikow N., Le Hir P. & Bassoullet P., 1992. Campagne "crème de Loire 1991": résultats bruts.*Rap. int. IFREMER DEL 92.22 (in French).*

Kirby R., 1988. High concentration suspension (fluid mud) layers in estuaries. *In J.Dronkers & W. Van Leussen (eds.): Physical Processes in Estuaries. Springer-Verlag, pp463-487.*

Kusuda T., Watanabe R., Futawatari T. and Yamanishi H., 1993. Fluid mud movement on an inclined bed. *In A. Mehta (ed.) Nearshore and Estuarine Cohesive SedimentTransport, Coastal and Estuarine Studies N°42, A.G.U.*

Le Hir P. & Karlikow N., 1991. Balance between turbidity maximum and fluid mud in the Loire estuary. Lessons of a first mathematical modelling. *Proc. of the Int. Symp. on "The transport of suspended sediments and its mathematical modelling", Florence, pp449-466.*

Le Hir P. & Karlikow N.,1992. Sediment transport in a macrotidal estuary : do we need to account for consolidation processes ? *Proc. 23rd Int. Conf. on Coastal Engineering, Venice.*

Mehta A., Hayter E., Parker R., Krone R and Teeter A., 1989. Cohesive sediment transport. I: process description. *Journal of Hydraulic Engineering, Vol.115, N°8.*

Odd N. & Owen M., 1972. A two-layer model of mud transport in the Thames Estuary. *Proceedings of ICE paper, 7517S, pp175-205.*

Sanchez M., Grovel A., 1994. Le tassement des vases comme processus sédimentaire. *Comptes rendus des 3èmes journées nationales Génie civil-Génie côtier, Sète 2-4mars (in french).*

Schiffmann R.L., Pane V. & Sunara V., 1985. Sedimentation and consolidation. *Proc. of the Eng. Foundation Conf., Sea Island, GA, pp57-121.*

Teisson C., Simonin O., Galland J.C. & Laurence D., 1992. Turbulence and mud sedimentation : a Reynolds stress model and a two-phase flow model. *Proc. 23rd Int. Conf. on Coastal Engineering, Venice.*

Teisson C. & Simonin O., 1993. Simulating turbulent vertical exchanges of mud concentrations and bed consolidation with a two-phase flow model. *Euromech 310, Le Havre, 13-17 sept.*

Toorman E., 1992. Modelling of fluid mud flow and consolidation. *Ph D thesis, Katholieke Universiteit Leuven.*

# 31 Equilibrium hypsometry of fine-grained shore profiles

**S.-C. LEE[1] and A. J. MEHTA[2]**
[1]*National Hydraulic Research Institute, Malaysia*
[2]*Coastal and Oceanographic Engineering Department,
University of Florida, USA*

## INTRODUCTION

The manner in which water waves mold the shore profile is evidently important in predicting such phenomena as coastal flooding, impact of sea level change on wetlands and coastal turbidity transport. In the simplest physical setting, consider a beach of uniform slope that is subjected to steady wave action normal to the shoreline. Assuming that the waves have sufficient fluid power to erode and transport the beach material, the profile will change with time until an equilibrium shape is attained. Thus, with respect to prediction of profile response, the time-rate of change of the profile and its ultimate equilibrium shape are matters of evident interest. The focus here is on profile hypsometry in the equilibrium state.

## COMPARISON BETWEEN COARSE-GRAINED AND FINE-GRAINED SHORE PROFIELS

As a starting point, the field behavior of the two generic, size-based classes of shore profiles is compared to illustrate the differences in the modeling perspective needed to describe the behavior of fine-grained profiles.

### Coarse-grained Shore Profile Hypsometry
Based on extensive field evidence, a generic equilibrium profile shape of the power form:

$$h(y) = Ay^n \tag{1}$$

where h is the water depth at an offshore distance y from the shoreline, A is a profile scale parameter and n is an empirical exponent, has been identified (e.g., Dean, 1977). To provide a physical basis for Eq. 1, various phenomenological models have been advanced. For example, Dean (1977) heuristically correlates, through the equation for the conservation of wave energy, the dissipation of energy due to wave breaking to profile dynamics, hence its equilibrium shape, according to which the mean value of $n = 2/3$ and A depends on the particle settling velocity.

### Fine-grained Shore Profile Hypsometry
The natural occurrence of stable profiles composed of fine-grained material suggests that an equilibrium shape also occurs for materials ranging in size from silt to clay. Yet, field observations point to two noteworthy differences between coarse-grained and fine-grained profile behaviors:

*Cohesive Sediments.*
Edited by Neville Burt, Reg Parker and Jacqueline Watts
©1997 John Wiley & Sons Ltd.

a) The fine-grained equilibrium profile shape differs from that for a sandy beach.

b) The comparatively much more compliant bottom composed of fine-grained material absorbs energy more readily than at a sandy beach, thus making it necessary to consider modes of wave energy dissipation in addition to, or other than, wave breaking.

However, there has not been any systematic examination of the hypsometry of field profiles composed of fined-grained sediments. In this respect, exceptions include work on glacial till around the Great Lakes area (Bishop et al., 1992), and on intertidal mudflat morphology (Osterthun and Partenscky, 1991; Friedrichs, 1993). Of particular interest here is the work of Friedrichs (1993) that links profile concavity and convexity along a wave-dominated coast to shoreline curvature. While a 2/3 power form (i.e., Eq. 1 with n = 2/3) results for a straight shoreline, a lobate shoreline only increases the profile concavity slightly relative to that for a straight shoreline. However, an embayed shoreline is found to decrease profile concavity substantially such that at a high enough embayed shoreline curvature, the profile becomes convex. The implication is that profile convexity may not appear along either straight or lobate shorelines. Inasmuch as most fine-grained shorelines fronting the ocean are generally straight or slightly lobate, for example in the vicinity of prograding river mouths, the presence of convex profiles cannot be fully explained by this analytical approach. It will be shown in the following that based on an alternative formulation that incorporates wave-bottom interaction, the resulting profile shape exhibits both concavity and convexity. In that context, results of a field data analysis are presented next to support the formulation of the alternative framework.

Figure 1 shows the results of regression fits using Eq. 1 to various profile data gleaned from the literature. Except for clay, the other data relate to sandy shorelines. The histogram for clay is based on fine-grained profiles (45 in number), the majority of which are from the Chenier Plain of Western Louisiana. Notwithstanding the fact that the analysis for clayey profiles is based on considerably fewer profiles than for sandy ones, three observations are pertinent. Firstly, convex profiles (n greater than 1) do occur on both sandy and fine-grained shoreline, albeit with much less frequency than their concave counterparts. Secondly, the median value of n for most sand data lie within the range 0.6 - 0.7 (the average for sand data ranges from 0.63 to 0.68). On the other hand, the value of n for clay is in the range 0.5 - 0.6 (average 0.51). Thirdly and more importantly, the distribution of n for sand is generally bell-shaped, and hence the use of a median/average value of n, e.g., 2/3, is not wholly inappropriate. On the other hand, while a single peak does show up in the histogram for clay, the distribution is more rounded and skewed, implying that the value of n has a wider spread and an uneven distribution than that given by the normal distribution. The implication is that the adoption of a single median/average value of n to characterize fine-grained profile geometry becomes tenuous.

The change in the exponent due to the observed variation in sediment size ranging from sand to fine-grained sediments is likely a reflection of the basic difference in their response to waves, and, therefore, unless this difference in process response is built into the approach, it is unlikely that the response of fine-grained shore profiles to waves can be adequately described. In this regard, it is recognized that the hypsometry and evolution of fine-grained shore profiles in general are not adequately described by available approaches based on sandy beach profiles, either in terms of fundamental processes or field behavior. Hence, an alternative framework must consider additional dominant processes, especially wave-soft bottom mud interaction.

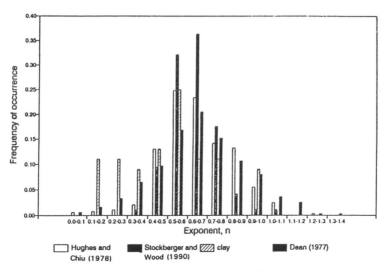

**Figure 1.** Histograms of occurrence for the exponent n in Eq. 1.

## WAVE ENERGY DISSIPATION MECHANISMS

The starting point of most existing expressions derived to support Eq. 1 applicable to coarse-

$$\frac{\partial E}{\partial t} + \nabla_h \cdot (E\vec{C}_g) = -\epsilon_D \tag{2}$$

grained sediments is the conservation of wave energy:
where $E$ = incident wave energy, $t$ = time, $\nabla_h$ = horizontal gradient operator, $\vec{C}_g$ = group velocity vector and $\epsilon_D$ = rate of energy dissipation per unit bed area. Frequently, a steady state is assumed by averaging over some suitable time-scale such as the wave period to remove time dependency. This is justified on the ground that the primary emphasis is on full equilibration, and not the transient response in profile evolution. Further considering steady wave incidence normal to the shoreline, Eq. 2 reduces to:

$$\frac{d}{dy}(EC_g) = -\bar{\epsilon}_D \tag{3}$$

where $y$ = shore normal coordinate positive landward and $\bar{\epsilon}_D$ = wave-mean rate of energy dissipation per unit bed area. The use of Eq. 3 is not contingent upon sediment size; however, its application in fine-grained profile dynamics must entail a different formulation of the wave energy dissipation mechanism than for coarse-grained profiles, for which wave breaking is identified as the primary energy sink (Dean, 1977).

The significant wave attenuation exhibited by mud bottom, as observed both in the field (e.g., Tubman and Suhayda, 1976) and in the laboratory (e.g., Maa and Mehta, 1987), is a consequence of the dissipative nature of the sedimentary medium under wave forcing. Viscous dissipation due to the wave-induced bottom motion can account for the observed wave damping that is beyond that attainable over a rigid bottom. Constitutive models proposed for describing soft bottom rheology include viscous fluid, viscoelastic or viscoplastic medium, poroelastic medium as well as those obtained empirically.

While breaking is nearly ubiquitous along sandy beaches, waves may exhaust themselves at the shoreline without breaking along mud shores, as observed by, among others, Wells and Coleman (1981), due to viscous dissipation. Since rheological parameters bear directly on the dissipation process associated with wave-soft bottom interaction, these parameters feature prominently in the phenomenological framework considered here.

## A MODEL OF PROFILE HYPSOMETRY

Equation 3 is used with an assumed uniform wave-mean rate of dissipation of wave energy ($\bar{\epsilon}_{D*}$) for the right-hand-side to obtain an analytical expression describing equilibrium shore profile shape as follows. The premise is that the profile adjusts in a manner consistent with the dissipation of incoming wave energy when equilibrium is reached. First, wave height decay is represented by the following exponential function (e.g., MacPherson, 1980):

$$H(y) = H_0 e^{-k_i y} \tag{4}$$

where $H_0$ is the incident wave height and $k_i$ is the wave attenuation coefficient. From shallow water linear wave theory, the wave group velocity is given by $C_g = (gh)^{1/2}$, where g is the acceleration due to gravity. Substituting Eq. (4) into Eq. 3 leads to:

$$\frac{\rho g^{3/2} H_0^2}{8} \frac{d}{dy}(e^{-2k_i y} h^{1/2}) = -\bar{\epsilon}_{D*} \tag{5}$$

where $\rho$ is the density of water. Integrating in the shoreward direction:

$$\int_{h_0,0}^{h,y} d(e^{-2k_i y} h^{1/2}) = \int_0^y \frac{-8\bar{\epsilon}_{D*}}{\rho g^{3/2} H_0^2} dy \tag{6}$$

which yields:

$$e^{-2\bar{k}_i y} h^{1/2} - h_0^{1/2} = \frac{-8\bar{\epsilon}_{D*}}{\rho g^{3/2} H_0^2} y \tag{7}$$

where $\bar{k}_i$ is now the wave attenuation coefficient averaged over the profile length and is related to $\bar{\epsilon}_{D*}$. Satisfying the boundary condition at the shoreline location ($y = y_0$) where h vanishes, Eq. 7 reduces to:

$$h_0^{1/2} = \frac{8\bar{\epsilon}_{D*} y_0}{\rho g^{3/2} H_0^2} \implies \frac{\sqrt{h_0}}{y_0} = \frac{8\bar{\epsilon}_{D*}}{\rho g^{3/2} H_0^2} \tag{8}$$

Substituting into Eq. 7, the following profile geometry that includes mud bed rheology by way of $\bar{k}_i$ is obtained:

$$h = h_0 e^{4\bar{k}_i y} \left(1 - \frac{y}{y_0}\right)^2 \tag{9}$$

where ($y_0$, $h_0$) locates the offshore terminus of the profile. Transforming the offshore coordinate such that y is now positive seaward with $y = 0$ being at the shoreline, Eq. 9 becomes:

$$h = h_0 e^{4\bar{k}_i(y_0-y)} \left(1 - \frac{(y_0-y)}{y_0}\right)^2 = h_0 e^{4\bar{k}_i(y_0-y)} \left(\frac{y}{y_0}\right)^2 \tag{10}$$

As seen in Fig. 2, depending on the value of $\bar{k}_i$, the resulting profile shape can be either concave- or convex-upward. If the bottom were rigid, $\bar{k}_i$ would be $O(10^{-5})$ m$^{-1}$, whereas for mud much higher values, $O(10^{-4})$ to $O(10^{-3})$ m$^{-1}$, are common (Jiang and Mehta, 1992). Hence, the $\bar{k}_i$ values used in Fig. 2 can be considered as representing a realistic range.

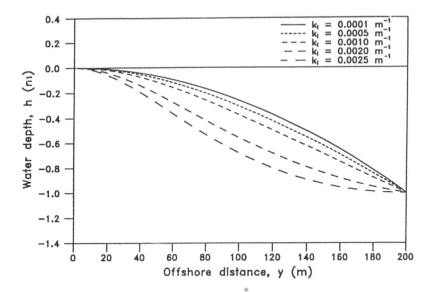

**Figure 2.** Theoretical shore profiles, given $h_0 = 1$ m and $y_0 = 200$ m.

The value of $\bar{k}_i$ can be calculated from the solution of the equations of motion describing wave-bed interaction. For the commonly assumed linear viscoelastic material of the Voigt type, it can be shown that the equations of motion reduce to the form of the linearized Navier-Stokes equations for a Newtonian fluid (MacPherson, 1980). By writing the general constitutive relation for a linear viscoelastic material in a differential form, a similar form for the momentum equation representing the behavior of the material under cyclic loading can be derived (Mehta et al., 1994):

$$\rho\frac{D\vec{u}}{Dt} = -\nabla p + \mu^*\nabla^2\vec{u} \tag{11}$$

where $\vec{u} =$ velocity vector, $p =$ pressure, $\mu^* = \mu' + i\mu''$ is defined as the complex dynamic viscosity, $\mu' =$ real part, or the dynamic viscosity and $\mu'' =$ the imaginary part, or second viscosity. In this case, the relevant constitutive equation is given by:

$$\tau_{ji}' = (\mu' + \mu'')\dot{e}_{ji}' = u^*\dot{e}_{ji}' \tag{12}$$

where $\tau_{ji}' =$ deviatoric shear stress component, $e_{ji}' =$ deviatoric component of shear strain and the dot denotes time-derivative. For any particular theoretical constitutive model, the associated mechanical analogue can be used to derive the relations linking $\mu^*$ and model parameters. On the other hand, in general $\mu^*$ does not have to be constrained to any particular theoretical model and can be left to be determined empirically using rheometric techniques.

433

It is noted that given the exponential decay in wave height per Eq. 4, $\bar{k}_i$ seemingly does not depend on wave height (e.g., Jiang and Mehta, 1992). However, a wave-height dependence is implicit by way of the change in bed rheological properties, including fluidization, due to the change in wave-induced shear strain, which is a function of wave height. Through this feedback mechanism, $\bar{k}_i$ varies with wave height via the solution of the equations of motion for wave-soft bottom interaction mentioned above.

Since increasing $\bar{k}_i$ values are associated with increasingly viscous nature of the soft bed, and with increasing wave height (Yamamoto et al., 1983), Fig. 2 can be qualitatively interpreted as supporting the tendency for a convex (accretionary) profile shape under low wave condition and conversely, concave (erosioinal) profile shape under high wave condition. This is consistent with observational evidence by, among others, Osterthun and Partenscky (1991).

A drawback that is apparent from Fig. 2 is that the profile slope becomes zero at the shoreline. While this may be acceptable for a convex profile, such is not the case for a concave profile where an "escarpment" due to vertical downcutting is frequently observed in the foreshore area. This limitation is inherent in the formulation of Eq. 10 because the approach fails to account for the effect of non-zero slope at the waterline. Thus, while a gently sloping beach predicted by Eq. 10 is appropriate for the seaward portion, clearly a different mechanism governs the profile hypsometry near the water's edge due to the presence of the steeply sloped face. This steepening can be considered to be the manifestation of gravity effect whereby the downslope component of the buoyant weight of the material above a potential failure plane (as a result of the over-steepening tendency by the remolding action of waves) is just balanced by the resisting force mobilized by the sediment. Hence, it seems logical to implement some type of geometric control based on field experience (Dean, 1991). In this regard, a modified form of Eq. 10 that incorporates the dominant influence of gravity near the shoreline where waves are substantially attenuated is obtained by heuristically including an exponential term in Eq. 10:

$$ h = \frac{FS}{\beta}(1 - e^{-\beta y}) + \left[ h_0 - \frac{FS}{\beta}(1 - e^{-\beta y}) \right] e^{4\bar{k}_i(y_0 - y)} \left( \frac{y}{y_0} \right)^2 \tag{13} $$

where FS = foreshore slope at the shoreline and $\beta$ is a profile-specific coefficient. The interpretation of Eq. 13 is that near the shoreline the first (gravity) term dominates while further seaward, the second term becomes dominant. The appearance of the gravity term in the second term is to ensure that the same seaward terminal point $(h_0, y_0)$ is recovered. It is easily shown that by retaining only the leading order term involving y in the series expansion of the first term, Eq. 13 reduces to a linear equation with a constant slope specified by FS, a condition frequently encountered in the field.

## COMPARISON BETWEEN THEORETICAL AND MEASURED PROFILES

A qualitative comparison with a limited field data set is shown in Fig. 3. The measured profiles were taken from a site along the Chenier Plain in the vicinity of Cheniere au Tigre of Western Louisiana (Kemp, 1986). The shoreline condition was described as having a mudflat morphology, sediment size in the silt-clay range, and dominated by fluid mud in the inshore area (mean sea level to 100 m offshore). The major clay mineral groups identified were illite, smectite and kaolinite. Viscometric measurement showed that the shear stress-strain rate plot for the mud (at a density of 1,260 kg/m³) could be approximated by a Bingham plastic with an upper Bingham yield strength of 17.1 Pa and an apparent viscosity of 5.2 Pa.s. However, this single viscosity value only permits calculation of one $\bar{k}_i$ value. For example, using the inviscid water-viscid mud model of Jiang and Mehta (1992) and taking the apparent viscosity to be the dynamic viscosity, $\bar{k}_i$ is found to lie within the range $10^{-4}$ m$^{-1}$ to $10^{-3}$ m$^{-1}$ depending on the thickness of the mud layer relative to the water depth. Additional guidance in this respect is provided by the measured value of $\bar{k}_i$ in East Bay, Louisiana (Tubman and Suhayda, 1976). At this site, two wave gages stationed 3.35 km apart (in 19.2 m and 4.6 m water depths, respectively) registered a percentage reduction of 48% in the root-mean-square wave height (from 0.53 m to 0.28 m), which translates to $\bar{k}_i$ = 0.0002 m$^{-1}$. This observation was made in May, 1976, a period of relatively low wave energy based on extrapolation of hindcast wave data in the Gulf of Mexico for the years 1955 - 1975 (Hubertz and Brooks, 1989). From the wave data at Station 16, which was the closest station to the above site, the mean significant wave height based on the 20-year statistics was found to range from 0.6 m to 0.9 m in the months May - October and from 1.0 m to 1.1 m in November - April. Hence, the $\bar{k}_i$ value measured in East Bay was taken as representative of the low wave energy period and transposed to the site at Chenier Plain. Thus, $\bar{k}_i$ = 0.0002 m$^{-1}$ has been used to simulate the convex curve (10/10/81) in Fig. 3. For the concave profile, a higher value of $\bar{k}_i$ = 0.0018 m$^{-1}$ has been used to obtain a good fit with the measured profile (1/14/81). An implicit assumption in the comparison is that the measured profiles are typical of the profiles that are in quasi-equilibrium with the prevailing wave condition. While this may be reasonable for the convex profile (10/10/81), which approximately marked the transition from low wave activity into high wave activity based on the 20-year statistics, it is very likely that the profile measured on 1/14/1981 may have been advancing toward equilibration, given that the next transition did not occur until April/May.

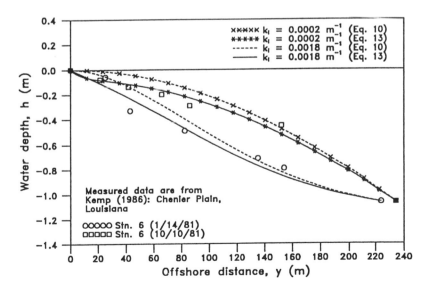

**Figure 3.** Comparison of theoretical and measured shore profiles.

435

The results of both Eq. 10 and Eq. 13 are superposed in Fig. 3 for comparison. it is seen that Eq. 10 only matches the offshore portion of the measured profile. However, the addition of the gravity term does improve the agreement considerably, especially in the case of the convex profile. In this case, the values of FS and $\beta$ are 0.01, and 0.125, respectively. For the concave profile, the agreement is not as good even though the improvement over Eq. 10 is apparent. The corresponding values of FS and $\beta$ are 0.001 and 0.04, respectively. From Fig. 3, it is clear that the overall effect of the gravity term is to lower the profile. While the values of FS have been selected to fit the measured foreshore slopes, $\beta$ is used here as a tuning parameter to achieve goodness-of-fit. Physically, $\beta$ can be interpreted as characterizing the transition between the first term and the second term in Eq. 13. More field data are required to test the generality of the proposed profile model.

## CONCLUSIONS

Based on a phenomenological approach whereby the shore profile is considered to adjust in a manner consistent with the dissipation of incoming wave energy at equilibrium such that the wave-mean rate of energy dissipation per unit bed area of the soft bed becomes uniform in the cross-shore direction, an analytical expression for the equilibrium profile hypsometry has been obtained. The profile equation depends on mud rheology, which follows from the premise that wave-soft bottom interaction leading to bed motion is the primary energy sink. Subject to additional assumptions, the resulting profile shapes are able to reproduce salient features of fine-grained shore profiles, including the transition from a convex to concave shape due to a change in the prevailing wave condition from low wave energy to high wave energy environment.

## ACKNOWLEDGMENT

Funds for this study were provided by the U.S. Army Engineer Waterways Experiment Station, contract DACW39-93-K-0008.

## REFERENCES

Bishop, C.T., M.G. Skafel and Nairn, R.B., 1992. Cohesive profile erosion by waves. *Proceedings of the 23rd Coastal Engineering Conference*, 3, ASCE, New York, 2976-2989.

Dean, R.G., 1977. Equilibrium beach profiles: U.S. Atlantic and Gulf Coasts. *Ocean Engineering Report No. 12*, Department of Civil Engineering, University of Delaware, Newark, 45p.

Dean, R.G., 1991. Equilibrium beach profiles: characteristics and applications. *Journal of Coastal Research*, 7(1), 53-84.

Friedrichs, C.T., 1993. Hydrodynamics and Morphodynamics of shallow tidal channels and intertidal flats. *Ph.D. dissertation*, Woods Hole Oceanographic Institution/Massachusetts Institute of Technology, Cambridge, 214p.

Hubertz, J.M. and Brooks, R.M., 1989. Gulf of Mexico hindcast wave information. *WIS Report 18*, U.S. Army Engineer Waterways Experiment Station, Vicksburg, MS, variously paginated.

Hughes, S.A. and Chiu, T.Y., 1978. The variation of beach profiles when approximated by a theoretical curve. *Report No. UFL/COEL-TR/39*, Department of Coastal and Oceanographic Engineering, University of Florida, Gainesville, 136p.

Jiang, F. and Mehta, A.J., 1992. Some observations on fluid mud response to water waves. In: *Dynamics and Exchanges in Estuaries and the Coastal Zone*, D. Prandle ed., American Geophysical Union, Washington DC, 351-376.

Kemp, G.P., 1986. Mud Deposition at the Shoreface: Wave and Sediment Dynamics on the Chenier Plain of Louisiana. *Ph. D. dissertation*, Department of Marine Science, Louisiana State University, Baton Rogue, 148p.

Maa, P.-Y. and Mehta, A.J., 1987. Mud erosion by waves: a laboratory study. *Continental Shelf Research*, 7(11/12), 1269-1284.

MacPherson, H., 1980. The attenuation of water waves over a non-rigid bed. *Journal of Fluid Mechanics*, 97(4), 721-742.

Mehta, A.J., Lee, S.-C. and Li, Y., 1994. Fluid Mud and Water Waves: Processes and Feedbacks. *Report UFL/COEL/MP94-01*, Department of Coastal and Oceanographic Engineering, University of Florida, Gainesville, 87p.

Morgan, J.P., VanLopik, J.R. and Nichols, L.G., 1953. Occurrence and Development of Mudflats along the Western Louisiana Coast. *Trafficability and Navigability of Louisiana Coastal Marshes Technical Report No. 2*, Louisiana State University, Baton Rouge, 34p.

Osterthun, M. and Parthenscky, H.-W., 1991. Changing salt-marshes along the southern North Sea coast. *Proceedings of the Third Conference on Coastal and Port Engineering in Developing Countries*, Mombassa, Kenya, 1093-1109.

Stockberger, M. and Wood, W.L., 1990. Application of equilibrium beach concepts to sandy Great Lakes profiles. *Proceedings of the 22nd Coastal Engineering Conference*, 3, ASCE, New York, 2291-2303.

Tubman, M.W. and Suhayda, J.N., 1976. Wave action and bottom movements in fine sediments. *Proceedings of the 15th Coastal Engineering Conference*, 2, ASCE, New York, 1168-1183.

Wells, J.T. and Coleman, J.M., 1981. Physical processes and fine-grained sediment dynamics, Coast of Surinam, South America. *Journal of Sedimentary Petrology*, 51(4), 1053-1068.

Yamamoto, T., Takahashi, S. and Schuckman, B., 1983. Physical modeling of sea-seabed interactions. *Journal of Engineering Mechanics*, 109(1), 54-72.

# 32 Simulating turbulent vertical exchange of mud with a Reynolds stress model

J.-C. GALLAND, D. LAURENCE, and C. TEISSON
*Electricité de France, Laboratoire National d'Hydraulique,*
*6 Quai Watier, 78401 Chatou Cedex, France*

## INTRODUCTION

When comparing vertical profiles of mean velocity or concentration produced numerically and measured in the field or in experiments, some adjustments are often required in order that numerical results fit measures. As far as the mean velocity profile is for example concerned, the most often used techniques mainly consist in varying the Von Karman constant, the roughness parameter and/or the wake function. These procedures yield best fit with experiments but are not predictive, e.g. see McLean (1991) for a critical review of this phenomenon and modelling procedures.

The aforementioned adjustments are needed due to the sediment-induced stratification which affects turbulence, the actor responsible for maintaining sediment in suspension since suspension results from the balance of the upward diffusive and the downward settling flux of particles. A careful attention should therefore be paid to the modelling of turbulent mechanisms in cohesive sediment transport models. However, turbulence in practical tools is still roughly described: it is often modelled through the eddy viscosity concept and the Reynolds analogy (identity between the eddy viscosity and diffusivity), assuming most of the time that viscosity and diffusivity are not affected by the sediment load (parabolic "clear water" profiles). Such hypothesis give for example a logarithmic velocity profile even for heavily laden flows although it is recognised that such a profile does not match many field or experimental data (Gust, 1976 ; Soulsby and Wainwright, 1987 ; Myrhaug et al., 1993), leading to some difficulties when estimating the bottom shear stress under the presence of sediment.

The discrepancies between calculations and measurements, the shortcomings in predicting the bottom friction velocity (which then fully determines erosion and deposition rates in numerical models) let suspect a stronger interaction between hydrodynamics and sediments than often assessed and have pleaded for the use of higher accuracy turbulence models. K-ε models (Celik and Rodi, 1988), Algebraic Stress Models (Hanjalic et al., 1982 ; Sheng and Villaret, 1989) and full Reynolds Stress Models (Brors, 1991 ; Teisson et al., 1992 ; Laurence et al., 1993) were then successfully applied.

By treating individually turbulent stresses and fluxes in the 3 spatial directions, the Reynolds Stress Model (RSM) enables to describe precisely the transfer mechanisms between

*Cohesive Sediments.*
Edited by Neville Burt, Reg Parker and Jacqueline Watts
© 1997 John Wiley & Sons Ltd.

the 3 components of the turbulent kinetic energy (which is important at the bottom and the free surface where turbulence tends to become 2 dimensional, decreasing thus the vertical component responsible for the generation of the upward sediment flux), and between turbulent and sediment fluxes. Furthermore no assumptions are prescribed on the eddy viscosity and diffusivity, the shear stress or the velocity profiles, which appear as results of the model naturally including stratification effects.

The standard Reynolds Stress model (Launder et al, 1975) is used here, slightly adapted in order to better treat the free surface, via a modified wall echo term (Gibson and Rodi, 1989) and some adjustments on the $\varepsilon$ equation to increase dissipation for highly anisotropic turbulence (Craft and Launder, 1989). The RSM is applied to the schematic case of a steady plane open channel flow under equilibrium conditions (that is a 1DV flow without any deposition or erosion rate).

This paper is not aimed at presenting the Reynolds stress model itself, but rather some features of the vertical processes highlighted by the fine modelling of this second order turbulence closure. The RSM is first compared to the classical zero-equation turbulence closure assumption. In a second step, the Reynolds stress equations are used to derive similarity laws for suspended sediment transport, which are used in the third paragraph for a tentative evaluation of the laboratory experiments procedure.

## RESULTS OF THE REYNOLDS STRESS MODEL

The procedure used to run the model is as follows. Variables are first computed without taking into account the effect of suspended sediment on hydrodynamics, by setting the gravity constant to zero (decoupled calculation). The bottom friction velocity $u_*$ and an homogeneous mean concentration profile are imposed as initial conditions. After convergence, a second calculation is initiated from the previous, taking that time gravity into account (coupled calculation). The flow rate is calculated according to the initial bottom friction velocity and is then kept constant throughout the procedure. The bottom friction velocity is a constant during the decoupled calculation, but then adapts to the sediment load during the coupled one: it is an output of the model. Also, the settling velocity $w_s$ is kept constant throughout the procedure.

The first numerical application (referred to as Case 1) is the following: $u_* = 1.9 \ 10^{-2}$ m/s (corresponding to a bulk velocity $U_0 \approx 0.5$ m/s), $C_0 = 1$ g/l, $w_s = 0.1$ cm/s, the water depth being h = 1 m. Shear stress and mean concentration profiles, for decoupled and coupled calculation, are presented on fig. 1. From these figures, it is noticeable that the influence of suspended sediment cannot be omitted for a correct prediction of concentration profile and hydrodynamic characteristics of the flow. When compared to the decoupled case, the bottom shear stress is reduced by more than 15 %, resulting in a 8% lower bottom friction velocity ($u_* = 1.76 \ 10^{-2}$ m/s instead of $u_* = 1.90 \ 10^{-2}$ m/s), and the mean relative concentration $C/C_0$ dies out in the upper part of the water column while it is 35% higher at the bottom ($C/C_0 = 2.28$ instead of $C/C_0 = 1.68$). Also the velocity profile for the coupled calculation strongly deviates from the logarithmic one, with a steeper gradient throughout the channel height and a maximum velocity at the free surface being 10% higher.

Such results are supported both by experimental investigations, from which it has been stated that the bottom friction velocity could be reduced by as much as 40 % in the presence of sediment (Gust, 1976), and by previous numerical simulations showing the same difference between decoupled and coupled bottom friction velocity $u_*$ (Sheng and Villaret, 1989 ; Huynh-Tanh et al, 1991). Comparison is also made with the Rouse profile, giving the concentration distribution through the water column as a function of the concentration

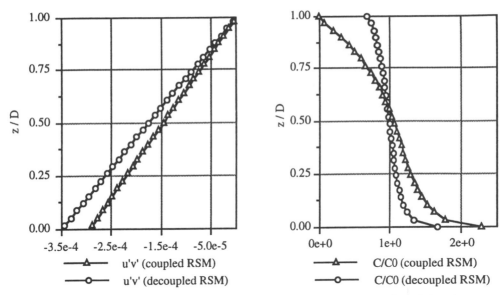

**Figure 1** Comparison between decoupled and coupled RSM simulations. Case 1: h = 1m, u* = 1.9 cm/s, $C_0$ = 1 g/l, $w_s$ = 0.1 cm/s. Left: shear stress, right: mean relative concentration.

$C_a$ at a reference level $z_a$, the bottom friction velocity and the settling velocity:

$$\frac{C}{C_a} = \left( \frac{z_a(h-z)}{z(h-z_a)} \right)^Z, \text{ Z being the Rouse parameter defined as } Z = \frac{w_s}{\kappa u_*}.$$

This formula results from the analytical integration of the sediment momentum equation

$$\frac{\partial C}{\partial t} + \frac{\partial w_s C}{\partial z} = -\frac{\partial \overline{w' c'}}{\partial z}$$ (neglecting molecular diffusivity), under the following

assumptions:
- the modelling of the upward turbulent sediment flux $\overline{w' c'}$ is based on an eddy diffusivity concept: $\overline{w' c'} = -K_t \frac{\partial C}{\partial z}$,
- the sediment eddy diffusivity $K_t$ is identical to the fluid eddy viscosity $v_t$ (the turbulent Schmidt number $\sigma_t$ is set to 1),
- the eddy viscosity $v_t$ is assumed to follow a parabolic profile (arising from a mixing length and linear shear stress hypothesis): $v_t = \kappa u_* z(1 - z/h)$.

As a consequence, this approach is, from a formal point of view, identical to the decoupled simulations performed with the RSM.

On figure 2, the mean relative concentration $C/C_a$ (for $z_a/h$ = 0.03) calculated with the coupled Reynolds stress model is compared with the Rouse profile, the Rouse parameter being calculated first with $u_*$ derived from the decoupled RSM and then with $u_*$ derived from the coupled RSM. It can be seen that the RSM profile strongly deviates from both: this is an indication that some of the above assumptions could be wrong.

As an illustration, the profiles for the eddy diffusivity and viscosity, calculated with the coupled RSM from their definitions as:

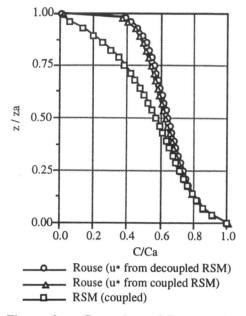

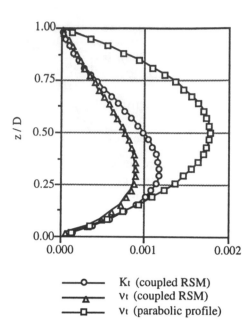

- ──○── Rouse (u* from decoupled RSM)
- ──△── Rouse (u* from coupled RSM)
- ──□── RSM (coupled)

- ──○── $K_t$ (coupled RSM)
- ──△── $v_t$ (coupled RSM)
- ──□── $v_t$ (parabolic profile)

**Figure 2**  Comparison of Rouse profile with RSM relative mean concentration profile (Case 1).

**Figure 3**  Comparison of RSM eddy diffusivity and viscosity with the parabolic profile (Case 1).

$$v_t = \frac{-\overline{u'w'}}{\frac{\partial U}{\partial z}} \quad \text{and} \quad K_t = \frac{-\overline{w'c'}}{\frac{\partial C}{\partial z}},$$

are compared with the theoretical parabolic profile calculated with $u_*$ derived from the coupled RSM (fig. 3). Although there is some agreement between both profiles in the lower quarter of the water column, they next significantly differ and neither the Reynolds analogy ($K_t = v_t$) nor the hypothesis of a parabolic eddy diffusivity or viscosity profile are verified. The most noticeable point is the difference in the order of magnitude: the parabolic profile gives values about twice those found by the RSM for $K_t$ and $v_t$ in the upper half of the channel. Now it can be seen from the Rouse approach that the use of a $K_t$ twice lower than predicted by the parabolic profile would have led to a concentration profile corresponding to a twice higher Rouse parameter Z (that is $Z \approx 1/3.75$ instead of $Z \approx 1/7.5$): stratification would then have been much stronger, and much more like the RSM prediction.

This implies that not only the bottom shear stress reduction should be taken into account for a realistic prediction of the vertical suspended sediment profile, but also processes occurring throughout the water column. As a matter of fact, from the two factors having a damping effect on turbulence that are the reduction of the bottom friction velocity and the reduction of the eddy diffusivity, the latter is seen to be the most important here (Case 1): when compared to clear water values, $u_*$ is reduced by only 8% when $K_t$ is reduced by as much as 100% in the upper half of the water column.

As $u_*$ is the only degree of freedom to adjust $K_t$ and $v_t$ for the above zero-equation turbulence model, a dependency on a Richardson number seems to be at least required for such a closure. A possible way is the use of the Munk-Anderson approach, developed for thermally stratified flows but which has already proved to accurately predict strongly sediment-stratified flows such as lutocline and fluid mud layer (Ross and Mehta, 1989 ; Wolanski et al, 1988). It consists in expressing $K_t$ and $v_t$ as functions of the gradient Richardson number $Ri_g$:

$$v_t = v_{t0}(1 + 10Ri_g)^{-0.5} \text{ and } K_t = K_{t0}(1 + \frac{10}{3}Ri_g)^{-1.5}, \text{ with } Ri_g = -\frac{\beta g \frac{\partial C}{\partial z}}{(\partial U/\partial z)^2} \text{ and } \beta$$

the relative mass density of the sediment.

$K_{t0}$ and $v_{t0}$ are the eddy diffusivity and viscosity for neutral conditions, and can be defined in a first time by the parabolic profile ($v_{t0} = K_{t0} = \kappa u_* z(1 - z/h)$).

Applied to the above case, with $Ri_g$ derived from the RSM results, this method gives better order of magnitude and shape for both $K_t$ and $v_t$ than the pure parabolic profile, but is not fully satisfactory because it predicts a turbulent Schmidt number ($\sigma_t = v_t / K_t$) greater than 1. This is in disagreement with RSM results in the region of high diffusivity and viscosity ($z/h < 0.75$) where $\sigma_t$ is about 0.7, as well as with fitting of k-ε results on experimental measures (Celik and Rodi, 1988) which indicates that the better choice for $\sigma_t$ is 0.5 for that model, or extrapolation from measurements by Coleman which gives $\sigma_t$ always lower than 1 (Van Rijn, 1989).

As suggested by the RSM results, the neutral turbulent Schmidt number $\sigma_{t0}$ is then set to 0.7. The eddy diffusivity and viscosity are then in good agreement with the RSM predictions (fig. 4), although they are still

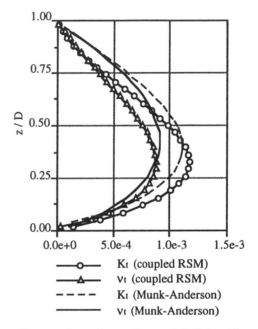

**Figure 4** Comparison of RSM eddy diffusivity and viscosity with the Munk-Anderson approach (Case 1).

slightly overestimated in the upper part ; this is an indication that the coefficients in the Munk-Anderson formulae should be fitted.

## DIMENSIONAL ANALYSIS

We have however to keep in mind that the results presented above have been obtained for an only slightly loaded flow ($C_0 = 1$ g/l, $Z \approx 7.5$), for which stratification effects have however been seen important. This is in contradiction with many laboratory experiments (carousels or flumes tests ), from which the damping effect of suspended sediment is indeed thought to be small for low concentrations, say $C_0 < 2$g/l (Berlamont et al, 1993). This is why attention was paid to analytical developments (Teisson et al, 1992 ; Laurence et al, 1993) and particularly to the dimensional analysis of the Reynolds stress equations.

Under the assumptions presented previously (1DV steady flow), the equations for the mean velocity U and concentration C read (neglecting molecular effects):

$$\begin{cases} \dfrac{\partial U}{\partial t} = -\dfrac{1}{\rho_w}\dfrac{\partial P^*}{\partial x} - \dfrac{\partial \overline{u'w'}}{\partial z} \\[2mm] \dfrac{\partial C}{\partial t} + w_s\dfrac{\partial C}{\partial z} = -\dfrac{\partial \overline{w'c'}}{\partial z} \end{cases} \qquad \text{with } w_s < 0 \text{ (constant)}$$

For simplicity, let us focus here on the equations for the mean concentration and the turbulent flux of concentration. Developing only the (exact) production terms, this later reads:

$$\dfrac{\partial \overline{w'c'}}{\partial t} = -\overline{w'w'}\dfrac{\partial C}{\partial z} - g\beta\overline{c'^2} + \Phi_{w,c} + D_{w,c}, \text{ where } \beta = -\dfrac{(\rho_s - \rho_w)}{\rho_w}, \text{ and } \Phi_{w,c},$$

$D_{w,c}$ are modelled terms respectively for pressure strain correlations and turbulent diffusion.

Use of the bottom friction velocity $u_*$, the channel depth $h$ and the bulk concentration $C_o$ to make the variables non-dimensional leads to:

$$z \to \hat{z} = \dfrac{z}{h}, u \to \hat{u} = \dfrac{u}{u_*}, t \to \hat{t} = \dfrac{u_*}{h}t, C \to \hat{C} = \dfrac{C}{C_0} \text{ and then to (dropping the symbol ' $\hat{}$ '):}$$

$$\begin{cases} \dfrac{\partial C}{\partial t} + \dfrac{w_s}{u_*}\dfrac{\partial C}{\partial z} = -\dfrac{\partial \overline{w'c'}}{\partial z} \\[2mm] \dfrac{\partial \overline{w'c'}}{\partial t} = -\overline{w'w'}\dfrac{\partial C}{\partial z} - \dfrac{g\beta h C_0}{u_*^2}\overline{c'^2} + \Phi_{i,c} + D_{i,c} \end{cases}$$

Two non-dimensional numbers appear to make the full set of equations non-dimensional (except of course terms including the shear, for which a Reynolds number is required):

- a bulk Richardson number: $Ri_b = \dfrac{g\beta h C_0}{u_*^2}$,

- and a non-dimensional settling velocity: $\dfrac{w_s}{u_*}$, which is a Rouse like parameter.

These two numbers fully characterise the sediment mixing under given conditions, and should therefore be scaled in an experiment aimed to reproduce stratification effects observed in the field.

## SIMILARITY EXPERIMENTS

To check if the RSM correctly reproduces the influence of the above derived non-dimensional numbers, and also to try an evaluation of the procedure used to perform laboratory experiments, calculations were conducted for several cases with two different water depths: the deeper ($h = 1$ m) could represent the field and the shallower ($h = 0.2$ m) a flume aimed to be a small scale model of the field situation. Such an experiment could be a deposition test for the evaluation of the critical bottom shear stress for deposition to be used in a 2DH model for the prediction of the morphological evolution in a given field area.

The usual procedure to perform laboratory experiment has been identified as:

*In the field:*
- to make measurements of the velocity profile and of the sediment concentration and characteristics,

- to evaluate the bottom friction velocity, by fitting a log-law on the velocity profile.
*In the laboratory:*
- to adjust the clear water flow rate to get the same $u_*$ as estimated for the field,
- then to set the mean concentration at the same value as in the field,
- to work with the same sediment as in the field.

To fulfil the last point, the settling velocity is set at the same value for the field and the flume ($w_s = 0.1$ cm/s), and is also kept constant in the following.

In a first run, we have reproduced exactly the described procedure: the decoupled calculations for both h = 1 m and h = 0.2 m were initiated with the same bottom friction velocity $u_* = 1.9$ cm/s and mean concentration $C_0 = 1$ g/l.

Results for the mean concentration and the shear stress are presented on fig. 5. The influence of suspended sediment is seen to be nearly negligible in the flume: the concentration profile looks much more like the decoupled one (fig. 5.a), which is identical for h = 1 m and h = 0.2 m, and the shear stress profile is similar to the clear water one (reduction at the bottom of about 1% instead of 15% in the field, see fig. 5.b).

Going now back to the dimensional analysis, it is seen that neither $Ri_b$ nor Z is scaled in the flume. Both are indeed underestimated, which is consistent with fig. 5:

$$-\hat{Z} = \frac{1}{\hat{u}_*} = 0.93 \text{ and } \hat{Ri}_b = \frac{\hat{h}}{\hat{u}_*} = 0.18.$$

To get identical profiles for the field and the flume, a second run is performed, by just setting the initial ("clear water") $u_*$ in the flume at the previous coupled field value (that is now $u_*$ [flume] = 1.76 cm/s instead of $u_*$ [flume] = 1.90 cm/s, as an initial condition).

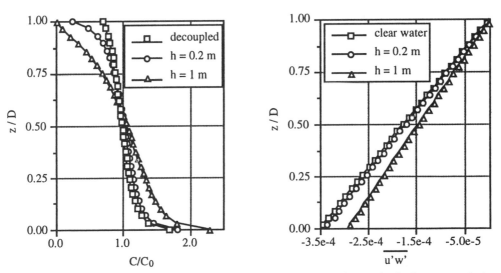

**Figure 5**    RSM simulations - $u_* = 1.9$ cm/s, $C_0 = 1$ g/l, $w_s = 0.1$ cm/s. Left: mean relative concentration, right: shear stress.

Shear stresses (not shown here) are now identical, so that $u_*$ is scaled and $\hat{Z} = 1$, but the flume concentration profile remains similar to the previous one (cf. fig. 6) and stratification is again underpredicted. This is because $Ri_b$ is still not correctly reproduced: now $Ri_b$ is scaled as h, so that stratification effects are 5 times lower in the flume ($\hat{Ri}_b = \hat{h} = 0.2$).

A third computation is then run with identical initial clear water bottom friction velocities ($u_*$[flume] = $u_*$[field] = 1.90 cm/s), but with an initial bulk concentration 5 times higher in the flume than in the field ($C_0$[flume] = 5 g/l, $C_0$[field] = 1 g/l).

As a result, concentration and shear stress profiles in the field and the flume are now identical (cf. fig. 7).

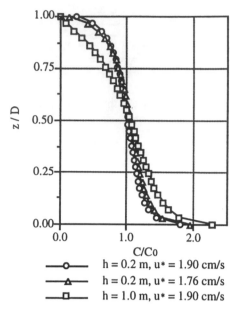

$h = 0.2$ m, $u_* = 1.90$ cm/s
$h = 0.2$ m, $u_* = 1.76$ cm/s
$h = 1.0$ m, $u_* = 1.90$ cm/s

**Figure 6**    Effect of adapting the clear water $u_*$ in the flume.

The most important in these results, is that they confirm the above dimensional analysis and the influence of the two non-dimensional numbers derived. That way, provided that the sediment load $hC_0$ is scaled, a flume test conducted with the same clear water bottom friction velocity $u_*$ and the same sediment settling velocity $w_s$ as in the field would yield (under our assumptions) the same concentration and shear stress profiles as in the field.

## CONCLUSIONS

The dimensional analysis of the Reynolds stress equations has highlighted that two non-dimensional numbers are to be considered and scaled to catch stratification effects when studying sediment transport in laboratory experiments:

- the bulk Richardson number:  $Ri_b = \dfrac{g\beta hC_0}{u_*^2}$

- and the Rouse parameter:  $Z = \dfrac{w_s}{ku_*}$.

As a matter of fact, this is a slightly different criteria than the one derived previously (Teisson et al, 1992), based on the scaling of the flux Richardson number, defined as:

$$Ri_f = \dfrac{g\beta w_s hC_0}{u_*^3}.$$

However, the later number also reads $Ri_f = Ri_b\ Z$: the new criteria is only more severe than the first one, as $Ri_b$ and $Z$ have to be scaled separately and not only their product.

The developments presented here have been derived under very simplifying assumptions (equilibrium flow, constant settling velocity), but they however could be of some help to answer some of the questions arising when working on suspended sediment.

446

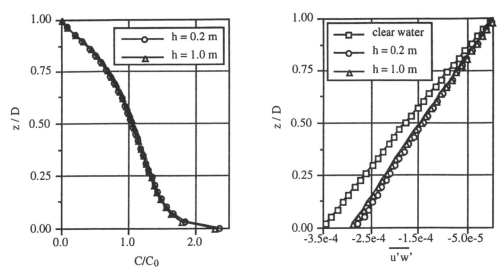

**Figure 7** RSM simulations - $u_* = 1.9$ cm/s (clear water), $hC_0$ = constant. For $h = 1$ m: $C_0 = 1$ g/l, for $h = 0.2$ m: $C_0 = 5$ g/l. Left: mean relative concentration, right: shear stress.

The analysis, with respect to $Ri_b$ and $Z$, of the laboratory procedure to perform tests for suspended sediment transport lets suspect that their results should be carefully interpreted, particularly when making reference to the equivalent field situation. Working in the laboratory with the same sediment, for the same range of concentration and flow rate as in the field has been shown to produce stratification effects lower in the flume than in the field, the ratio being defined as the scale of $Ri_b$ (which is here the one of h): what has to be scaled to catch stratification effects is the mass of sediment over the water column, $hC_0$.

So, laboratory experiments conducted with cohesive sediment in flumes, that are usually rather shallow (typically of the order 0.2 m), do not often show as strong stratification as can be observed in the field. As a matter of fact, it is seen from the similarity criteria derived here that such experiments should be conducted with very high sediment concentrations to reproduce marked stratification, so high that they should anyhow be unrealistic because of physical processes (flocculation, hindered settling) which are impossible to scale.

Carousels, where are determined erosion and deposition laws, present a second shortcoming compared to flumes: even if such a facility is equipped with a rotating bottom (and not only with a rotating roof), turbulence is partly generated at the roof, unlike the field where it is produced nearly completely at the bottom and has to fight against high concentration gradients to reach the upper part of the water column. Therefore vertical concentration profiles are most often seen homogeneous in carousels.

Numerical models for the transport of suspended sediment are known to sometimes strongly underestimate deposition and overestimate erosion, requiring critical shear strength for erosion and deposition up to 2 order of magnitude higher than those determined from field or laboratory measurements to work properly (Costa and Mehta, 1990). The reason for that shortcoming could be that, neglecting the effect of suspended sediment on hydrodynamics, these models systematically overestimate the bottom shear velocity $u_*$.

Comparison of the Reynolds stress model with the zero-equation model (used in particular to derived the so-called Rouse profile) also shows that the prediction of the vertical concentration profile could be much improved by using a Munk-Anderson like approach, which makes not only the eddy diffusivity but also the eddy viscosity dependent on the gradient Richardson number.

## ACKNOWLEDGEMENTS

This work was part of the MAST G8 - Coastal Morphodynamics Programme and funded partly by the Commission of European Communities, Directorate General for Science, Research and Development under MAST Contract n° 0035-C and by the French Sea State Secretary (STCPMVN).

## REFERENCES

Berlamont J., Ockenden M., Toorman E., Winterwerp J, 1993. The characterisation of cohesive sediment properties. Coastal Eng., n°21, pp. 105-128.

Brors B., 1991. Turbidity current modelling. Dr. Ing. Thesis; Univ. of Trondheim, Norway.

Celik I., Rodi W., 1988. Modeling suspended sediment transport in nonequilibrium situations. J. Hydr. Eng.,vol. 114, n°10, pp. 1157-1191, October.

Costa R. G., Mehta J. M., 1990. Flow-fine sediment hysteresis in sediment-stratified coastal waters. Proc. 22nd ICCE, Delft (The Netherlands), pp. 2047-2060.

Craft T., Lauder B.E., 1989. A new model for pressure/scalar-gradient correlation and its application to homogeneous and inhomogeneous free shear flows. Proc. 7th Symp. on Turb. Shear flows, Paper 17-1, Stanford.

Gibson N. M., Rodi W., 1989. Simulation of free surface effects on turbulence with a Reynolds stress model. J. Hydr. Res., vol. 27, pp. 233-244.

Gust G., 1976. Observations on turbulent-drag reduction in a dilute suspension of clay in sea-water. J. Fluid Mech., vol. 75, part 1, pp. 29-47.

Hanjalic K., Ivanovic M., Vujovic V., Milisic V., 1982. Buoyancy effects and mutual interaction of suspended particles and velocity field in turbulent shear flows. Proc. ICMH, Structure of turbulence in H. & M. transf., pp. 529-544, Hemisphere Ed.

Huynh-Thanh S., Hamm L., Temperville A., 1991. Analysis of the erosional behavior of cohesive sediments with a stratified turbulent transport model. Int. Symp. on the Transport of Suspended Sediments and its Mathematical Modelling, Florence (Italy), pp. 433-466.

Launder B. E., Reece G. J., Rodi W., 1975. Progress in the development of Reynolds stress turbulence closure. J. Fluid Mech., n° 118, pp. 341-377.

Laurence D., Maupu V., Galland J.-C., Teisson C., 1993. A sediment laden open channel flow simulation with recent Reynolds stress-flux transport models. Proc. 2nd Int. Symp. on Eng. Turbulence Modelling and Measurements, pp. 73-82. Florence (Italy).

McLean S. R., 1991. Stratification effects due to suspended sediments. Int. Symp. on The Transport of Suspended Sediments and its Mathematical Modelling. Florence (Italy), pp. 145-164, Elsevier.

Myrhaug D., Slaattelid O. H., Lambrakos K. F., 1993. Suspended sediment concentrations near the seabed in North Sea storms. MAST G8-M, Overall Workshop, Grenoble (France).

Ross M. A., Mehta A. J., 1989. On the mechanics of lutoclines and fluid mud. J. Coastal Res. Special issue n° 5, pp. 51-61.

Sheng P., Villaret C., 1989. Modelling the effect of suspended sediment stratification on bottom exchange processes. J. Geophysical Res., vol. 94, n°C10, pp. 14, 429 -14, 444.

Soulsby R. L., Wainwright B.L.S.A., 1987. A criterion for the effect of suspended on near-bottom velocity profile. J. Hydr. Research, vol. 25, n°3, pp. 341-356.

Teisson C., Simonin O., Galland J.-C., Laurence D., 1992. Turbulence and mud sedimentation: a Reynolds stress model and a two-phase flow model. 23nd ICCE, Venise (I), pp.2853-2866.

Van Rijn L. C., 1989. Handbook. Sediment transport by currents and waves. Delft Hydraulics Report H 461.

Wolanski E., Chappell J., Ridd P., Vertessy R., 1988. Fluidization of mud in estuaries. J. Geophysical Res., vol. 93, n° C3, pp 2351-2361.

# 33 An intercomparison exercise of cohesive sediment transport numerical models

L. HAMM[1], T. CHESHER[2], M FETTWEIS[3],
K. P. P. PATHIRANA[4] and E. PELTIER[5]

[1]*SOGREAH Ingénierie, Grenoble, France*
[2]*HR Wallingford Ltd, Wallingford, UK*
[3]*International Marine & Dredging Consultants N.V., Antwerp, Belgium*
[4]*Katholieke Universiteit te Leuven, Leuven, Belgium*
[5]*Laboratoire National d'Hydraulique, Chatou, France*

## INTRODUCTION

The aim of the paper is to report first results of an intercomparison exercise involving four cohesive sediment transport numerical models versus a common database. This dataset considers transport on a flat bottom (1D case) under steady and alternating unsteady (estuarial regime) currents. Current velocities and suspended sediment concentrations have been measured over a vertical profile (water depth of 0.32 m). Eight tests with steady currents (from 0.25 to 0.85 m/s) and four tests with unsteady currents have been carried out on deposited beds of various ages (between 0.25 and 7.6 days). The paper is divided into fourth parts: firstly, the four models are briefly described, then the experimental data are summarized. The third and four parts describe the simulations carried out in steady and unsteady currents respectively. The result obtained so far from this collaborative research is the identification of the aspects of the physics that required improvement in each model and in the database.

## BRIEF DESCRIPTION OF THE NUMERICAL MODELS

The numerical models used by the four institutes for the present exercise are all two-dimensional depth-averaged (2DH) models. Cohesive sediment transport processes involved include erosion of deposited or placed beds, advection and diffusion of suspended sediment, deposition and consolidation of the deposits. Erosion laws proposed by Ariathurai and Arulanandan(1978, eq.1), Thorn and Parsons (1980, eq. 2) and Parchure and Mehta (1985, eq. 3) have been used together with deposition laws proposed by Krone (1962, eq. 4) and Mehta and Partheniades (1975, eq.5).

(1)     $E = M_1 (t - t_e)/t_e$ when $t > t_e$
(2)     $E = M_2 (t - t_e)$ when $t > t_e$
(3)     $E = E_0 \exp [a(t - t_e)^{1/2}]$ when $t > t_e$
(4)     $D = - C W_s(C) (1 - t/t_d)$ when $t < t_d$

*Cohesive Sediments.*
Edited by Neville Burt, Reg Parker and Jacqueline Watts
©1997 John Wiley & Sons Ltd.

(5)    $D = $ function ( C, h, t, $t_{50}$, $s_1$, $s_2$) when $t \le t_d$

where E is the erosion rate (kg/m$^2$/s), D is the deposition rate (kg/m$^2$/s), t is the bed shear stress (N/m$^2$), $t_e$ is the critical shear stress for erosion, $t_d$ is the critical shear stress for deposition, $M_1$ (kg/m$^2$/s), $M_2$ (kg/N/s), $E_0$ (kg/m$^2$/s), a (m/N$^{1/2}$), t, $t_{50}$ (s), $s_1$, $s_2$ are empirical constants, C is the depth-averaged suspended sediment concentration, $W_s$(C) is the settling velocity (m/s) and h is the water depth (m). The choice of the empirical constants is quite essential and will be detailed in the following parts.

## Description of MUDFLOW-2D

The transport equation solved in the HR model (MUDFLOW-2D see Miles,1985) reads:

(6)    $\dfrac{\partial}{\partial t}(hC) + \alpha[\dfrac{\partial}{\partial x}(huC) + \dfrac{\partial}{\partial y}(hvC)] =$

$\dfrac{\partial}{\partial s}(hK_s\dfrac{\partial C}{\partial s}) + \dfrac{\partial}{\partial n}(hK_n\dfrac{\partial C}{\partial n}) + E + D$

where (x, y) are the Cartesian co-ordinates in the horizontal plane, (s, n) are the natural co-ordinates (parallel with and normal to the mean flow), t is the time, (u, v) are the depth-averaged components of the velocity (m/s), $K_s$ is the longitudinal (shear flow) dispersion coefficient (m$^2$/s), $K_n$ is the lateral (turbulent) diffusivity (m$^2$/s), h is the water depth (m), a is an advection factor to recover the true sediment flux from the product of depth-averaged quantities. The settling velocity is considered to increase linearly with concentration and is computed as:

(7)    $W_s$(C) = maximum (CW, $W_{smin}$)

where W is a constant of proportionality and $W_{smin}$ is a constant minimum value corresponding to very little interaction between particles.

The initial deposit is considered as a layer with an initial thickness and a continuous vertical density profile. When erosion occurs, the thickness of this layer decreases and the surface critical shear stress for erosion is increased consequently. Deposition of new sediment is assumed to give rise to a bed density equal to the initial surface dry density. By this means the process of resuspension of unconsolidated mud is achieved.

## Description of SUBIEF

The equation of suspended sediment solved in the LNH model (SUBIEF see Peltier and Chausson, 1991) is:

(8)    $\dfrac{\partial C}{\partial t} + u\dfrac{\partial C}{\partial x} + v\dfrac{\partial C}{\partial y} = \dfrac{\partial}{\partial x}\left(K_x\dfrac{\partial C}{\partial x}\right) + \dfrac{\partial}{\partial y}\left(K_y\dfrac{\partial C}{\partial y}\right) + \dfrac{E+D}{h}$

where $K_x, K_y$ are dispersion coefficients. In this equation, hydrodynamic conditions (u, v, h) are known, being previously computed by the model TELEMAC. Equation (8) is numerically solved with a finite element technique in two steps: a. advection without source term (calculated with the characteristic method) and b. treatment of the dispersion and of the source term (equation solved implicitly). Consolidation of the bed is simulated at each node in the numerical model by a pile of several layers of increasing concentration (Teisson and Fritsch, 1988), each layer of various thickness being characterized by a maximum residence duration (after which mud of the layer goes into the underlocated more concentrated layer), a critical shear stress for erosion, which could be related to the layer density and an erosion rate (constant $M_1$ of equation (1)). When deposition occurs, the sediment goes into a storage upper layer where the dry density is kept at a low value. In such a case, its thickness increases consequently. When the bed shear stress exceeds its critical value for erosion, erosion occurs in the first layer. Each successive layer is similarly tested for potential erosion when the preceding layer has been totally eroded.

### Description of PATHIRANA'S model

The transport model of Pathirana (1993) is very similar to SUBIEF but the modelling of the deposit is rather different. Pathirana considers two layers simulating the initial deposit and new soft deposits respectively. The initial deposit layer can only be eroded and its dry sediment density is continuously varying over the depth following empirical data. The upper layer is made of fresh deposits and could be easily resuspended. No consolidation is taken into account in both layers.

### Description of TRASIL

The IMDC (TRASIL) model (Praagman, 1986) solves in a Lagrangian way the advection equation in the following form:

$$(9) \quad \frac{\partial C}{\partial t} + \frac{\partial u C}{\partial x} + \frac{\partial v C}{\partial y} = \frac{E + D}{h}$$

It is assumed that the velocity vector (u, v) is given at nodal points of the domain of interest. The structure of the domain is either a finite element or a finite difference type of grid. Using these discrete velocity nodal points, the velocity vector in each internal point is calculated by interpolation. An initial concentration field is created by covering the whole domain of interest by small squares and by assigning to each the representative amount of the species. Each square keeps it identity as long as it remains within the domain. Once the field of squares has been created the transport of each of these "particle" squares is computed by numerically integrating the velocity field to obtain trajectories of the particles using the classical fourth order Runge-Kutta method. During the computational process the right hand side functions u and v are being adjusted each substep, using interpolated values of the velocity field. Since the fourth order Runge-Kutta method has very good amplification and phase error properties the advection of the squares will almost be optimal and positiveness and mass conservation are guaranteed. At the inflow boundary new particles with a well defined concentration and volume are entering the model. The

451

particles which have been transported outside the model are deleted. Erosion and sedimentation change the amount of mud in the particles.

A disadvantage of the method is encountered if the results are needed on an Eulerian grid. Several smoothing effects are present due to the fact that the concentrations are given at the squares and not at the nodal points. Once the new position of the particle has been calculated erosion and sedimentation are determined. The structure of the deposit is modelled according to Pathirana(1993).

## SUMMARY OF THE EXPERIMENTAL DATA

The measurements were carried out in a racetrack flume 1 m wide and 0.4 m high consisting of two straight sections 30 m long and two 180° bends, 12 m long giving a total length of 84 m. Currents were generated by an Archimedean screw submerged in the central part of one of the straight sections. The water level was kept constant during each test. Five optical turbidity meters and six micro propellers were used for the measurements. The mud used for the tests was taken from the river Garonne at the mouth of Bordeaux wet lock. It has been treated chemically prior to the tests in order to eliminate the organic matter.

Preliminary hydraulic tests without mud were carried out to measure the vertical profile of the velocity for seven mean flow regimes. Then, eight tests were performed in uniform flow conditions with a movable bed. The initial water depth over the bed was 0.32 m. An amount of roughly 800 kg of mud was introduced into the flume and a deposited bed was built by putting the mud into suspension prior each test and then allowing it to consolidate naturally for a variable period ranging from 0.25 to 7.6 days corresponding to a mean dry density range of the deposit of 98 to 309 kg/m$^3$ and a thickness of 97.2 to 32.4 mm. Then, the current speed was increased in successive stages from 0.15 up to 0.85 m/s, with each stage lasting for more than 40 minutes. Current speeds and turbidity were measured at four depths namely 2, 5, 15 and 25 cm above the bed in the centreline of the flume. No vertical concentration gradient was observed during the tests.

In the second part of these experiments, four tests were carried out with alternating current regimes and consolidation periods of the bed ranging from 0.7 to 7.8 days (mean density of 193 to 310 kg/m$^3$ and thickness of 40 to 25 mm). For each test, five "tides" were reproduced successively with a period of 2 h 20 mn and a constant water depth of 0.32 m. Current speeds and turbidity were measured at five depths namely 2, 5, 19, 15 and 25 cm above the bed in the centre line of the flume. All the data have been provided in digitalized form and in a report (Viguier et al., 1994). For the purposes of presentation, all graphs show depth-averaged quantities of the measured turbidity.

Data related to the yield stress of the deposit and its global consolidation rate (evolution of the depth-averaged density profile with time) were also provided as well as two density profiles for consolidation times of 0.25 and 1 day. These two density profiles were normalized relative to the depth of the initial deposits and the depth-averaged density in order that the density profile (and hence yield stress profile) of the mud for each test could be indirectly ascertained

The analysis of the vertical velocity profiles near the bed provided an estimate of the friction coefficient.

## SIMULATION OF THE EROSIVE STEADY CASES

The steady cases have been used to adjust the empirical constants to be introduced into the different models. The participants have been encouraged to use the same input data as much as possible. Some values have nevertheless been modified in order to better fit the data or the model requirements. $t_e$ was estimated from the yield stress curve related to the the bed dry density ($T_s$). The measured bed density profile was used with linear interpolation between measured points. The magnitude of the erosion rate was adjusted to fit the data. A constant value of $M_2$ in equation (2) gave good results for most tests. On the other hand both $M_2$ and $t_e$ were adjusted for each layer in SUBIEF in order to closely follow the data. Table 1 summarizes the input coefficients used for the simulations. Numerical results of four tests (namely tests 2, 6, 7 and A1) are presented in figures 1 to 4 together with the experimental data. In the SUBIEF model, the choice made of only one layer per velocity step induces a linear increase of the suspended sediment concentration during each step which does not follow exactly the observed trend. The use of a continuous density profile in the other models gives a better qualitative agreement. Quantitatively, results obtained with a unique value of $M_2$ look reasonable. Major discrepancies appear during test 6 with a well consolidated bed and for test A1 with the IMDC model. These factors remain presently unexplained.

### Table 1: List of Inputs

| | | MUDFLOW-2D HR | SUBIEF LNH | PATHIRANA KUL | TRASIL IMDC |
|---|---|---|---|---|---|
| 1. | Friction coefficient | Darcy: 0.015 | Strickler: 90 | Chezy: 69 | Chezy: 69 |
| 2. | Critical shear stress for erosion (N/m²) | $1.26 \ 10^{-9} \ T_s^{3.56}$ | fit to steady cases | $7.95 \ 10^{-10} \ T_s^{3.64}$ | $7.95 \ 10^{-10} \ T_s^{3.64}$ |
| 3. | Erosion law | eq 2 | eq 1 | eq 2/ eq 1 | eq 2/ eq 1 |
| 4. | Erosion rate (constant $M_i$) | 0.009 | $9.73 \ 10^{-8} \ T_s^{1.8}$ | 0.009/ 0.003 | 0.009/ 0.003 |
| 5. | Structure of the initial deposit: . Density gradient . Consolidation | discretized density profile No | multiple layers No | discretized density profile No | discretized density profile No |
| 6. | Critical shear stress for deposition (N/m²) | 0.05 | 0.05 | 0.05 | 0.05 |
| 7. | Settling velocity (mm/s) | 0.1 C (C in kg/m³) | 0.2 mm/s | 0.11 mm/s | 1.5 mm/s |
| 8. | Deposition law | eq 4 | eq 4 | eq 4 + eq 5 | eq 4 + eq 5 |
| 9. | Resuspension of fresh deposit | Yes | Yes | Yes | Yes |
| 10. | Erosion rate of fresh deposits | eq 2 | eq 3 | eq 3 | eq 3 |
| 11. | Time step used for the simulations | 24 s | 4s | 60 to 120 s | 0.2 to 1 s |

## SIMULATION OF THE UNSTEADY CASES

The four unsteady test cases have been run using almost the same input data described in the preceding section. The modifications are as following: In the TRASIL model, a linear relationship of the form $H/H_o = 0.355 \ T_s/T_{smean} + 1.125$ (H and $H_o$ are the actual and initial thicknesses of the deposit) has been used to model the dry density profile of the deposits.

Pathirana and IMDC used the erosion law given by equation (1) in order to better fit the data. Simulation results for three tests are given in figures 5 to 7 together with the experimental data. From these results it is concluded that all the model results look good except for the observed overestimation in deposition. They all show hysteresis, as with the data between the first and last tide and appear to represent the basic physics adequately. The quantitative differences observed between the simulated results could be explained by the different schematizations of the properties of the initial and fresh deposits over the vertical. An analysis of the thickness of the eroded part of the deposit should also be carried out in order to assess the relative importance of new and initial deposits.

## CONCLUSIONS

The aim of the tests was for each institute to run their most appropriate model versus the measured data, as a means of identifying qualitatively the model's capability of representing the basic physics associated with each test. Thus, transport modelling was not really an issue in this first step. It should be stressed that standard parameter settings were defined where appropriate, and that the exercise was not one of which model gives the most "accurate" results, since this would lead to a (much less worthwhile) process of "curve fitting" by each institute. On the contrary, any differences between model and data were used to identify aspects of the physics (eg description of the bed structure) that required improvements.

At the present stage of the study, the following conclusions could be drawn up: Available erosion and deposition laws seem to adequately model the basic processes in uniform flows. Results from the alternating current cases are adequate for all models. Quantitative differences observed tend to indicate that the structure of the bed and the resuspension of soft material having previously settled are the key parameters to be considered. It should be noted that a full assessment of the data is being carried out in parallel with these studies, and as a consequence, further refinements to, or additional data may arise.

## ACKNOWLEDGEMENTS

This work was carried out as part of the G8 Coastal Morphodynamics research programme, which is funded partly by the Commission of the European Communities, Directorate General for Science, Research and Development, under contract No. MAS2-CTP2-0027 and by national agencies including the Service Technique Central des Ports Maritimes et des Voies Navigables (Compiègne, France) and the UK Ministry of Agriculture, Fisheries and Food (Flood and Coastal Defence Division). Data presented in this report are taken from a study commissioned by the Service Technique Central des Ports Maritimes et des Voies Navigables to the Laboratoire Central d'Hydraulique de France in 1985. Permission to publish these data is gratefully acknowledged.

## REFERENCES

**Ariathurai, R. and Arulanandan, K. (1978).** Erosion rate of cohesive soils. J. of the Hyd. Div., ASCE, vol.104, HY2, 279-283

**Krone, R.B. (1962).** Flume studies of the transport of sediment in estuarial shoaling processes. Final report. Hydraulic Eng. Lab., Univ. of California, Berkeley, California

**Mehta, A.J. and Partheniades, E. (1975).** An investigation of the depositional properties of flocculated fine sediments. J. of Hyd. Research, IAHR, vol.13, 14, 361-381

**Miles, G.V. (1985).** Numerical Simulation of the Erosion of Marine Mud - HR Wallingford Report SR 31.

**Parchure, T.M. and Mehta, A.J. (1985).** Erosion of soft sediment deposits. J. of Hyd. Engng, ASCE, vol.111, 10, 1308-1326.

**Pathirana, K.P.P. (1993).** Cohesive sediment transport in estuaries and coastal environments. Ph. D. Thesis. Civil Eng. Dept., K.U. Leuven, Leuven, Belgium

**Peltier, E. and Chausson, P. (1991).** Code Subief de Calcul de Transport de Sédiments en Suspension.Rapport EDF/HE-RE/91.23

**Praagman, N. (1986).** TRASIL: A numerical simulation model for the transport of silt. Internal report, Svasek B.V., Rotterdam

**Teisson, Ch. and Fritsch, D. (1988).** New advanced in Numerical Modelling of Suspended Sediment Transport in the Loire Estuary -. IAHR Symposium on Mathematical Modelling of Sediment Transport in the Coastal Zone.Copenhague 30 May 1st June 1988.

**Thorn, M.F.C. and Parsons, J.G. (1980).** Erosion of cohesive sediments in estuaries: an engineering guide. in H.S. Stephens (ed.), Proc. third Int. Conf. on Dredging Techn., BHRA, Bordeaux, France, 349-358

**Viguier, J., Gallissaires, J.M. and Hamm, L. (1994).** Flume measurements of mud transport on a flat bottom under uniform and alternating currents. Internal report MAST-G8M Project 4 topic E., SOGREAH 52184R1, Grenoble, France

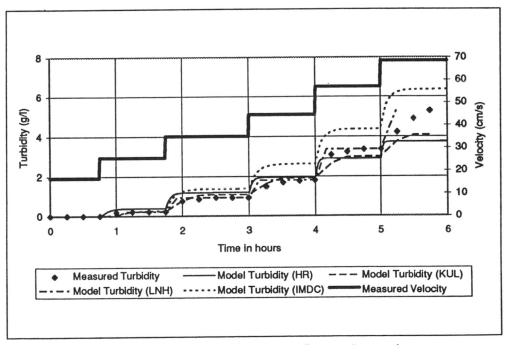

**Figure 1.** Uniform flow test: turbidity and flow speed versus time
Test 2 (consolidation time= 4.9 days, $T_{smean}$ = 291 kg/m$^3$, H$_o$ = 32.7 mm)

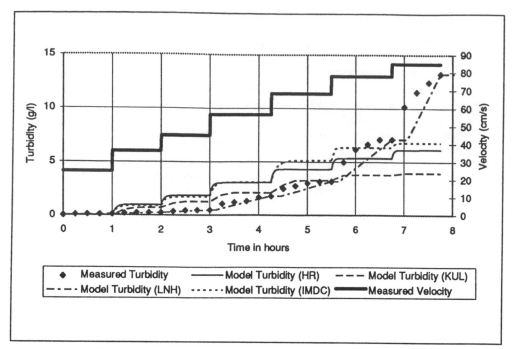

**Figure 2.** Uniform flow test: turbidity and flow speed versus time
Test 6 (consolidation time= 7.6 days, $T_{smean} = 309$ kg/m$^3$, H$_o$ = 32.4 mm)

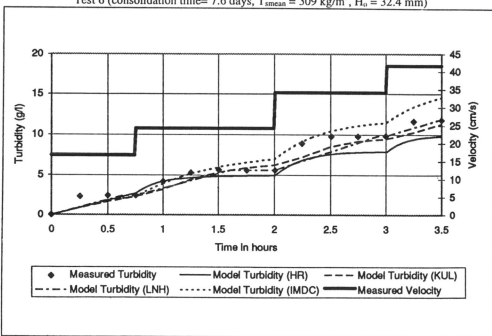

**Figure 3.** Uniform flow test: turbidity and flow speed versus time
Test 7 (consolidation time= 0.25 days, $T_{smean} = 98$ kg/m$^3$, H$_0$ = 97.2 mm)

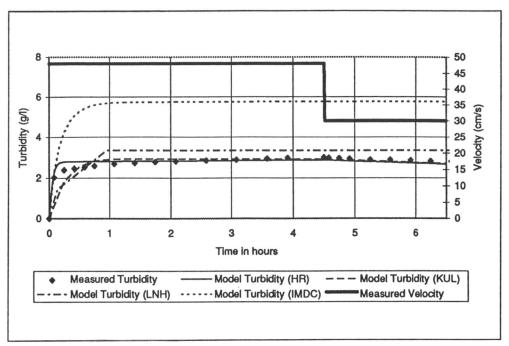

**Figure 4.** Uniform flow test: turbidity and flow speed versus time
Test A1 (consolidation time= 2.9 days, $T_{smean}$ = 259 kg/m$^3$, $H_o$ = 36.8 mm)

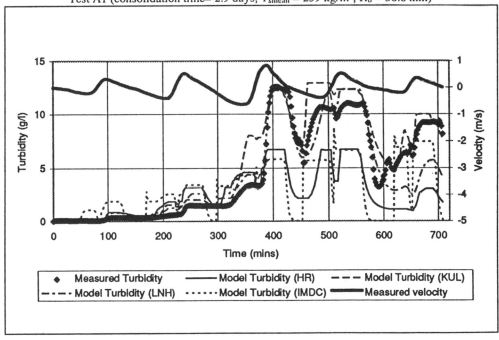

**Figure 5.** Alternating current test: turbidity and flow speed versus time
Test 1 (consolidation time= 4.6 days, $T_{smean}$ = 287 kg/m$^3$, $H_o$ = 29.0 mm)

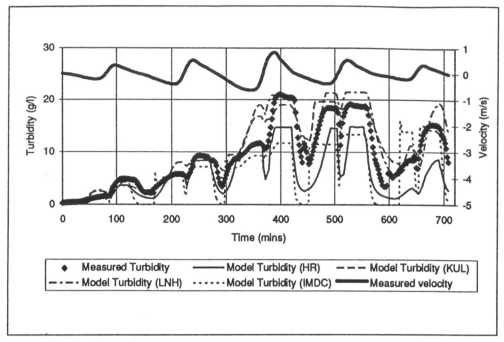

**Figure 6.** Alternating current test: turbidity and flow speed versus time
Test 3 (consolidation time= 0.75 days, $T_{smean}$ = 193 kg/m$^3$, $H_0$ = 40.0 mm)

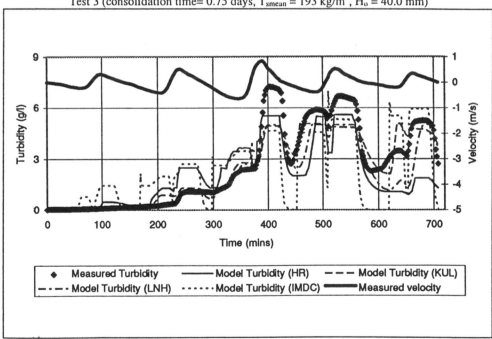

**Figure 7.** Alternating current test: turbidity and flow speed versus time
Test 4 (consolidation time= 7.8 days, $T_{smean}$ = 310 kg/m$^3$, $H_0$ = 25.0 mm)